The Changing Flow of Energy through the Climate System

Kevin Trenberth emphasizes the fundamental role of energy flows in the climate system and anthropogenic climate change. The distribution of heat, or more generally, energy, is the main determinant of weather patterns in the atmosphere and their impacts. The topics addressed cover many facets of climate and the climate crisis. These include the diurnal cycle; the seasons; energy differences between the continents and the oceans, the poles and the tropics; interannual variability such as El Niño; natural decadal variability; and ice ages. Human-induced climate change rides on and interacts with all of these natural phenomena, and the result is an unevenly warming planet and changing weather extremes. The book emphasizes the need to not only slow or stop climate change, but also to better prepare for it and build resilience. Students, researchers, and professionals from a wide range of backgrounds will benefit from this deeper understanding of climate change.

Dr. Kevin E. Trenberth is a Distinguished Scholar at the National Center for Atmospheric Research (NCAR). He was a Lead Author of the 1995, 2001, and 2007 Scientific Assessment Reports of the Intergovernmental Panel on Climate Change (IPCC), and shared the 2007 Nobel Peace Prize which went to the IPCC and Al Gore. He served from 1999 to 2006 on the Joint Scientific Committee of the World Climate Research Programme (WCRP). He chaired the WCRP Observation and Assimilation Panel from 2004 to 2010 and the Global Energy and Water Exchanges (GEWEX) Scientific Steering Group from 2010 to 2013 (member 2007–2014). He has also served on many US national committees. He is a Fellow of the American Meteorological Society (AMS), the American Association for the Advancement of Science (AAAS), the American Geophysical Union (AGU), and an honorary fellow of the Royal Society of New Zealand Te Apārangi. In 2000 he received the Jule G. Charney award from the AMS; in 2003 he was given the NCAR Distinguished Achievement Award; in 2013 he was awarded the Prince Sultan Bin Abdulaziz International Prize for Water, and he received the Climate Communication Prize from AGU. In 2017 he was honored with the Roger Revelle medal by the AGU.

The Changing Flow of Energy through the Climate System

KEVIN E. TRENBERTH
National Center for Atmospheric Research

CAMBRIDGE
UNIVERSITY PRESS

University Printing House, Cambridge CB2 8BS, United Kingdom

One Liberty Plaza, 20th Floor, New York, NY 10006, USA

477 Williamstown Road, Port Melbourne, VIC 3207, Australia

314–321, 3rd Floor, Plot 3, Splendor Forum, Jasola District Centre, New Delhi – 110025, India

103 Penang Road, #05–06/07, Visioncrest Commercial, Singapore 238467

Cambridge University Press is part of the University of Cambridge.

It furthers the University's mission by disseminating knowledge in the pursuit of
education, learning, and research at the highest international levels of excellence.

www.cambridge.org
Information on this title: www.cambridge.org/9781108838863
DOI: 10.1017/9781108979030

First published 2022

Printed in the United Kingdom by TJ Books Limited, Padstow Cornwall

A catalogue record for this publication is available from the British Library.

Library of Congress Cataloging-in-Publication Data
Names: Trenberth, Kevin E., author.
Title: The changing flow of energy through the climate system / Kevin E. Trenberth.
Description: Cambridge ; New York, NY : Cambridge University Press, 2022. | Includes bibliographical references and index.
Identifiers: LCCN 2021019452 (print) | LCCN 2021019453 (ebook) | ISBN 9781108838863 (hardback) | ISBN 9781108972468 (paperback) | ISBN 9781108979030 (epub)
Subjects: LCSH: Climatology. | Hydrometeorology. | Force and energy. | Climatic changes. | BISAC: SCIENCE / Earth Sciences / Meteorology & Climatology
Classification: LCC QC866 .T74 2022 (print) | LCC QC866 (ebook) | DDC 551.6–dc23
LC record available at https://lccn.loc.gov/2021019452
LC ebook record available at https://lccn.loc.gov/2021019453

ISBN 978-1-108-83886-3 Hardback
ISBN 978-1-108-97246-8 Paperback

Contents

Foreword

We have reached a tipping point in humanity's struggle to fully understand the magnitude of the threat posed to the survival of our civilization by the climate crisis and to accelerate implementation of a global plan to begin solving it. The good news is that in 2021 it is now obvious to the majority of women and men throughout the world that if we do not act with greater urgency, this crisis could be the source of our demise. And the many new pledges of impressive action are encouraging.

But the bad news is that still, every day, we continue to release 162 million tons of heat-trapping global warming pollution into our Earth's atmosphere, as if it were an open sewer. The extra heat energy trapped by all of the man-made greenhouse gas pollution already accumulated in our atmosphere is equal to what would be released by 600,000 Hiroshima-class atomic bombs detonating on Earth every single day.

The consequences of this extraordinary amount of additional heat energy are all around us, and they are spelled out in startling detail in *The Changing Flow of Energy through the Climate System*: the disruption of the hydrological cycle which is evaporating much more water vapor from the oceans into the sky, where the warmer air holds more of it, filling the atmospheric rivers that fuel stronger storms and more extreme downpours and floods. We're seeing deeper and longer droughts, increased water stress, declining crop yields, the spread of tropical diseases poleward, refugee crises and the resulting political instability in several regions. Concurrently, excessive pollution is contributing to the melting and fracturing of the cryosphere, accelerating sea-level rise, threatening coastal cities and freshwater aquifers; and increased outgassing of both CO_2 and methane from thawing permafrost in the Arctic.

Kevin Trenberth is one of the world's premier climate scientists, and along with his colleagues he has been warning us about this mounting crisis for decades. Kevin has an extraordinary ability to take the complicated scientific dynamics of global warming and communicate what's happening in a clear and compelling way. I have learned a great deal from him over many years, and I am grateful for Kevin's patience as a teacher and his ability to convey complexity in simple and understandable language. He's taught me about quite a few of the intricate ways that we humans are disrupting the balance of important ecological systems. And I highly value the way he not only informs, but also motivates action.

His new book gives his readers an opportunity to learn from Kevin's expertise before it's too late. As of this writing, the outlook is bracing, with a global pandemic shaking our way of life, our coastal cities being battered by storm after unrelenting storm, and record wildfires destroying enormous areas of our forest land. Yet, because of champions like Kevin and his dedication to communicating scientific reality, I remain hopeful that we will ultimately solve this crisis and usher in a new era of clean solutions.

Due in large part to action from private and public sector leaders, spurred by an increasingly galvanized citizenry, we now have the solutions available to us. Over the past five years, solar energy jobs grew five times faster than average job growth. The fastest-growing job in the United States over the next decade is projected to be wind turbine technician, and solar panel installer is projected as the third fastest growing job. Natural climate solutions like sustainable forestry and regenerative agriculture are also developing rapidly. These are just a few of the exciting advancements that represent major opportunities for pollution reduction, enabled by new policies and technologies that can speed their adoption in response to the rising public awareness of the crisis and its impacts.

Deepening our understanding of the climate crisis is central to solving it – it helps us understand why we need to implement the solutions so rapidly, it connects us with the natural world and the web of biodiversity that we desperately need to protect, and it prepares us for the accelerating climate impacts that are already part of our reality. This is why I find the clear language and accessible science education represented in this book so very important. *The Changing Flow of Energy through the Climate System* is an essential read to understand the underlying scientific dynamics of the climate crisis.

Put most simply, we need a full and widespread understanding of the problem before we can act in contributing to its solutions. The old adage, "knowledge is power," applies. Otherwise, we will be navigating our future blindly and will face unacceptably dangerous consequences. Beyond the obvious target audiences for this book, which include policymakers, scientists, and even the local solar panel installer, everyone from all walks of life must be bought in. Ultimately, we share the same home, and there is no alternative other than fighting to save it.

Former Vice President Al Gore
January 4, 2021

Preface

This book is all about the flow of energy through the climate system, where the energy is sequestered or resides, and the consequences. It is also about the changes in energy as a result of climate change. This also means tying together the fundamental underpinnings of weather and climate with climate change. The human-induced disruption in our life-sustaining thin atmospheric layer is on a runaway course toward unprecedented rapid change unknown to any previous human civilization.

The distribution of heat or, more generally, energy is the main determinant of subsequent weather patterns in the atmosphere and their impacts. The allocation of energy has always been unequal, mostly because of the Sun–Earth orbit and Earth's rotation, and on Earth from the distribution of land. The movement around of heat, mainly by the winds in the atmosphere and the currents in the ocean, is primarily a response to the gradients or contrasts. The atmosphere is always trying to get rid of the big contrasts, but the Sun–Earth geometry insistently restores them. "Warm air rises" is fundamental and leads to clouds and rain. The equatorial regions, or the tropics more generally, are much hotter than the polar regions, the Arctic and Antarctic, and as a result the atmosphere experiences huge contrasts in temperature which it is always trying to alleviate. Consequently, in middle latitudes large-scale storms develop, called cyclones or depressions, along with cold and warm fronts, and warm air is moved polewards while cold outbreaks on the other side of the storm bring cooler air to lower latitudes. In the tropics, the movement of energy polewards is more often in the form of a large-scale overturning, as in monsoon circulations or the Hadley Circulation, in which the relatively warmer air moves polewards above the cooler but moister near-surface flow towards low latitudes.

But now climate change enters the fray. Primarily caused by human activities that interfere with the natural flow of energy through the climate system, especially because of changes in composition of the atmosphere, an energy imbalance has formed for the planet. Radiation to space is somewhat trapped by the increasing greenhouse gases that form a blanket in the atmosphere so that more radiation is coming in from the Sun than escapes back to space. The result is a warming planet.

This is very fundamental, yet often ridiculed by so-called climate deniers who go so far as to question the basic science.

It has been difficult to accurately assess Earth's Energy Imbalance because the values are very small compared with the natural flows of energy through the climate system, but since they are always of one sign the effects accumulate. Accordingly, rather than measuring the imbalance directly from satellites in space, the best method is to perform a comprehensive bookkeeping of the inventory of energy, its distribution, and its changes over time. In turn, this information has consequences for the future weather and climate. Even a decade ago, this assessment was difficult owing to inadequate observations, but it is now possible and can be done reasonably well globally, and even locally on space scales of 1000 km. The main revolution making this possible has been the advances in the ocean observing system through thousands of expendable profiling floats.

Hence, the observational evidence is mounting, and year after year the ocean is found to be at its warmest state ever. Land and sea ice are melting, and heatwaves and wildfires are a consequence on land. Ironically, both droughts and heavy rains, with their associated flooding risk, are increasing too. Therefore, beyond the basic understanding and the science knowledge, it becomes more and more apparent that climate change has consequences. These are big and costly. Hundred billion–dollar disasters have become all too common. But as well as the dollar expense, it is the cost in terms of lives lost, houses burned, lives disrupted, structures destroyed by floods or strong winds, and people displaced that really matters. Nonetheless, many economists and politicians and the general public have not put two and two together, and accordingly underestimate the impacts and costs. The purpose of this book is to help rectify these oversights, and better inform decision-makers at all levels of what is happening and why, and what it means for the future.

As discussed, it is in everyone's personal interest to exploit the environment, but there are now so many people that this is actually changing and destroying the global commons of the atmosphere and oceans that we all share. This is *the tragedy of the commons*. While each of us can do our part in reducing our carbon footprint and trying to live more sustainably, this is a global challenge. But there is no global government. The nations on Earth have yet to make serious inroads into addressing the problems that are mounting. Governments have a key role to play in setting the stage to encourage the private sector and individuals to change their ways of operating. Once a sea change happens, with a real framework to encourage the decarbonization of our energy and electrical systems, we may be surprised at how much change can occur in a decade or two and help put our spaceship Earth back on track to a more sustainable state. The COVID-19 (coronavirus) pandemic effects on the peoples of the planet illustrate the threat and how poorly it has been dealt with in many places. The climate change issue perhaps represents a similar but longer-term threat – except there will be no vaccine to cure it.

Although this book provides a somewhat novel focus on how to view the climate change problem, the topics addressed are general and cover many facets of climate. These include the diurnal cycle, the seasons, the continents versus the oceans, the ice regions versus the tropics, interannual variability such as El Niño, natural decadal variability, and even ice ages. Human-induced climate change rides on and interacts with all of these natural phenomena but the result can be new records set and thresholds crossed, as a result of which things break, and people die. This book briefly touches on the need to not only slow or stop climate change, but also to better prepare for it and build resilience, in part with a better information system.

The aim here is to produce a book that details the physical processes but mostly without mathematics. In several places, relatively simple mathematics or models are introduced in sidebars in a different color, and the intent is that these can be skipped by most readers. Several short boxes are also introduced in the text to call out an aside. A short statement is given at the start of each chapter to outline the main content. References, especially to sources for figures, are listed along with some further reading at the end of each chapter, but the main sources, not called out, are in the bibliography.

Enjoy!

Kevin E. Trenberth
Distinguished Scholar
National Center for Atmospheric Research
Auckland
New Zealand
November 5, 2020

Acknowledgments

Although I was a professor at the University of Illinois for seven years or so, most of the material in this book has been developed for the hundreds of lectures I have given since becoming employed at NCAR in 1984. In addition to the many talks at conferences, from 1990 until the end of 2019 (when the coronavirus pandemic upset things), I gave an estimated 366 invited or keynote talks of a scientific nature, and 157 invited public general lectures. Some of this material has been made freely available through PowerPoint presentations at my website www.cgd.ucar.edu/staff/trenbert/.

I especially thank my close collaborators in recent years who helped develop many of the graphics and figures: John Fasullo, Yongxin Zhang, and Adam Phillips from NCAR, and Lijing Cheng from the Institute of Atmospheric Physics, Beijing, China. Special thanks also to Tom Karl, Aiguo Dai and Jim Hurrell, and many colleagues at NCAR. I also benefitted greatly from interactions with colleagues in NOAA, WCRP, and IPCC.

I gratefully thank Richard Rosen, Neil Gordon and Tom Karl for reading through a draft of the book and providing useful comments that have helped to improve the manuscript.

I dedicate this book to my family, especially my wife Gail, my daughters Annika and Angela, and the next generation, led by Ngaira and Maeve. This book may help them understand why I spend a lot of time in our study, and why I was on travel so much in the past.

1 Earth and Climate System

This introduction to the book briefly describes the climate system and flows of energy, and the challenges presented by climate change and climate modeling.

1.1 The Climate System and a New Approach

The Sun is the center of our climate system. It provides a fairly steady stream of radiant energy to Earth, and in order to keep from heating up, Earth in turn radiates energy back out into space. In between the incoming and outgoing radiation are all of the rich complex processes involved in the climate system including all of the weather systems, the entire *hydrological cycle*,[1] the ocean, land and ice, and the multitude of forms of heat and energy on the planet. The internal interactive components in the climate system (Fig. 1.1) include the *atmosphere*, *oceans*, *ice*, and *land*.

This book is focused on all of these forms of energy, how they vary and change, and interact; how energy flows through the climate system to maintain a stable climate. This also provides a framework for dealing with climate change. Our recent climate is clearly favorable to life on the planet, and humans have thrived, even as some dwell in the tropics while others prefer cooler climes at higher latitudes. We are adapted to our current climate. While we can relocate and change the climate we experience, climate change is potentially disruptive because it threatens the food, water, and other resources used to directly sustain human life.

Natural climate changes have occurred in the past as Earth has gone from ice ages (*glacials*) to *interglacials*, but on very long timescales of millennia to hundreds of thousands of years. Now human activities are causing changes of comparable magnitude on decadal to century timescales, creating Earth's Energy

[1] Italicized words are defined in the Glossary at the back of the book.

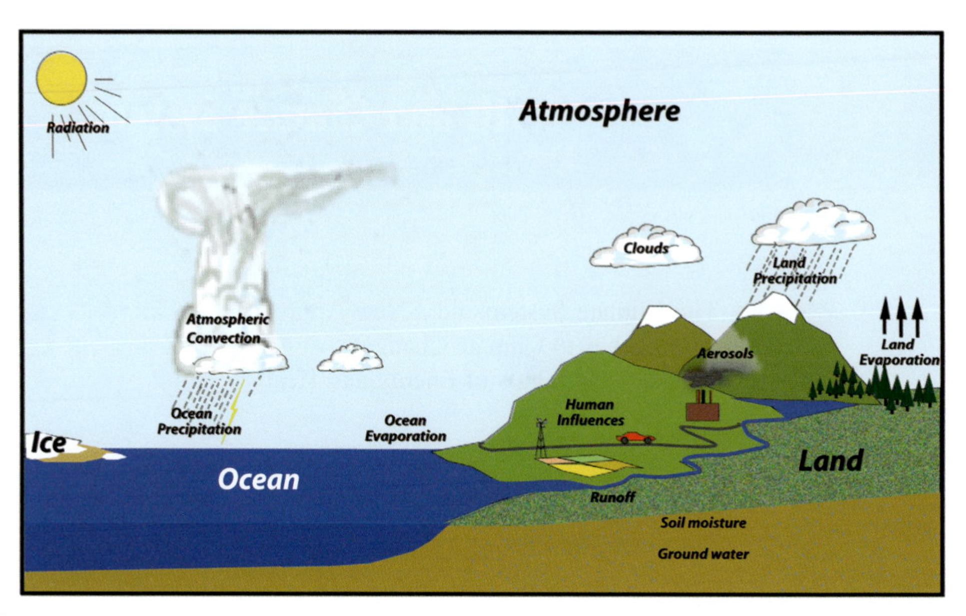

Fig. 1.1 The main components of the climate system are indicated: atmosphere, ocean, land, ice, and the complexity on land of vegetation, water and human influences. Adapted from Trenberth et al. (1996)

Imbalance (EEI).[2] This disequilibrium and its consequences in terms of what it is doing to the flows of energy through the climate system, which become manifested as climate changes of various sorts, are also the focus of this book.

The climate of Earth fundamentally depends on the Sun–Earth geometry and the flows of energy from the Sun to Earth and back out to space. To understand and predict the climate variations and changes, it is important to examine the processes involved and follow the absorption, reflection, storage, flows, exchanges, and release of energy.

Changes in climate, whether natural or anthropogenic, over a range of timescales can be identified and studied from different climate variables. Because humans live in and breathe the atmosphere, it is natural to focus on the atmospheric changes where phenomena and events are loosely divided into "*weather*" and "*climate*." The large fluctuations in the atmosphere from hour to hour, or day to day, constitute the weather; they occur as weather systems move, develop, evolve, mature and decay as forms of atmospheric turbulence. These weather systems arise because of temperature and heating contrasts which set up atmospheric instabilities. Their evolution is governed by nonlinear "*chaotic*" dynamics, so that they are not predictable in an individual deterministic sense for very long: perhaps hours for thunderstorms, or beyond a week or two into the future for extratropical cyclones. Unstable temperature gradients may guarantee that a disturbance develops, but the answer to the

[2] A list of acronyms is given at the back of the book.

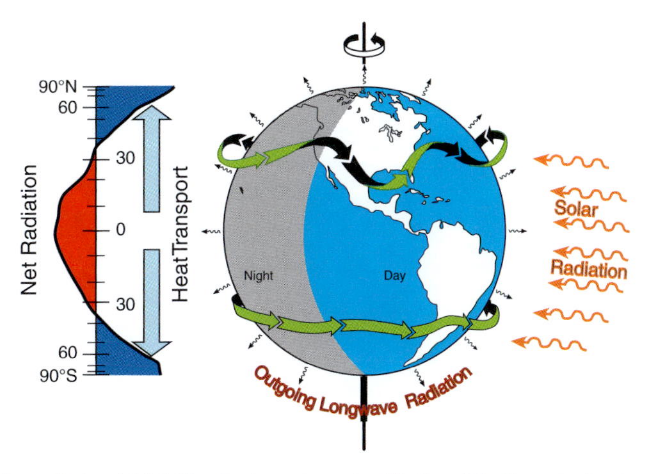

Fig. 1.2 The incoming solar radiation (right) illuminates only part of Earth, while the outgoing longwave radiation is distributed more evenly. On an annual mean basis, the result is an excess of absorbed solar radiation over the outgoing longwave radiation in the tropics, while there is a deficit at middle to high latitudes (left), so that there is a requirement for a poleward heat transport in each hemisphere (broad arrows) by the atmosphere and the oceans. The radiation distribution results in warm conditions in the tropics but cold at high latitudes, and the temperature contrast results in a broad band of westerlies in the extratropics of each hemisphere in which there is an embedded jet stream (shown by the green ribbon arrows) at about 10 km above Earth's surface. The flow of the jet stream over the different underlying surfaces (ocean, land, mountains) produces waves in the atmosphere and geographic spatial structure to climate. (The excess of net radiation at the equator is 68 W m^{-2} and the deficit peaks at −100 W m^{-2} at the South Pole and −125 W m^{-2} at the North Pole.). Adapted from Trenberth et al. (1996)

question of "which one and how quickly?", or maybe "instead of one big event, is it two smaller events?", can be sensitive to very minor details.

Weather and climate on Earth are determined by the amount and distribution of incoming radiation from the Sun. The distribution of solar radiation absorbed on Earth is very uneven and is largely determined by the Sun–Earth orbit and its variations (Fig. 1.2). Climate is usually defined to be average weather, described in terms of the mean and other statistical quantities that measure the variability over a period of time and possibly over a certain geographical region. Climate involves variations in which the atmosphere is influenced by and interacts with other parts of the climate system, and "*external*" forcings.

For an equilibrium climate, *outgoing longwave radiation* (OLR) necessarily balances the incoming *absorbed solar radiation* (ASR), although there is a great deal of fascinating atmosphere, ocean and land phenomena that couple the two. The balance is not instantaneous but takes place over weeks, months, or years. Although incoming radiant energy may be scattered and reflected by clouds and aerosols, and some is absorbed in the atmosphere, most is transmitted through clear

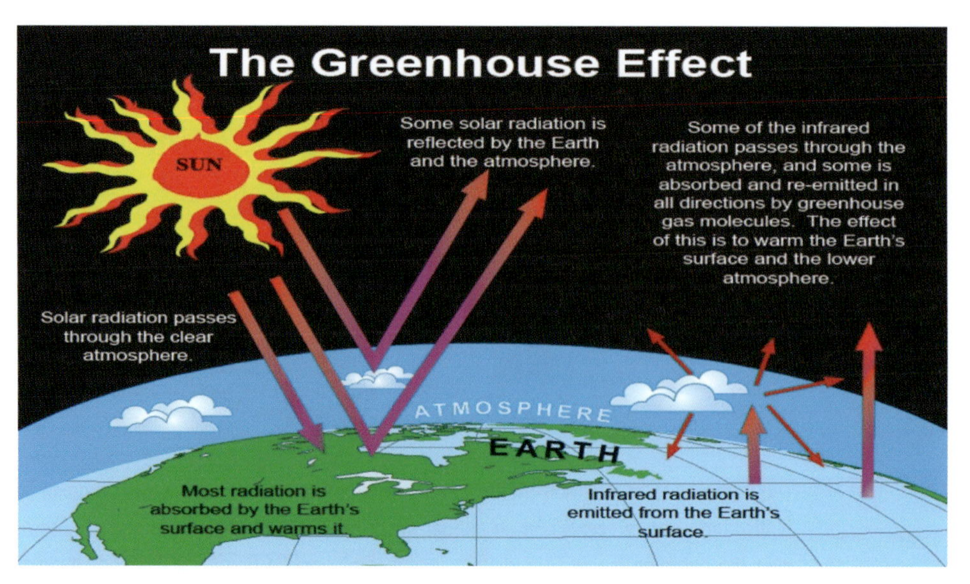

Fig. 1.3 Schematic of the flow of energy through the climate system illustrating the greenhouse effect, reflection and scattering of radiation.

skies in the atmosphere. The transmitted radiation is then either absorbed or reflected at Earth's surface. Radiant solar (shortwave) energy is transformed into sensible heat (related to temperature), latent energy (involving different water states and thus changes in phase), potential energy (involving gravity and altitude, or depth in the ocean), and kinetic energy (involving motion) before being emitted back to space as infrared (longwave) radiant energy. The bulk of the outgoing radiation from the surface is intercepted and re-emitted both up and down by clouds and certain *greenhouse gases* (GHGs) in the atmosphere. These then provide a coverlet for the planet and keep it warmer than it otherwise would be, and thus the blanket is called the *greenhouse effect* (Fig. 1.3). Because OLR has a strong dependency on the temperature of the emitting body, it increases if the planet warms until a new balance is achieved.

Energy may be stored for some time, transported in various forms, converted among the different types, and redistributed among the climate system components, giving rise to a rich variety of weather or turbulent phenomena in the atmosphere and ocean. Moreover, the energy balance can be upset in various ways, changing the weather and associated temperature and moisture fields. Repartitioning energy among oceans and atmosphere, say, can cause climate variability on timescales of years to centuries. Trends in certain influences on the climate system may lead to climate change.

The *climate system* has several internal interactive major components. The *atmosphere* does not have very much *heat capacity* but is very important as the most changeable and dynamic component of the climate system. Winds move heat, moisture, and energy around; they change continuously and wind speeds can exceed 50 m s^{-1} (112 mph) in the jet streams. The *oceans* have enormous heat capacity and, being fluid, can also move heat and energy around in important ways. Ocean currents are typically up to a few cm s^{-1} at the surface but may be >1 m s^{-1} in strong currents like the Gulf Stream. The land is where we live but, in addition to urban areas, it is covered by a rich tapestry of vegetation, fields, and mountains, as well as rivers, lakes and surface and subsurface water. These include a huge variety of plants, biomass and ecosystems with varied reflective character, called *albedo.* The *cryosphere* embraces sea ice, snow cover, land ice (including glaciers and the semi-permanent ice sheets of Antarctica and Greenland), and permafrost, and may be on land or ocean. These components determine the mean climate, which may vary from natural causes. Their role in energy storage and the energy balance of Earth is addressed below and throughout this book.

The climate system is subjected to influences from external *forcings*; effects from outside, including especially the Sun, but also internal upheavals from tectonics and volcanoes, and humans. Usually, the following factors are taken as given and imposed externally: the basic Sun–Earth geometry and the slowly changing orbit; the physical components of the Earth system, such as the distribution of land and ocean, the geographic features on the land and the ocean bottom topography and basin configurations; the mass and basic composition of the atmosphere and ocean; and major ice sheets. However, many of these aspects can change on very long timescales as volcanoes erupt and change the composition of the atmosphere, and continents drift. Further, human activities are changing the land surface through the destruction of forests and the building of roads, cities and farms, as well as changing the composition of the atmosphere, especially through fossil fuel burning. For the most part, these are also external influences on the climate system.

Changes in any of the climate system components, whether internal and thus a natural part of the system, or from the external forcings, cause the climate to vary or change. Thus, climate can vary because of alterations in the internal exchanges of energy or in the internal dynamics of the climate system. A change in the average net radiation at the top of the atmosphere due to perturbations in the incident solar radiation or the emergent infrared radiation leads to what is known as *radiative forcing* of the system. An example of natural variability is *El Niño–Southern Oscillation* (ENSO), whose events arise from natural coupled interactions between the atmosphere and the ocean centered in the tropical Pacific (Chapter 12). Climate may also vary from changes in the Sun, such as the sunspot cycle (Chapter 3). The greatest variations in the composition of the atmosphere involve water in various phases in the atmosphere, as water vapor, clouds containing liquid water and ice crystals, and rain, snow and hail. However, other constituents of the atmosphere and the oceans can also change, thereby bringing in considerations of atmospheric chemistry, marine biogeochemistry, and land surface exchanges. *Paleoclimate* deals with effects of the usually slowly changing land surface, oceans, or the Sun, and these are usually referred to as part of natural climate change. Such interactions and factors are also discussed below from the standpoint of energy.

However, climate is now changing quite rapidly compared with the historical record from human activities. The main way humans influence the climate is by interfering with the natural direct flows of energy through the climate system by changing the composition of the atmosphere. In particular, the build-up of various GHGs from human activities is dominated by increasing carbon dioxide from the burning of fossil fuels. Atmospheric pollution from aerosols also plays a role. With increasing greenhouse gases, the planet warms until the OLR increases to match the ASR. There are many feedbacks and complexities, but the case can be made that the most fundamental measure that the climate is changing is Earth's Energy Imbalance. A major advantage of EEI is that it is the net result of all the complicated processes and feedbacks.

The climate change because of human activities has been detailed through many global scientific consensus reports from the Intergovernmental Panel on Climate Change (IPCC) – see Chapter 16. The IPCC issued its first report in 1990, and assessments have been issued about every six years since then. These assessment reports survey the published scientific literature, and typically cover observational aspects of both changes in forcings and observed state variables, such as temperature, how well we understand the links between external changes and observed changes, called attribution, and projections of climate into the future. Necessarily, these involve comprehensive climate models or Earth system models, and an assessment is usually included of how well these perform, their limitations, and also their use. In addition to assessing the physical climate system, the biogeochemistry (that is, all of the biological, geological and chemical processes in the atmosphere and ocean and on land) has increasingly been included. A second working group under IPCC has assessed adaptation, including the climate impacts, vulnerabilities, resilience, and how best to deal with climate change; while a third working group has dealt with mitigation, which includes how to slow or stop the problem, mainly by assessing strategies and the economics of cutting emissions and decarbonizing the economy. These are not intended to be prescriptive or to make recommendations, but rather the reports provide advice and many examples of "what if" outcomes, that help decision-makers craft policy.

It is not the purpose of this book to reproduce the IPCC assessments, but rather to adopt a new approach focused on providing an overview of the flow of energy through the climate system, and the perspective of the EEI. Where the excess energy goes and its consequences provide a complementary way of viewing and understanding changes in climate and associated weather based upon physical principles.

1.2 Approaches to Climate Change

The standard approach to understanding and projecting climate change is to analyze all of the observations and determine what is going on, put all of the understanding into comprehensive climate system models, and use the models to make projections of the future. A key focus then is to encapsulate the knowledge into a computer-based numerical model. This then depends on the degree of understanding of the processes involved, and how well we can represent them by mathematical equations, code them into numerical form, and solve them. The mathematical equations represent the laws of physics and thermodynamics, and are well established; for instance, Newton's laws of motion (after Sir Isaac Newton). There is no analytic solution to the governing physics and biogeochemistry, some aspects of which we do not know, although some we do. The result is the need to break up and discretize the model into specified elements on a grid that will be explicitly resolved, simulated, and predicted and, in turn, that depends hugely on the size and speed of the available computer. The size of the grid then determines what phenomena can be explicitly included and hopefully simulated. Computers have become more and more sophisticated and capable, but still fall short of the real needs. Necessarily, a number of approximations have to be made to simplify the problem somewhat, and over time, as increased knowledge has accrued, some of these have been relaxed.

This is a major challenge. The best example of the success of this approach is numerical weather prediction, which uses a high-resolution computer-based model of the atmosphere to make forecasts. But first the models need a starting point: the state of the atmosphere today. These models have improved enormously, as has the observing system, including data and imagery from satellites that are combined through a four-dimensional (three space dimensions plus time) data assimilation of all observed data. Together they allow the state of the atmosphere at any time to be determined to a known level of accuracy through a combination of past predictions of the model, which essentially carry forward in time all of the past observed data, blended and adjusted with the new observations. This observed state is then used as the starting point for new predictions for up to about 2 weeks. The predictability of each analyzed situation can be determined by making imperceptible perturbations in the fields, and watching how well the resulting predictions based on slightly different starting points group or spread. Because the atmosphere is chaotic, these predictions become completely different by about two weeks out. Hence it is common to use an *ensemble* of predictions to explore these aspects.

For climate simulation, the major components of the climate system must be represented in sub-models (atmosphere, ocean, land surface, cryosphere, and biosphere), along with the processes that go on within and among them (Fig. 1.4). In the late 20th century, computers had advanced to enable atmospheric

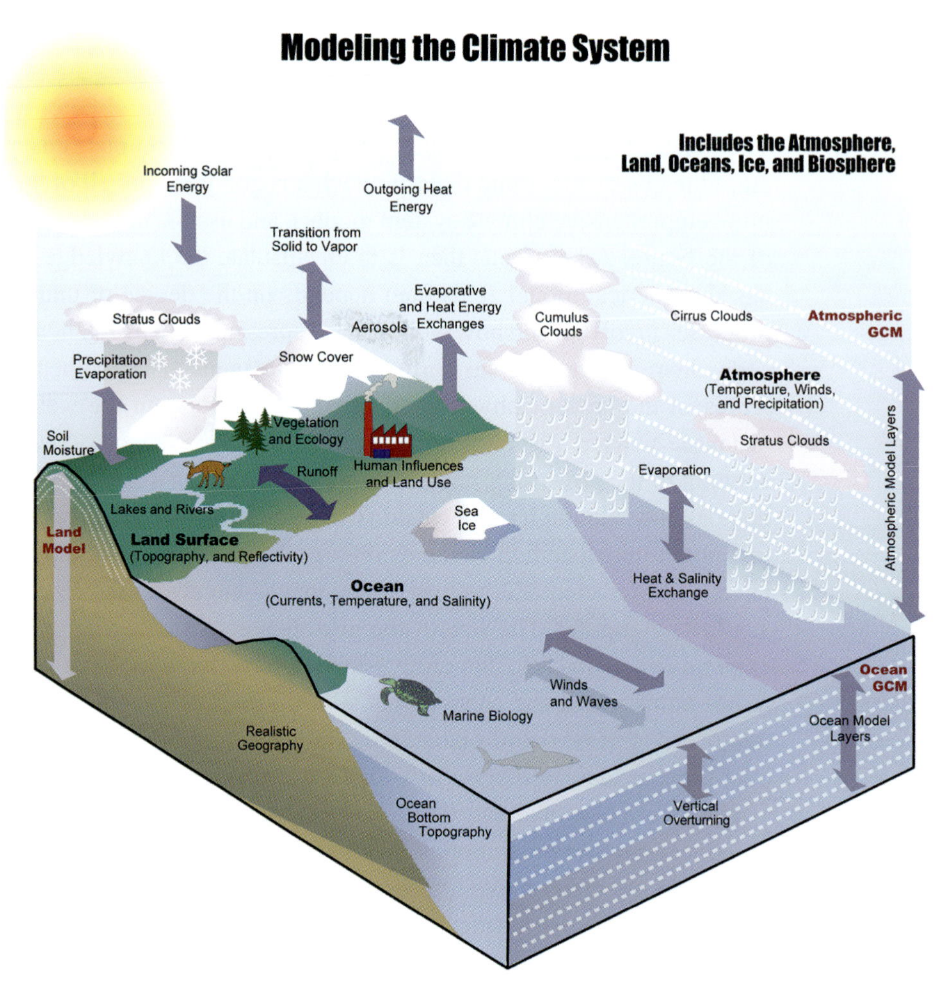

Fig. 1.4 Schematic of the main components and processes involved in the climate system that must be modeled. The discrete vertical levels for the atmospheric, oceanic and land models are indicated. From Karl and Trenberth (2003). Reprinted with permission from the American Association for the Advancement of Science

models with order 1 km resolution in the vertical and horizontal resolutions of about 250 km. Although some atmospheric chemistry and pollution had been introduced, biogeochemistry was mostly specified or diagnosed. Since then, the carbon cycle has been included along with approximations to biology and other aspects of chemistry, so that climate models have become Earth System Models, and their horizontal resolution has improved to perhaps 25 km in the atmosphere. Even so, there are many important processes that cannot be explicitly resolved and which therefore have to be included as a parametric representation (parameterization). These may be empirical or somewhat physical relationships of the small-scale processes with resolved phenomena.

At 25 km, crude but somewhat realistic hurricanes can be simulated although hurricane simulations are much better with less than 4 km resolution because spiral arms can then be resolved. Even then thunderstorms and clouds are not explicitly included, yet they play a vital role in climate and their effects have to be included through some form of parameterization. Boundary layer turbulence, convection, topographical complexity (gravity wave drag), radiative transfer, clouds, precipitation, and many land surface (lakes, rivers, roads, cities, wetlands, vegetation, soils, etc.) processes, including those involving the carbon cycle, chemistry and aerosols, all have to be parameterized. The microphysics of clouds have to be included in some form, and the complexity of clouds and their interactions with pollution (aerosols) makes this especially challenging.

The computing demands are enormous and require the biggest supercomputers available with thousands of processors. The deep ocean takes millennia to reach equilibrium, and century-long runs are routinely needed. Because the chaotic weather varies in each run and each model, large ensembles of runs have to be made to get reliable averages of the climate signal. High-resolution atmospheric cloud-resolving models can be run for short times or limited areas and alleviate many of the issues of the coarse-resolution models, but remain too computationally demanding for climate change simulations for the foreseeable future. At the same time, Earth system models generate huge amounts of output data on the climate system components, including the model weather. The volumes can easily exceed petabytes.

The aim is to model as much as possible of the Earth's climate system so that all components can interact, and properly consider the effect of feedbacks among components, then make predictions of future climate change. This firstly requires careful evaluations of model performance, biases and other shortcomings, using analyses of the observations. What is often done is that first a "control" climate simulation is run with the model. Then, the climate change experiment simulation is run; for example, with increased carbon dioxide in the model atmosphere. Finally, the difference is taken to provide an estimate of the change in climate due to the perturbation. The differencing technique removes most of the effects of any biases and artificial adjustments in the model, as well as systematic errors that are common to both runs. However, complicated feedbacks do take place, because comparison of different model results makes it apparent that the nature of some errors influences the outcome.

While there have been huge advances in understanding and modeling the climate system, and simulations by models are remarkably realistic, there are also some chronic problems and uncertainties that compromise their utility. The biggest climate variation routinely experienced is the annual cycle of the seasons. Most, but not quite all, of this is very well simulated by climate models and hence this provides a strong test of processes important for up to a year. Perhaps the main outstanding problem is the ability of models to simulate precipitation and clouds. Not only do the patterns of precipitation contain large errors, such as a double

Inter-Tropical Convergence Zone (ITCZ), but the intensity and frequency of precipitation are considerably in error, which means the lifetime of moisture in the model atmosphere is typically too short. Clouds are continually cited as the biggest uncertainty with regard to feedback processes and this relates also to their interactions with aerosols (see Chapter 13). The power consumption of high-intensity supercomputing is a real issue. For example, running an exascale (a billion billion – i.e., a quintillion – calculations per second), cloud-resolving climate model is estimated to consume 0.5 GW (billion watts), the equivalent of a small nuclear power plant. (See Sidebar 1.1.)

These sorts of problems, along with the enormous requirements for data processing, suggest that a hard look at other complementary approaches is desirable. Accordingly, in addition to the use of first principles in modeling, another approach is to make more complete use of observations as to what the outcome has been, and especially the consequences of the EEI. That is the emphasis of this book.

Sidebar 1.1: Units of Energy and Heat

In dealing with climate, there is a huge range of scales, from the microphysics associated with a tiny cloud droplet, to circulations in the atmosphere the size of Earth, to distances to the Sun, and beyond. In mathematics, to accommodate a shorthand for all of the zeros that come in, various prefixes and mathematical notation are used. For instance, 1000 meters has a kilo prefix and thus kilometer, and represented as 10^3. That is to say the superscript is the number of zeros after the 1. For a millimeter, 0.001 m, the prefix is milli, which is 10^{-3}. In this case the exponent is negative and counts the numbers to the right of the decimal point.

Here is the summary:

Prefix	Symbol	Base 10	Decimal	English
yotta	Y	10^{24}	1 000 000 000 000 000 000 000 000	septillion
zetta	Z	10^{21}	1 000 000 000 000 000 000 000	sextillion
exa	E	10^{18}	1 000 000 000 000 000 000	quintillion
peta	P	10^{15}	1 000 000 000 000 000	quadrillion
tera	T	10^{12}	1 000 000 000 000	trillion
giga	G	10^{9}	1 000 000 000	billion
mega	M	10^{6}	1 000 000	million
kilo	k	10^{3}	1 000	thousand
hecto	h	10^{2}	100	hundred
deca	da	10^{1}	10	ten
		10^{0}	1	one
deci	d	10^{-1}	0.1	tenth
centi	c	10^{-2}	0.01	hundredth
milli	m	10^{-3}	0.001	thousandth
micro	μ	10^{-6}	0.000 001	millionth
nano	n	10^{-9}	0.000 000 001	billionth
pico	p	10^{-12}	0.000 000 000 001	trillionth

Energy Units

The main source of energy on Earth is from the Sun. Consequently, a few facts about our Sun and the solar system are in order. To consider magnitudes of heat and energy, the units commonly used are briefly introduced. As will be discussed in detail later, heat is one form of energy. In the past, a measure of heat in common use was the calorie, which is still used to measure the energy content of food. In industry, the British thermal unit (BTU) has often been used. However, in science under the international system of units (SI) the unit of energy is the joule (J). The rate of energy transfer or use is a watt (W) which is a joule per second. Consequently, using 1 kilowatt (1000 W) for 1 hour is a kilowatt-hour (kWh) and a unit of energy. We may also find a watt-hour, and kilocalorie (kcal) in use.

- 1 BTU $= 1055.06$ J $= 2.931 \times 10^{-4}$ kWh $= 252$ cal $= 0.293$ watt-hours.

Sidebar 1.1: *(cont.)*

- An item using one kilowatt-hour of electricity generates *3412 BTU.*
- 1 kcal = 1000 cal = 4186.8 J = 1.163×10^{-3} kWh = 3.9683 BTU.

For common everyday use, a 100 W lightbulb is familiar, or a one-bar electric heater which uses 1 kW. The power generated by a power station is measured in kW, megawatts (10^6 W) or gigawatts (10^9 W). Power stations vary greatly in capacity depending on the type of power plant and other factors. The Sun's energy is often measured in W per unit area, such as watts per square meter (W m^{-2}), and so the total energy depends on the surface area of Earth (5.1×10^{14} m^2).

In 2018 the total US energy consumption was 101.27 quadrillion BTUs, according to the US Energy Information Administration. A quadrillion is 10^{15}, or in SI units "peta," as in petawatts (PW). Hence 101.27 QBTUs is 1.068×10^{20} joules over 1 year (3.156×10^7 seconds). As a zettajoule is 10^{21} J, this is also 0.1068 ZJ. Thus, the energy rate of US consumption is 3.385×10^{12} watts or 3.385 TW (terawatts), or 0.0066 W m^{-2} averaged over Earth.

As discussed below, this may be compared with the input from the Sun, averaged over all regions and day and night, of 340.3 W m^{-2}. Because some of that energy is reflected back to space, the net solar input is about 240 W m^{-2}, or over a factor of 36 000 times the total US energy consumption, which clearly demonstrates that it is not the direct heating or energy from human activities that threatens us with climate change, but rather it is the interference with the natural flows of energy by changing the greenhouse effect or the brightness of the planet that matters. These numbers are discussed further below.

References and Further Reading

Intergovernmental Panel on Climate Change, 2013: *Climate Change 2013. The Physical Science Basis*, ed. Stocker, T. F., et al. Cambridge: Cambridge University Press.

Karl, T. R., and K. E. Trenberth, 2003: Modern global climate change. *Science*, **302**(5651), 1719–1723. Doi: 10.1126/science.1090228.

Lorenz, E. N., 1967: *The Nature and Theory of the General Circulation of the Atmosphere*. Vol. **218**. Geneva: World Meteorological Organization, 161pp.

National Academy of Sciences and The Royal Society, 2020: *Climate Change: Evidence and Causes: Update 2020*. 36pp. Doi: 10.17226/25733.

Oreskes, N., and E. M. Conway, 2010: *Merchants of Doubt*. London: Bloomsbury Press, 355pp.

Trenberth, K. E., ed., 1992: *Climate System Modeling*. Cambridge: Cambridge University Press, 788pp.

Trenberth, K. E., 1997: The use and abuse of climate models in climate change research. *Nature*, **386**, 131–133.

Trenberth K. E., 2001: Stronger evidence of human influence on climate: the 2001 IPCC Assessment. *Environment*, **43**(4), 8–19.

Trenberth, K. E., 2018: Climate change caused by human activities is happening and it already has major consequences. *Journal of Energy and Water Resources Law*, **36**, 463–481. Doi: 10.1080/02646811.2018.1450895.

Trenberth, K. E., J. T. Houghton, and L. G. Meira Filho, 1996: The climate system: an overview. In: Houghton, J. T., L. G. Meira Filho, B. Callander, et al., eds., *Climate Change 1995. The Science of Climate Change*. Second Assessment Report of the Intergovernmental Panel on Climate Change. Cambridge: Cambridge University Press, 51–64.

2 Earth's Energy Imbalance and Climate Change

Global climate change is introduced. This also sets the stage for understanding the Earth's energy imbalance and direct and indirect human influences.

2.1 Climate Change

Climate change is a general term for long-term changes in climate of either sign. The term "global warming" was popularized by Wally Broecker in 1975 when he published a paper in *Science* magazine titled "Climatic change: are we on the brink of a pronounced global warming?"; although the term had been used as early as 1957 about Roger Revelle's research. It has most commonly been interpreted to be synonymous with rises in global mean surface temperature (GMST) and associated with increasing carbon dioxide in the atmosphere, as put forward by Broecker. However, the term is ambiguous because warming can also refer to "heating," which is more appropriate, and just one consequence of heating is an increase in temperature (another, for example, is "melting"). In fact, this would be a better way to use the term. Another ambiguity is whether it refers to all global temperature rise, for whatever reason, or whether it refers to only anthropogenic temperature increases. In any event, many people did not like it and climate change skeptics, in particular, preferred the term "climate change" to embrace both natural and anthropogenic sources, as well as the possibility that decreases could occur.

2.2 Carbon Dioxide

In 2020, there is no doubt whatsoever that global warming is occurring and it is caused by human activities, in particular those that change the composition of the atmosphere by adding more and more greenhouse gases, and especially carbon dioxide. Prior to 1958, CO_2 concentrations are determined from bubbles of air trapped in ice cores in the Greenland and Antarctic ice sheets. Pre-industrial values are estimated at 280 ppm. In 1958, Charles (Dave) Keeling began making measurements of CO_2 high on the north flank of Mauna Loa in Hawaii, well away from any sources, and the annual mean value was about 315 ppm. By 2020 the value had increased to 415 ppm and continues to trend relentlessly upwards.

The strong sawtooth character of the seasonal variations (Fig. 2.1) of 7 ppm or so relates to the onset of northern spring after the peak in May and increases in photosynthesis in plants that draw down the amounts in summer (minimum in September), while in fall and winter, leaves, twigs, senescent plants, and so forth, decay and put CO_2 back into the atmosphere to produce highest values after winter. Hence this annual cycle is a substantial large natural variation. In addition, it relates to the greater amount of land in the northern hemisphere.

Nevertheless, the annual cycle was soon dwarfed by the overall trend whose rate has systematically increased over time from about 1 ppm yr^{-1} to 2.8 ppm yr^{-1} after 2015. During the coronavirus pandemic, several reports have indicated lower emissions from human activities, but concentrations continue to rise apace, suggesting that effects of bushfires and wildfires may have compensated.

Figure 2.2 shows the longer perspective for CO_2 concentrations as well as estimates of the total emissions predominantly from burning fossil fuels. Note that

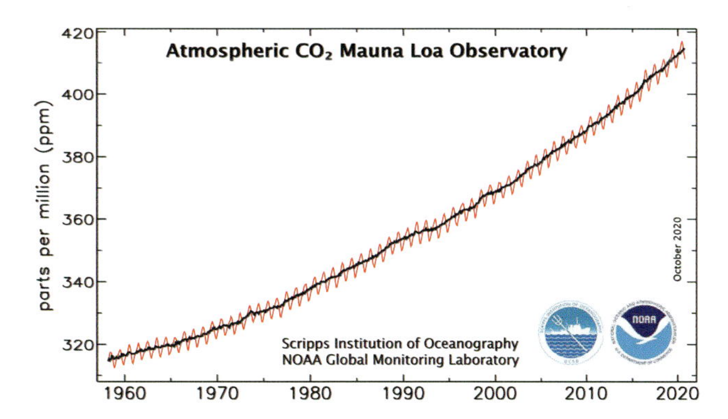

Fig. 2.1 The CO_2 record of combined Scripps and NOAA data in ppm. The red lines represent monthly mean values, and black lines are after removal of the average seasonal cycle based on a 7-year moving average. Courtesy NOAA, October 2020.

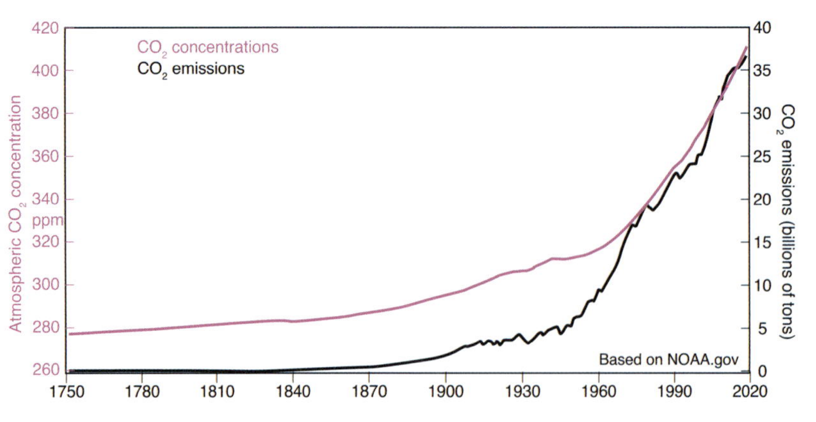

Fig. 2.2 Estimates of annual carbon dioxide concentrations in the atmosphere (violet line) along with human emissions (black line) since the approximate start of the industrial revolution in 1750. Emissions rose slowly to about 5 billion tons a year in the mid-20th century before skyrocketing to more than 35 billion tons per year by the end of the century (a ton is 0.91 tonnes). Adapted from NOAA Climate.gov; CO_2 emissions data https://ourworldindata.org/co2-and-other-greenhouse-gas-emissions#how-have-global-co2-emissions-changed-over-time

the concentrations are effectively the accumulated emissions, except for any losses. Accordingly, the increases in CO_2 in the atmosphere come from human activities. The largest contributors to emissions have been the Western countries, led by the United States and Europe until 2003, when Chinese emissions passed those of the USA whose emissions have declined modestly since then. However, because of the long lifetime, for reasons given below, a more relevant number is the accumulated emissions, which is still led by the United States. In 2019 the USA accounted for about 25% of cumulative emissions, the European Union 22%, China 13%, followed by Russia 7%, and Japan 4%. Of large countries, the USA and Canada lead the emissions per capita, with about 16 tonnes per capita vs. 12 for Russia, and 7 for China and the European Union (see Chapter 17 for more details).

The additional CO_2 that remains in the atmosphere is only about half of the total from the emissions. The rest is split approximately equally into uptake within the oceans and on land. Figure 2.3 provides the 2019 carbon budget assessment from the Global Carbon Project. Most of the emissions come from burning fossil fuels, complemented by changes in land use.

A primary removal process for carbon dioxide is through uptake and absorption by the oceans on about a 5- to 200-year time frame, leading to acidification of the ocean, which places many small organisms at risk owing to effects on bone and shell structures. Carbon may be used in the formation of eggs, corals, shells and bones of fish, and when they die this calcium carbonate may fall to the ocean floor

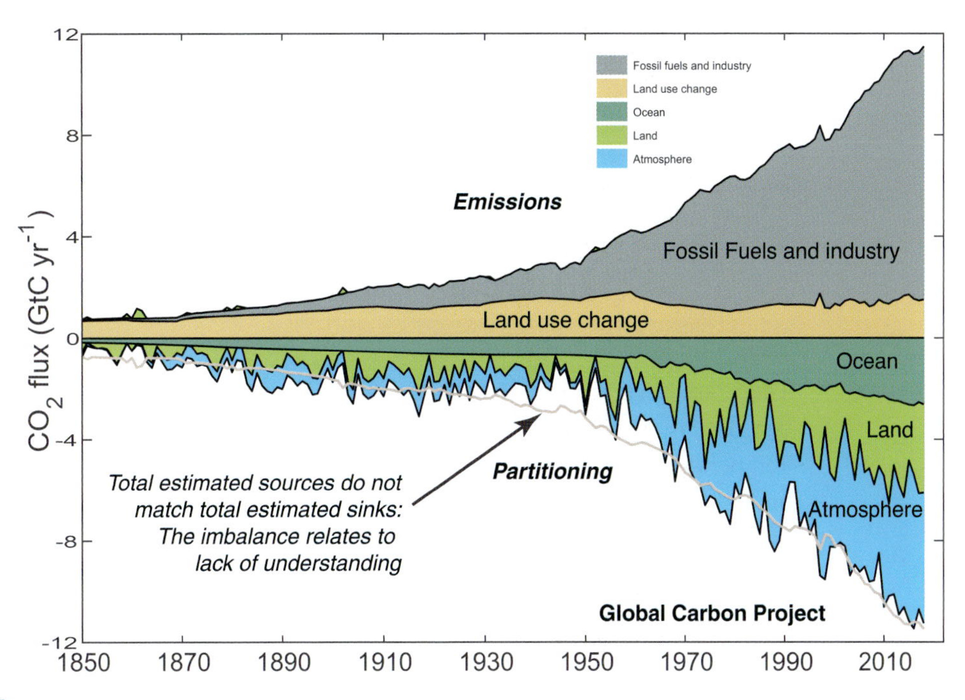

Fig. 2.3 Sources and sinks of carbon dioxide over time, as estimated from the Global Carbon Project (adapted from Friedlingstein et al., 2019). Units are gigatonnes of carbon per year.

and be essentially removed from the climate system. Limestone, chalk, and marble may result eventually.

The land sink (destination of carbon dioxide) is thought to be quite variable in both space and time owing to the large regional variations in drought, much of which is associated with El Niño. Carbon dioxide can be seen to help fertilize some vegetation, but drought and associated wildfire add carbon dioxide to the atmosphere through reduced productivity and *respiration*. Plant respiration involves taking in oxygen and releasing it as carbon dioxide into the atmosphere. In many ways, respiration is the opposite of *photosynthesis*. Plants take carbon from the atmosphere as they grow, but it goes straight back when they die and perhaps when harvested for short-term use (such as for paper), if not locked up in a semi-permanent structure. Slower removal processes also occur on land with the chemical weathering of rock on timescales of millennia. Because carbon dioxide is recycled through the atmosphere many times before it is finally removed, it generally has a long lifetime exceeding 100 years, and therefore emissions lead to a buildup in concentrations in the atmosphere.

Hence, there is an important difference between carbon uptake and actual carbon sequestration. However, as Earth warms, higher temperatures reduce the amount of

CO_2 absorbed by surface ocean waters and the amount of carbon sequestered in soils. They can also accelerate tree death. Wildfires are apt to increase along with drought and heatwaves. The uncertainties are substantial; lightly marked on Fig. 2.3 is the small residual, not accounted for by summing up the estimated sinks. However, most of the carbon sources and sinks are at least qualitatively understood.

2.3 Global Mean Surface Temperature

There are several estimates of the GMST, and here we use the NOAA record because it has been carefully adjusted and corrected for inhomogeneities. Temperature measurements can be quite quirky depending on the site, whether on a slope, shaded by a tree, on a hill or in a valley, and the nature of vegetation nearby. Care is taken not to contaminate readings by human-made structures including buildings and roads, etc. But most of these factors are systematic and hence anomalies of temperatures are used: values that are departures from a standard, usually 30-year, base period that also accounts for the mean annual cycle. Experience has shown that these anomalies are far more coherent spatially than absolute temperature values and therefore more reliable and easily analyzed.

Over the oceans, where sampling comes from transient ships and has often not been sufficient, it was realized that sea surface temperatures (SSTs) were usually more representative of the monthly mean than the few capricious instantaneous atmospheric values that measured the weather. Indeed, a reliable monthly mean anomaly can be computed from just three (or more) spaced out SST values. Then some allowance is needed to account for SST versus atmospheric values.

BOX 2.1: Temperature Scales.

At sea level, water freezes at 0°C and boils at 100°C where C refers to Celsius (used to be called centigrade). Absolute temperature is in kelvin.

$T°F = 32 + 1.8 \times T(°C)$	Fahrenheit
$T°C = [T(°F) - 32]/1.8$	Celsius
$T\,K = T(°C) + 273.15$	Kelvin

The development of thermometers and temperature scales dates from the early 18th century, when Daniel Fahrenheit (1686–1736) produced a mercury thermometer and scale. However, it took some time before instruments were standardized and reliable. A central England temperature is the longest monthly instrumental

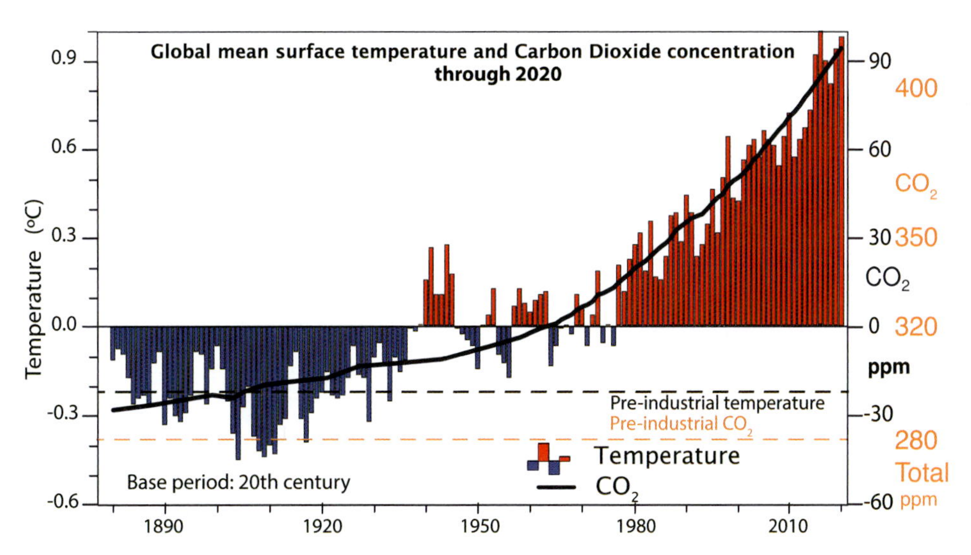

Fig. 2.4 Estimated changes in annual global mean surface temperatures (°C, color bars) and CO_2 concentrations (thick black line) over the past 150 years relative to 20th-century average values. Carbon dioxide concentrations since 1957 are from direct measurements at Mauna Loa, Hawaii, while earlier estimates are derived from ice core records. The scale for CO_2 concentrations is in parts per million by volume (ppm), relative to the 20th-century mean of 333.7 ppm, while the temperature anomalies are relative to a mean of 14°C. Also given as dashed values are the pre-industrial estimated values, where the value is 280 ppm, with the scale in orange at right for carbon dioxide. Updated from Trenberth (1997)

record and dates back to 1659; daily values begin in 1772. The quasi-global temperature record goes back to 1850, although coverage is spartan in the early decades, and usually values are used back to 1880. Prior to then, climate sensitive proxy indicators, such as tree rings, ice cores, coral cores, and sediment cores have been used to create the paleoclimate record, through carbon dating and other dating methods. It was not until the International Geophysical Year 1957–58 that measurements were systematically begun in Antarctica and thence over the southern oceans. Hence there is some spurious variability in the earlier years, before the 1950s, associated with calibration and inadequate sampling.

In Fig. 2.4 the annual values of global mean temperature anomalies are plotted, with the zero corresponding to the 20th-century mean. The exceptionally warm years in the early 1940s were influenced by a large prolonged El Niño event and hence most of that is real, but values were also influenced by changing ways of taking measurements and changing locations because of World War II. An estimate of the pre-industrial value is also given. The warmest year on record is 2016, which was a large El Niño year, and indeed all of the highest values tend to occur in the latter stages of large El Niños. The year 1998, an El Niño year, was the warmest prior to 2000.

There was a plateau of little or no increase in GMST from 1999 to 2013, called the "hiatus" or pause (see Chapter 15), but then 2014 came in as the warmest year on record, and was quickly surpassed by 2015, then 2016. The year 2020 is now the second warmest, which is notable as it was not an El Niño year. The seven warmest years on record are the last seven years. Global warming is clearly evident, although it only becomes clear after about 1976, when values exceed previous variability. Moreover, the oceans kept warming during the hiatus, and hence the extra heat was sequestered there.

Also shown in Fig. 2.4 is the annualized carbon dioxide record, with the zero corresponding to the mean for the 20th century. The scale on the right has been chosen to put the two curves together in a way that suggests they are related, because scientists can prove that they are by using climate models. Again, the pre-industrial value of 280 ppm is indicated and, as the most recent annual value is about 415 ppm, there is a 48% increase, over half of which has occurred since 1985.

As noted, the GMST in Fig. 2.4 is mostly a blended combination of surface air temperatures over land and SST information over the oceans. It has been estimated that the true surface air temperature has risen perhaps 5% more, by about 0.06°C, than the blended product because SSTs are not increasing as fast as air temperatures. This difference can be important in evaluating climate model results because it is vital to compare like with like.

There is no doubt whatsoever that the climate is warming; the warming is associated with increases in carbon dioxide and other greenhouse gases in the atmosphere and is caused by human activities. More insights are developed over subsequent chapters.

2.4 Earth's Energy Imbalance

It is well understood that carbon dioxide is a greenhouse gas (see Chapter 3) and traps outgoing radiation. Accordingly, there is a radiative imbalance for the planet. There are many complexities and feedbacks, both positive and negative (Chapter 13), but it is argued here that the most fundamental measure that the climate is changing is Earth's Energy Imbalance (EEI). The EEI is the net effect after all the complicated feedbacks (from clouds, aerosols, and water vapor, etc.) have operated.

Combustion of fossil fuels not only generates heat but also generates particulate pollution (e.g., soot, smoke), as well as gaseous pollution that can become particulates (e.g., sulfur dioxide, nitrogen dioxide; which get oxidized to form tiny sulfate and nitrate particles). Other gases are also formed in burning, such as carbon dioxide, and so the composition of the atmosphere is changing. Several other gases, notably methane, nitrous oxide, the chlorofluorocarbons (CFCs) and tropospheric ozone, are also observed to have increased from human activities (especially from biomass burning, landfills, rice paddies, agriculture, animal husbandry, fossil fuel use, leaky fuel lines, and industry), and these are all greenhouse gases. However, the observed decreases in lower stratospheric ozone since the 1970s, caused principally by human-introduced CFCs and halons, contribute to a small cooling in that region. All of these changes in atmospheric composition interfere with the natural flow of energy through the climate system and create an energy imbalance.

Nevertheless, EEI has been difficult to measure (Chapter 14), which is why it has not been widely used until now. In Chapter 3, we introduce the global mean energy flows through the climate system and suggest that the EEI is about 0.9 W m^{-2}, as an annual global mean average over the entire Earth in the past 15 years. The details on how EEI is estimated and uncertainties are given in Chapter 14. EEI necessarily includes all of the important seasonal variations and also the huge differences between equatorial and polar regions. For comparison, a small incandescent Christmas tree light is about 0.4 W, and therefore the EEI is about 2.5 such lights in every square meter over the entire planet. This is also in the context of the average flow of energy through the climate system of about 240 W m^{-2} (Fig. 3.1).

The EEI varies a bit over time, but 0.9 W m^{-2} is about 500 terawatts (500 000 000 000 000 = 500×10^{12} W), when totaled over the area of Earth. To get a sense of this, the largest power plants that exist are of order 1000 megawatts, and these service human needs for electricity in appliances and heating that use power in units of kilowatts. The 500 TW would be 500 000 of these huge power plants. It can also be compared with the 2018 US electricity consumption of about 0.5 TW (versus total energy consumption of 3.39 TW, see Sidebar 1.1), or the global electricity generation of about 2.6 TW (e.g., see https://www.eia.gov/energyexplained/index.php?page=electricity_in_the_united_states#tab2). Consequently, the EEI is order 200 times all of the electricity used in recent years.

There have been major changes in land use over the past two centuries. Conversion of forest to cropland, in particular, has led to a higher *albedo* and changes in evapotranspiration in places such as the eastern and central United States, both of which have probably cooled the region, in summer by perhaps 1°C, and in autumn by more than 2°C, although global effects are less clear. In cities, the building of "concrete jungles" allows heat to be soaked up and stored during the day and released at night, moderating nighttime temperatures and contributing to an *urban heat island*. Space heating also contributes to this effect. Urbanization changes also affect runoff of water, leading to dryer conditions unless compensated for by water usage and irrigation, which can cool the area. However, these influences, while real changes in climate, are quite local. Widespread irrigation on farms can have more regional effects.

Firstly, these numbers show that the direct effects of energy from humans are small compared with the Sun, and even with the EEI. Direct human influences are not the way global climate is changed, although such effects are much more important locally in big cities where an urban heat island is very evident, arising not just from space heating and energy usage, but also from the concrete jungle that greatly affects water availability. Direct global human influences are small. Instead, the way humans influence the climate is by interfering with the natural flows of energy through the climate system.

Secondly, the 0.9 W m^{-2} is very small compared with the 240 W m^{-2} energy flow. It is not possible to go outside and become aware of the global warming (heating) effect! So, this is not how global climate change is experienced. Instead, the 0.9 W m^{-2} has to accumulate. And it does so under a number of circumstances, because it is always in the same direction.

One obvious place that it accumulates is in the melting of ice in polar regions. At its minimum in late summer, the loss of Arctic sea ice is over 40% since 1979 (13% per decade). There is widespread evidence of retreating glaciers on land around the world. Further, the major ice sheets of Greenland and Antarctica are melting. All of these will be explored in much more detail in Chapter 14.

A second place the heat accumulates is in the oceans. The ocean heat content (OHC) is relentlessly increasing. It is only relatively recently that we have been able to quantify this reliably, with the advent of a new ocean observing system using autonomous profiling floats called Argo (Chapter 8). The increased knowledge and understanding have further enabled reconstructions of OHC back in time to 1958 (Chapter 14). Data are too few before then.

A consequence of both of these is that sea level is rising. Increased OHC causes expansion of the ocean, while melting of land ice puts more water into the ocean. Very reliable global estimates of sea level rise have been possible since 1992, when altimeters were first deployed on spacecraft. Global sea level is rising at over 3 mm yr^{-1} and has gone up by over 80 mm since these observations began. Sea level rise estimates are possible much further back in time from coastal and island tide gauges (see Chapter 14 for more details).

What about land? Land is a lot more complex because of the heterogeneity of the land surface (see Chapter 6, Section 6.3). However, the main issue is precipitation and surface water. As long as water is present, much of the extra heat goes into evaporating moisture, which is readily replenished by more rain (Chapter 10). The main times when EEI effects accumulate on land are during dry spells, and especially droughts. Firstly, there is drying, and then there is warming of the surface. The warmer air in turn demands more moisture from the vegetation and surface. Plants have rooting systems of various depths and can access soil moisture for quite some time (weeks or even months). Therefore, not only surface water, but also subsurface water is dried out, and then plants begin to wilt. As evaporative cooling of the land diminishes with water supply, more heat goes into raising temperatures. Temperatures creep up and the regions become vulnerable to prolonged heatwaves. Moreover, the warmer atmosphere increases drying by increasing its demand for moisture and increases wildfire risk (Section 10.6).

The energy amount of 0.9 W m^{-2} accumulated over 1 month is equivalent to 700 W for nearly 6 minutes for every square foot (there are 10 square feet in a square meter). This is about the size and power of a small microwave oven, and hence 1 month's accumulation of heat is equivalent to full power on a small microwave oven, every square foot (where the dry conditions occur), for almost 6 minutes. No wonder things catch on fire!

These concepts apply particularly to locations with a Mediterranean climate, in which there is a short winter wet season, followed by a long dry sunny summer season. As well as the Mediterranean, California and many subtropical regions, such as Australia, stand out. The combination greatly increases the risk of wildfire, or bushfire in Australia. Indeed, recent wildfires in Spain and Portugal, California and the whole western United States and Canada, and Australia have become more severe and widespread, with major consequences for society and wildlife. Hence, increased drought intensity, heatwaves and wildfires are widely observed.

The atmosphere is the other major component of the climate system. It does not have much heat capacity, but it nonetheless warms up as the ocean and land warm. But, more importantly, the extra heat over both land and ocean, and the extra drying, put more water vapor into the atmosphere, which can hold more moisture when it is hotter (Chapter 5). The result is more vigorous weather systems and heavier rains, where it does rain. Or heavier snows occur if conditions are right (Chapter 10).

All of these aspects are explored and documented in more detail in later chapters, with Chapter 14 devoted to EEI estimates.

References and Further Reading

Blunden, J. and D. S. Arndt, eds., 2020: State of the climate in 2019. *Bulletin of the American Meteorological Society*, **101**, Si–S429. Doi: 10.1175/2020BAMSStateofthe Climate.1.

Broecker, W., 1975: Climatic change: are we on the brink of a pronounced global warming? *Science*, **189**, 460–463.

Friedlingstein, P, M. W. Jones, M. O'Sullivan, et al., 2019: Global carbon budget 2019, *Earth System Science Data,* **11**, 1783–1838. doi: 10.5194/essd-11-1783-2019.

Hausfather, Z, and R. Betts, 2020: Importance of carbon-cycle feedback uncertainties. www.carbonbrief.org/analysis-how-carbon-cycle-feedbacks-could-make-global-warming-worse

Trenberth, K. E., 1997: The use and abuse of climate models in climate change research. *Nature*, **386**, 131–133.

3 Earth's Energy Balance

The focus is on the average flow of energy through the climate system and the greenhouse effect.

3.1 Global Energy Flows

The incoming energy to the Earth system is in the form of solar radiation and roughly corresponds to that of a *black body* at the temperature of the Sun of about 6000 K. The Sun's emissions peak at a wavelength of about 0.6 μm and much of this energy is in the visible part of the electromagnetic spectrum although some extends beyond the red into the infrared and some extends beyond the violet into the ultraviolet (Fig. 4.5). As noted earlier, because of the very nearly spherical shape of Earth, at any time half of Earth is in night (Fig. 1.2) and the average amount of energy incident on a level surface outside the atmosphere is one-quarter of the total solar irradiance, or 340 W m^{-2}. About 30% of this energy is scattered or reflected back to space by air molecules, tiny airborne particles (known as *aerosols*), clouds in the atmosphere, or by Earth's surface, which leaves about 240 W m^{-2} on average to warm Earth's surface and atmosphere (Fig. 3.1).

To balance the incoming energy, Earth must radiate on average the same amount of energy back to space (Fig. 3.1). The amount of thermal radiation emitted by a warm surface depends on its temperature and on how absorbing it is. For a completely absorbing (black) surface to emit 239 W m^{-2} of thermal radiation, it would have a temperature of about -19°C (254 K).

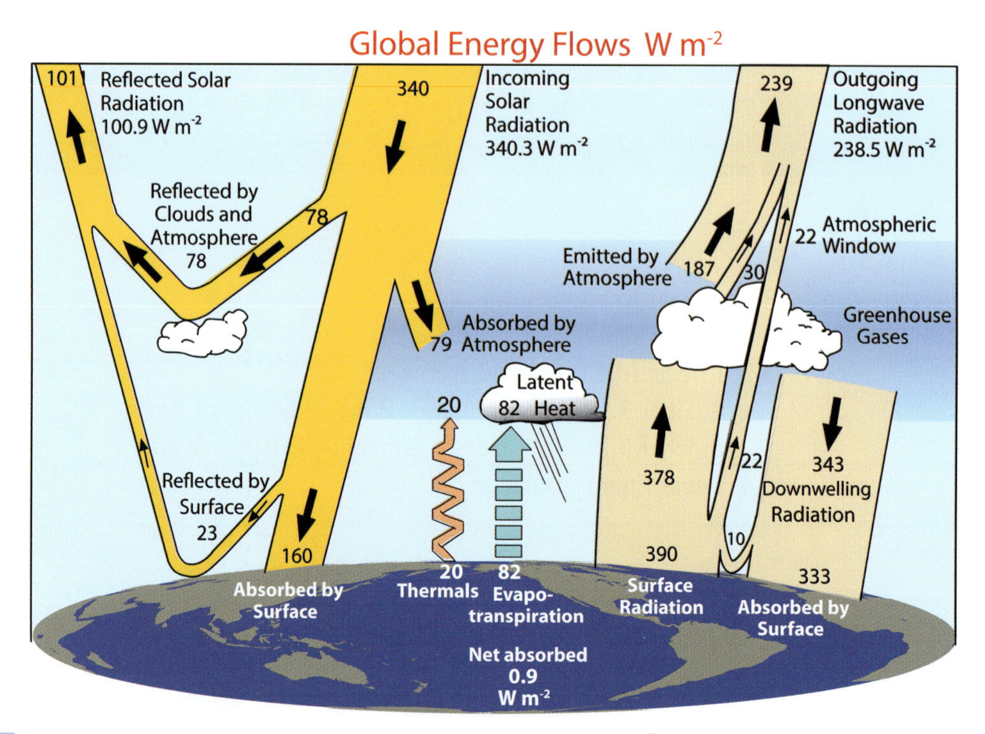

Fig. 3.1 Earth's radiation balance. The net incoming solar radiation of 340 W m^{-2} is partially reflected by clouds and the atmosphere, or at the surface, but 47% is absorbed by the surface. Some of that heat is returned to the atmosphere as sensible heating (thermals) and most as evapotranspiration that is realized as latent heat in precipitation. The rest is radiated as thermal infrared radiation. Most of that is absorbed by the atmosphere and re-emitted both up and down. A lot of the downwelling radiation comes from the lower atmosphere where most water vapor exists, and some (10 units) is reflected. The radiation lost to space comes from cloud tops and from parts of the atmosphere much colder than the surface, producing a greenhouse effect. The net absorbed at the surface is the Earth's Energy Imbalance. Adapted from Trenberth et al. (2009), Trenberth and Fasullo (2011), and Wild et al. (2015)

Planck's law describes the electromagnetic radiation emitted for such a body as a function of temperature (see also Sidebar 4.1). It therefore follows that the emitted thermal radiation occurs at about 10 μm which is in the infrared part of the electromagnetic radiation spectrum. Near 4 μm, radiation from both the Sun and Earth is very small, and hence there is a separation of wavelengths which has led to the solar radiation being referred to as *shortwave* radiation, while the outgoing terrestrial radiation is referred to as *longwave* radiation. Note that $-19°$C is much colder than the conditions that actually exist near Earth's surface, where the annual average global mean temperature is about 15°C. However, because the temperature in the lower atmosphere (*troposphere*) falls off quite rapidly with height (Section 6.1; Fig. 6.1), a temperature of $-19°$C is reached typically at an altitude of 5 km above the surface in mid-latitudes. This provides a clue about the role of the atmosphere in making the surface climate hospitable.

Sidebar 3.1 introduces a very simple zero-dimensional energy balance model of Earth. It simply requires the outgoing radiation, which is controlled by temperature, to match the incoming radiation from the Sun. The latter is affected by clouds and the surface which reflect (scatter) about 30% of the radiation back to space.

3.2 The Greenhouse Effect

Some of the infrared radiation leaving the atmosphere originates near Earth's surface and is transmitted relatively unimpeded through the atmosphere; this is the radiation from areas where there is no cloud and which is present in the part of the spectrum known as the atmospheric "window" (Fig. 3.1). The bulk of the radiation, however, is intercepted and re-emitted both up and down. The emissions to space occur either from the tops of clouds at different atmospheric levels (which are almost always colder than the surface), or by gases present in the atmosphere which absorb and emit infrared radiation. Most of the atmosphere consists of nitrogen and oxygen (99% of dry air), which are transparent to infrared radiation. It is the water vapor, which varies in amount from 0 to about 3%, carbon dioxide and some other minor gases present in the atmosphere in much smaller quantities that absorb some of the thermal radiation leaving the surface and re-emit radiation from much higher and colder levels out to space. These radiatively active gases are known as greenhouse gases (GHGs) because they act as a partial blanket for the thermal radiation from the surface and enable it to be substantially warmer than it would otherwise be, analogous to the effects of a greenhouse. Note that while a real greenhouse works somewhat in this way, the main heat retention in a greenhouse actually comes through protection from the wind. In the current climate, for clear sky conditions, water vapor is estimated to account for about 67% of the greenhouse effect, carbon dioxide 24%, ozone 6%, and other gases 3%; while with clouds included, water vapor is half, clouds a quarter, and carbon dioxide about one-fifth (Fig. 3.2).

Carbon dioxide, ozone, methane, nitrous oxide, and chlorofluorocarbons do not condense and precipitate as water vapor does. The result is orders of magnitude differences in the lifetime of these gases (decades to centuries) compared with about 9 days for water vapor. Hence, they serve as a stable structure of the atmospheric heating as the climate changes, and the resulting temperature is what enables the levels of water vapor (see Chapters 5 and 10), as a powerful feedback (Chapter 13). In this regard, carbon dioxide is the single most important greenhouse gas.

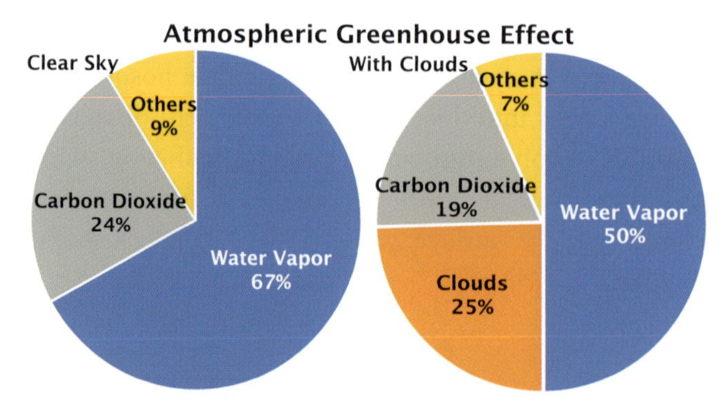

Fig. 3.2 The main components of the atmospheric greenhouse effect for clear sky (left) and all sky conditions, including clouds (right). Adapted from Kiehl and Trenberth (1997) and Schmidt et al. (2010)

3.3 Effects of Clouds and Aerosols

Clouds also absorb and emit thermal radiation and have a blanketing effect similar to that of greenhouse gases (Fig. 3.2). At nighttime the temperature does not drop nearly as fast or as much when clouds are present. But clouds are also bright reflectors of solar radiation and thus also act to cool the surface. While on average there is strong cancellation between the two opposing effects of shortwave cloud cooling and longwave cloud heating, the net global effect of clouds in our current climate, as determined by space-based measurements, is a small cooling of the surface. A key issue is how and where clouds will change as climate changes. If cloud tops get higher, the radiation to space from clouds is less since it is emitted from a colder temperature, and so this produces a relative warming. However, more extensive low clouds would be likely to produce relative cooling because of the greater influence on solar radiation being reflected. This issue is complicated by the fact that clouds are also strongly influenced by particulate pollution. Additionally, diurnal and seasonal aspects of cloud cover are also important. For example, if more clouds occur during the winter or at night, when there is no incoming solar radiation, they only serve to emit less radiation to space while not reflecting any incoming solar radiation.

Aerosols occur in the atmosphere from natural causes; for instance, they are blown off the surface of deserts or dry regions. The eruption of Mt. Pinatubo in the Philippines in June 1991 added considerable amounts of aerosol to the stratosphere which, for about 2 years, scattered incoming solar radiation leading to a loss of radiation and a cooling at the surface. Human activities also affect the amount of aerosol in the atmosphere. Aerosols are described in more detail in Section 6.1.3.

Sidebar 3.1: Simple Energy Balance Model of Earth

The rates of change of the global mean temperature must be weighted by the heat capacity C, related to the specific heat. The forcing consists of the incoming solar radiation Q, but with a certain fraction, α the albedo, reflected, leaving $(1-\alpha)Q$ absorbed. The outgoing longwave radiation is given by the Stefan–Boltzmann law, σT_e^4 as a response to the heating, where T_e is the emission temperature. However, if we are interested in the surface temperature T_s, rather than the temperature at the effective level from which the radiation to space occurs, T_e, then we have to also empirically relate the two temperatures: $T_e = \beta T_s$ and the β is an empirical value. Then the energy balance of this planet is governed by:

$$CdT_s/dt = (1-\alpha)Q - \sigma(\beta T_s)^4 \tag{3.1}$$

This is a first-order ordinary differential equation for T_s as a function of time t and it constitutes the simplest climate model of Earth: a global energy balance model. Under equilibrium conditions, the left-hand side is zero, and the result is

$$T_s = 1/\beta[(1-\alpha)Q/\sigma]^{1/4} \tag{3.2}$$

Plugging in the value of Q of 1361/4 W m^{-2}, and $\alpha = 0.3$, as observed, gives 254.58 K for the $[(1-\alpha)Q/\sigma]^{1/4}$ and, given that the observed value of T_s is 288 K, then β is 0.884. This tells us that seen from space, Earth appears to have an equilibrium emission temperature of 255 K, which it does nearly halfway up the atmosphere at an altitude of order 5 km (Fig. 6.1). The average height of the 500 hPa surface is about 5500 m, and this is about halfway through the atmosphere in terms of mass since the surface pressure is close to 1000 hPa.

If β is specified, then Eq. (3.1) can be solved as a time-varying temperature, for instance in response to changes in Q or α. Or by adding a bit more complexity, the equation may need to be solved numerically using a computer.

Sidebar 3.2: How Does a Greenhouse Effect Work?

Imagine Earth is just a plate in space with sunlight shining on it at a rate of 240 W m^{-2} (to simplify things we use round numbers and have accounted for the reflection). See panel 1 of Fig. 3.3. The Sun warms the plate, but because Earth is rotating (and is round) both sides get heated. To allow for that, we double the incoming radiation to 480 W m^{-2}. So the Sun warms the plate and the plate radiates until the radiated heat matches the heat being absorbed from the Sun (panel 2). The Stefan–Boltzmann law tells us that the total radiant heat power emitted from a surface is proportional to the fourth power of its absolute temperature. Hence, using the Stefan–Boltzmann law, we calculate the temperature of the plate when it reaches equilibrium:

$$480\ \mathrm{W\,m^{-2}} = 2\sigma T_{eq}^4$$

where σ is the Stefan–Boltzmann constant 5.67×10^{-8} W/(m^{2}K^4); the factor of 2 arises for a two-sided plate. Therefore $T_{eq} = 255.1$ K.

Now add another plate that we color green for greenhouse. It is heated by the first at a rate of 240 W m^{-2} (panel 3). Think of this as an analogy for adding GHGs to the atmosphere. It too has to heat up and reach an equilibrium temperature. As a first guess we might suppose something like panel 4. However, that is wrong because there are 240 W m^{-2} going into the two-plate system and only 120 coming out. At equilibrium an equal amount of energy has to be going in as coming out. Hence what happens is that the entire system has to heat up until it radiates more to space to reach the equilibrium condition. T_1 and T_2 are the equilibrium temperatures of the plates (panel 5).

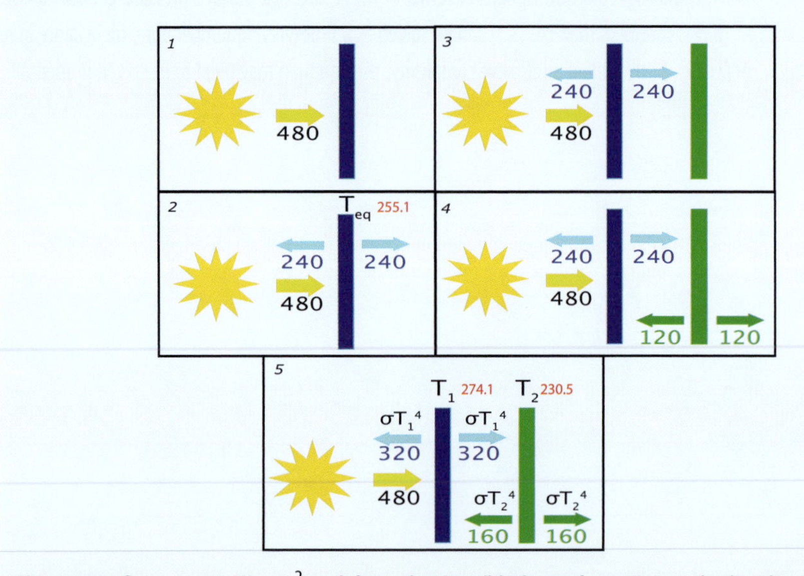

Fig. 3.3 The energy flows are in W m^{-2} and from the Sun (black numbers; 240 on both sides is 480), from the blue plate in dark blue numbers, and from the green plate in green numbers. The red numbers give the temperatures in kelvin. Adapted from Rabett (2017)

Sidebar 3.2: *(cont.)*

Looking at the two-plate system, the energy going in is 480 W m^{-2} and the energy going out is $\sigma T_1^4 + \sigma T_2^4$. Since these will be equal at equilibrium,

$$480 \text{ W m}^{-2} = \sigma T_1^4 + \sigma T_2^4.$$

Also there has to be an equilibrium for the energy going in and out of the green plate

$$\sigma T_1^4 = 2\,\sigma T_2^4.$$

Substitute for σT_2^4 back into the first equation:

$$480 \text{ W m}^{-2} = 1.5\,\sigma T_1^4.$$

Solving for T_1, the answer is: $T_1 = 274.1$ K, while $T_2 = 230.5$ K. Without the greenhouse plate T_1 was 255.1 K. Introduction of the second plate raised the equilibrium temperature of the first by 19 K.

Hence introducing an atmosphere (the green plate) firstly causes warming to occur and secondly, the equilibrium temperature is substantially higher than without the atmosphere.

3.4 Local Energy Balance

The global energy cycle is simplified by not having to worry about transports among and movement of energy within climate system components. But the outstanding research challenge has been to fully include those aspects, which means including the interactions among the atmosphere with land and ocean locally. Vertical movements of energy add further complications, and hence the approach mostly used here is to take the vertical integral in each domain, as in Fig. 3.1. That is to say, for the atmosphere, we sum over all layers and deal only with the radiation at top-of-atmosphere (TOA) and surface energy exchanges, called fluxes. Any heat or energy that goes out of the atmosphere into either the ocean or land must be equally accounted for in both, and thus several approaches are possible, detailed in Chapters 8 and 9.

The main fluxes of concern, given globally in Fig. 3.1, include

(i) net surface radiation, as the difference between the downward solar radiation and the upwelling longwave radiation;
(ii) surface heat fluxes which involve the direct exchanges of heat;
(iii) surface evaporation, E, which produces evaporative cooling of the surface but adds moisture and energy to the atmosphere in the form of latent heat energy, mostly tied up in the form of phase changes of moisture among water vapor, liquid water, and ice particles;
(iv) surface precipitation, P, which involves the phase of water, whether rain, snow or ice (e.g., hail), plus the temperature of the precipitation.

Because the global capacity to store moisture in the atmosphere is limited, there is a close correspondence between global evaporation and global precipitation amounts. Locally there are enormous differences between P and E, and that difference is often referred to as the freshwater flux $P - E$ (see Chapter 10). The effects of the temperature on rainfall or snow have generally been ignored since they are relatively small and, in any case, there are almost no measurements. However, recently estimates of these effects have been made as the effects are included in models and global analyses of the atmosphere.

Historically, a great deal of emphasis has been placed on estimating the surface fluxes in various ways. But every bit as important, because it can be done much more accurately, is measuring the energy of the atmosphere before and after an exchange, and similarly for the ocean or land. It is vital to ensure that what energy is lost from one domain is gained by the other. Depicting the exchange as the flux in a fully coupled framework is the only way to properly do this problem (see Figs. 8.8 and 8.9).

In both the atmosphere and ocean, the imbalances among the various vertical fluxes of energy in and out of each domain have to be compensated for by either a change in the energy storage in that domain or a transport of energy into or out of the domain. The heat storage in the atmosphere is a fairly small term, but changes in ocean heat content can be very large.

The transport is a vector quantity; it has both magnitude and direction. If the transport into a region is the same as that out, then there is no net contribution, and hence it is the divergence of the energy transport that balances the other terms.

This framework considers the entire energy budget: what comes in to heat or cool the region, what is moved or transported in or out, plus what the net change in energy storage is as a consequence. Often, in the past, just some of the terms have been estimated in an unconstrained manner, so that when they are added together, there may not be anything like a balance, indicating errors in one term or another.

All of these aspects are expanded and applied in Chapters 8 and 9.

References and Further Reading

Kiehl, J. T., and K. E. Trenberth, 1997: Earth's annual global mean energy budget. *Bulletin of the American Meteorological Society*, **78**, 197–208.

Lacis A. A., G. A. Schmidt, D. Rind, and R. A Ruedy, 2010: Atmospheric CO_2: principal control knob governing Earth's temperature. *Science*, **330**, 356-359. www.sciencemag.org/cgi/content/full/330/6002/356/DC1.

Rabett, E., 2017: The Green Plate Effect. http://rabett.blogspot.com/2017/10/an-evergreen-of-denial-is-that-colder.html.

Schmidt, G. A., R. A. Ruedy, R. L. Miller, and A. A. Lacis, 2010: Attribution of the present-day total greenhouse effect. *Journal of Geophysical Research,* **115**, D20106. doi: 10.1029/2010JD014287.

Trenberth, K. E. and J. T. Fasullo, 2011: Tracking Earth's energy: from El Niño to global warming. *Surveys in Geophysics*, doi: 10.1007/s10712-011-9150-2.

Trenberth, K. E., J. T. Fasullo, and J. Kiehl, 2009: Earth's global energy budget. *Bulletin of the American Meteorological Society*, **90**, 311–324. doi: 10.1175/2008BAMS2634.1.

Wild, M., D. Folini, M. Z. Hakuba, et al., 2015: The energy balance over land and oceans: an assessment based on direct observations and CMIP5 climate models. *Climate Dynamics,* **44**, 3393–3429.

4 The Sun–Earth System

The Sun's radiation and sunspot variations are introduced along with the changing Sun–Earth orbit and aspects of paleoclimate.

4.1 The Sun

Changes in the average net radiation at the top of the atmosphere due to perturbations in the incident solar radiation from the changes internal to the Sun or from changes in the orbit of Earth around the Sun lead to a change in heating, and climate change. There is no doubt about this physically, but the question is how large are these fluctuations?

The Sun is the star at the center of our solar system. For all practical purposes, the Sun is a tremendous continuous source of radiant heat. It is essentially inexhaustible but it does vary by small amounts that matter. The Sun's diameter is about 1.39 million km (864 000 miles), or 109 times that of Earth, and its mass is about 330 000 times that of Earth. It is about 149.6 million km (93 million miles) from Earth (called 1 Astronomical Unit) on average. There is no solid surface to the Sun, as the Sun is an extremely hot ball of gas with granules all over its surface forming the tops of convection cells. The visible surface of the Sun, called the photosphere, has a temperature of about 5700 K (kelvin, which is degrees Celsius + 273.15). There are very strong internal gyrations that occur within the Sun associated with strong magnetic fields and rotation within the Sun, so that internal convection may be inhibited or enhanced.

During active periods, sunspots form on about an 11-year cycle. Sunspots are dark cooler areas, but they are often surrounded by brighter spots called faculae. Temperatures in the dark centers of sunspots drop to about 3700 K. They typically last for several days, although very large ones may live for several weeks. While the sunspots tend to make the Sun look darker, the surrounding faculae make it

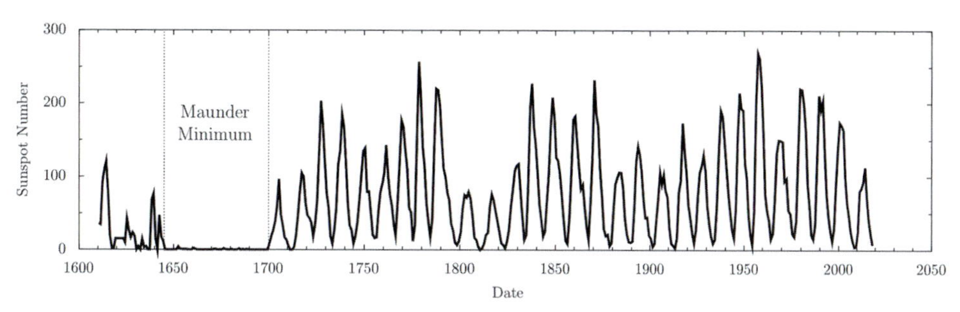

Fig. 4.1 The observed values of the sunspot number. The amplitude of the sunspot cycles from 1600 to the present varies substantially. This includes the Maunder Minimum, a period from 1645 to 1715 when the magnetic fields were too weak to produce sunspots. Reproduced from SolarCycleScience.com with permission

look brighter. During a sunspot cycle the faculae actually win out over the sunspots and make the Sun appear slightly (about 0.1%) brighter at sunspot maximum than at sunspot minimum. As discussed below, more profound changes occur at different wavelengths of radiation within the electromagnetic radiation spectrum.

Sunspot activity over the last 400 years has shown that the amplitude of the sunspot cycle varies from one cycle to the next (Fig. 4.1). The average cycle has a peak sunspot number of about 150. At times, as in the period known as the Maunder Minimum between 1645 and 1715, solar activity can become so weak that sunspots seem to disappear for several decades at a time.

Since 1979, satellites have borne instruments to allow the total and spectral solar irradiance to be estimated from measurements. It is important to have these instruments well above any atmospheric influences; ground- or air-based measurements before 1979 are compromised by atmospheric influences. However, the quality and calibration of the instrumentation and the understanding of how best to make these measurements have evolved over time, and the most accurate values have come from a mission called the Solar Radiation and Climate Experiment (SORCE) and an instrument called the Total Irradiance Monitor (TIM), launched in 2002. Fortunately, there has been overlap among the various satellites and instruments that have enabled a continuous time series after 1979 to be constructed (Fig. 4.2, adapted from Kopp 2016).

Our planet orbits the Sun once per year at an average distance of 1.50×10^{11} m. It receives from the Sun an average radiation of 1361.1 W m^{-2} at this distance, and this value is referred to as the *total solar irradiance* TSI. This used to be called the "solar constant" even though it does vary by small amounts with the sunspot cycle and related changes on the Sun. The TSI is correlated with the sunspot numbers, but the full extent of the variability in the past is quite uncertain.

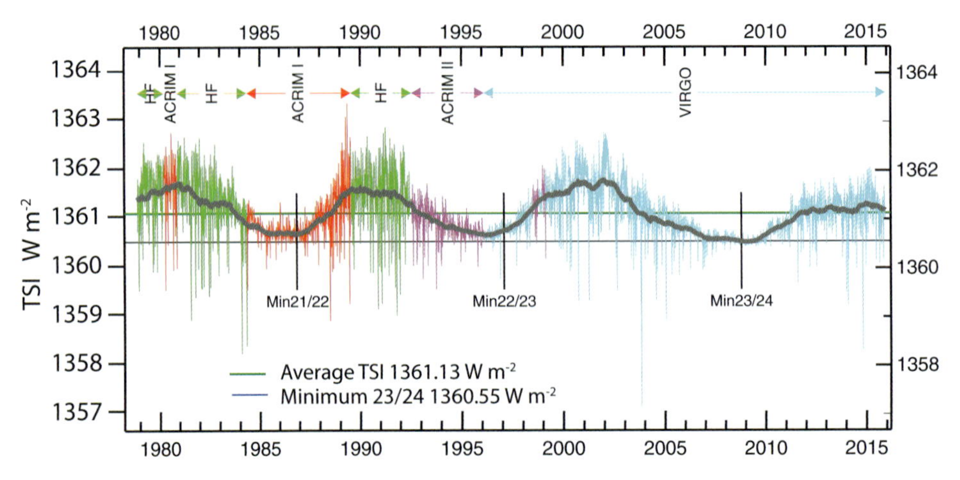

Fig. 4.2 Reconstructed TSI calibrated to the recent SORCE measurements. Adapted from Kopp (2016). The colors indicate the binary selections of different instruments used in the creation of the composite.

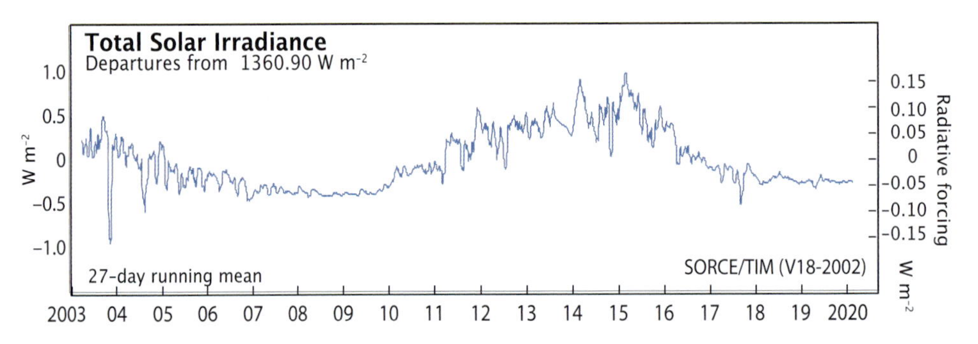

Fig. 4.3 The total solar irradiance for the last solar cycle dating from the beginning of the Total Irradiance Monitor (TIM) instrument on SORCE, with gaps filled in by interpolation and a 27-day running mean applied. The latter is approximately the rotation time in low latitudes of the Sun and therefore integrates over the much larger fluctuations from bright faculae and dark sunspots. The mean for this time is 1360.9 W m^{-2}. Data courtesy Greg Kopp, https://lasp.colorado.edu/home/tsis/data/tsi-data/

Nonetheless, the TSI is generally assumed to be lower in the Maunder Minimum in the 1600s to perhaps as low as 1360.1 W m^{-2}.

The most recent high-quality TSI values (Fig. 4.3) show the last solar cycle and include two sunspot minima. Variations can be large from day to day as sunspots and faculae point toward Earth, and the values in Fig. 4.3 are averaged over the average rotation rate near the Sun's equator of 27 days and normalized to the mean Sun–Earth distance. In reality, values vary with the orbit of Earth around the Sun. On the right side of this figure, the corresponding radiative forcing values are given. This divides by a factor of 4 (see the middle of Section 4.2) and accounts for the albedo (making a factor of 0.7). The mean radiative forcing varies within ± 0.1 W m^{-2}.

4.2 Earth's Orbit

Earth's shape is close to that of an oblate spheroid – a slightly flattened ball – with an average radius of 6371 km. It rotates on an axis with a tilt relative to the ecliptic (orbital) plane of 23½° around the Sun once per year in a slightly elliptical orbit that brings Earth closest to the Sun on January 3 (called perihelion); see Fig. 4.4.

Where the continual incoming radiation is absorbed, though, clearly depends on Earth's orbit around the Sun, the Sun–Earth geometry, and the rotation of the Earth on its axis. The shape of Earth means that incoming solar radiation varies enormously with latitude (Fig. 1.2). On timescales of tens of thousands of years, Earth's orbit slowly changes: the shape of the orbit is altered, the tilt changes, and Earth precesses on its axis like a rotating top, all of which combine to alter the annual distribution of solar radiation received at Earth.

It is the tilt of the axis, or "obliquity," and the revolution of Earth around the Sun that give rise to the seasons, as the northern hemisphere points more toward the Sun in June while the southern hemisphere points toward the Sun in December (Fig. 4.4). The tilt of the axis means that one hemisphere is favored to receive much more radiation, giving rise to the summer season, while the other hemisphere has a much shorter day and experiences winter. The peak value occurs at the solstice on or about June 21 and December 21, while 3 months on either side the heating of both hemispheres is about equal near the time of the equinoxes.

Earth also turns on its axis once per day, resulting in the day–night cycle (Fig. 1.2). A consequence of Earth's roughly spherical shape and the rotation is that the average energy in the form of solar radiation received at the top of Earth's atmosphere is the total solar irradiance divided by 4, which is the ratio of Earth's surface area ($4\pi a^2$, where a is the mean radius) to that of the cross section (πa^2). Because of the oblateness of Earth, the actual factor is 4.0034 instead of 4 and a spherical Earth assumption causes a +0.29 W m^{-2} bias in net TOA radiation flux.

In our current climate, the North Pole basks in continuous sunshine in midsummer while the South Pole experiences a long polar night. Six months later, the situation is reversed, as winter returns to the North Pole and the Arctic while Antarctica basks in continuous sunshine.

The rotation of Earth on its axis gives rise to the day–night cycle, and peak incoming radiation is at local noon (Fig. 1.2); the maximum occurs on the side of Earth directly facing the Sun. However, whether the radiation is absorbed or not depends upon whether clouds block and reflect the radiation back to space. Even in the absence of clouds, the underlying land may be bright (perhaps desert) or dark (forest) and in turn this affects how much is reflected and how much is absorbed.

As noted in Sidebar 4.1, the incoming radiation consists of many wavelengths, which become readily apparent when refracted, such as by raindrops to form a rainbow. The very short wavelengths are absorbed in the high thin atmosphere above 70 km altitude and ionize the molecules of air. Shortwave ultraviolet radiation (UV-C) is also absorbed in the upper atmosphere while UV-B radiation penetrates into the lower stratosphere where it plays a vital role in forming the ozone layer. UV-A radiation may pass through to the surface where it can have adverse effects, such as causing cancer; see Section 6.1.2 on ozone.

4.3 Earth's Changing Orbit around the Sun

Over geological time, the angle of Earth's obliquity cycles between 21.4° and 24.4° and has a period of about 41 000 years. If there were no tilt at all, then there would be no seasons and the polar regions would never receive any direct sunshine down to the surface. This situation is obviously one designed to promote very cold and icy conditions, as snow fails to melt at high latitudes. It can be readily understood then that the smaller tilt is one that favors ice ages, an idea put forward by Milutin Milankovitch in 1930.

In addition to the changes in tilt, the shape of Earth's orbit also changes from a nearly perfect circle to an oval shape on about a 100 000-year cycle, and this is referred to as a change in the eccentricity of the elliptical orbit. Finally, Earth wobbles on its axis as it spins, like the movement of a spinning toy top, over the span of 19 000–23 000 years, called precession, or precession of the equinoxes. These small variations in Earth–Sun geometry alter how much sunlight each hemisphere receives during Earth's year-long trek around the Sun, when in the orbit (the time of year) the seasons occur, and how extreme the seasonal changes are. Fig. 4.4 also depicts the situation about 9000 years ago, when perihelion – the closest point – was in July. However, the tilt was slightly greater than at present. The result is that the average radiation received by the planet in July at 9000 years ago was about 7% greater than today (equivalent to about 30 W m^{-2}). But in January, the reverse is true: the radiation was correspondingly less. More generally, Fig. 4.5 shows the changes in solar radiation received at Earth as a function of month for the past 21 000 years, which covers a precession cycle.

Earth is currently decreasing in obliquity, which sets the stage for more moderate seasons (cooler summers and warmer winters) while increases in obliquity create more extreme seasons (hotter summers and colder winters). However, glaciers tend to grow when Earth has many cool summers that fail to melt back the winter snows. If short, cool summers fail to melt all of the winter's snow then the snow would slowly accumulate from year to year, and its bright, white surface

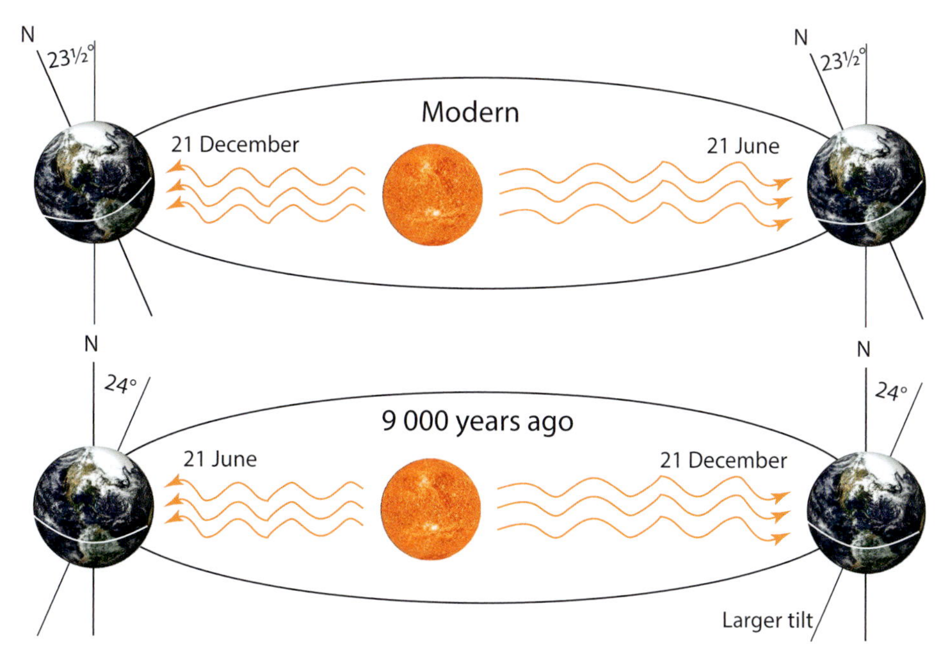

Fig. 4.4 Earth has an obliquity of about 23½° in modern times (top) and the southern hemisphere is tilted most toward the Sun (austral or southern summer solstice) on December 21, while the northern hemisphere is tilted most toward the Sun on June 21 (boreal or northern summer solstice). Nine thousand years ago (bottom) these were reversed owing to the precession of the equinoxes, and the tilt (obliquity) was greater at 24°.

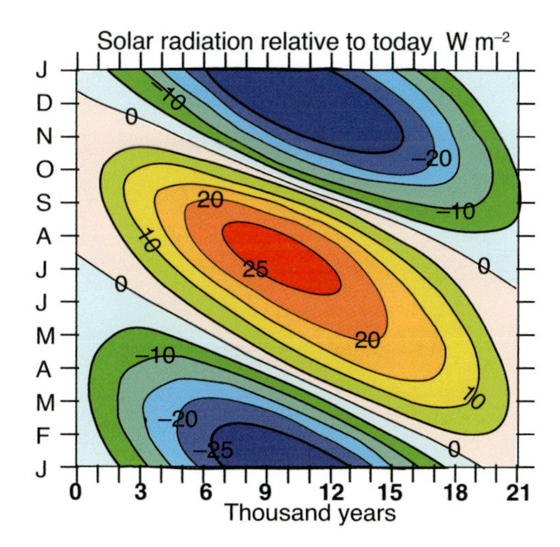

Fig. 4.5 Variations in solar radiation received at Earth relative to present day in W m^{-2} as a function of time of year (J is January, F is February, etc.) and for thousands of years in the past.

would reflect more radiation back into space. Temperatures would drop even further, and eventually an ice age would be in full swing.

Based on the orbital variations, Milankovitch predicted that the ice ages would peak every 100 000 and 41 000 years, with additional perturbations every 19 000–23 000 years. Paleoclimate studies over many millennia have indeed shown all of these periodicities to be prominent in the climate record.

It is generally thought that solar forcing has contributed minimally to climate change over the past 50 years. Two main mechanisms are thought possibly important. The first involves absorption of anomalous shortwave radiation in the upper atmosphere. Then the responses in the stratosphere would have to propagate downwards somehow. This is a little like the tail wagging the dog, and assessments suggest that although very important in the stratosphere, its effects are small for the climate system as a whole. In contrast, the second mechanism relates to changes in the total solar irradiance, which, even if small, can accumulate if they persist long enough and thereby contribute significantly to climate change through changes in the ocean and sea surface temperatures.

Sidebar 4.1: Radiation

All bodies emit radiation in amounts that depend upon their temperature. This is the *Stefan–Boltzmann law* which describes the energy radiated from a *black body* in terms of its temperature. A *black body* is idealized because it absorbs all incident electromagnetic radiation, regardless of wavelength or angle of incidence. It also emits radiation according to *Planck's law* which has a wavelength dependence determined by temperature alone. Actual bodies or substances do not absorb all incident radiation, and are sometimes called gray bodies. They are approximated in terms of the black body equivalent value by introducing an *emissivity* ε which has a value between 0 and 1. The Stefan–Boltzmann law states that the total energy radiated per unit area is directly proportional to the fourth power of the black body's absolute temperature T (in kelvin): $E = \varepsilon\sigma T^4$. The constant of proportionality σ, called the Stefan–Boltzmann constant, can be derived from other known physical constants, and has a value of $\sigma = 5.670 \times 10^{-8}$ W m^{-2} K^{-4}. The result is that E is in watts per square meter.

Moreover, radiation comes in many "colors." The nature of radiation is determined by its wavelength, and the peak radiation from the Sun is in the visible part of the electromagnetic spectrum (Fig. 4.6). Very short wavelengths (gamma rays and X-rays) play an important role only in the thin upper atmosphere where they contribute to the breakdown of atoms and molecules into ionized particles (hence the ionosphere) above about 70 km altitude. When the Sun is disturbed and more active, with solar flares and sunspots, the potential exists for ionospheric disturbances, adversely affecting radio signal propagation and possibly causing radio blackouts, and contributing to auroras at high latitudes. Ultraviolet radiation is mostly absorbed in the stratosphere in the ozone layer. Earth radiates at temperatures that give rise to mostly infrared radiation. Because the Sun's rays and associated energy are mainly in the band near and encompassing the visible, they are called shortwave radiation, while the infrared rays are often referred to as longwave or thermal radiation. For the most part these two bands do not overlap very much, and hence they can be dealt with separately.

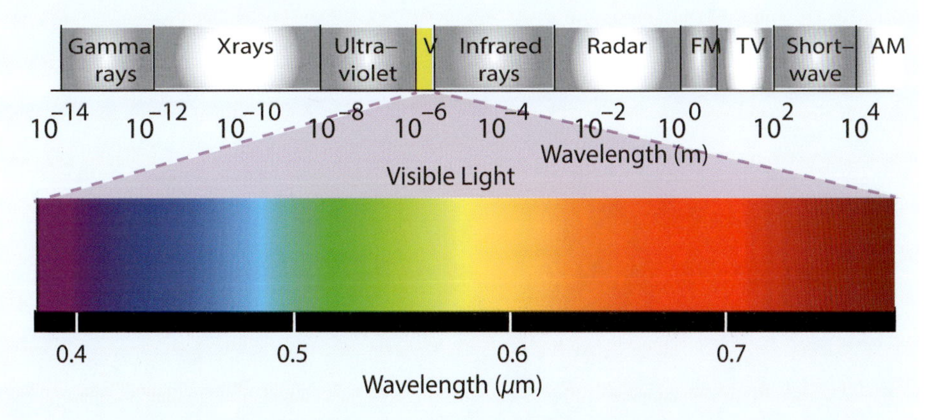

Fig. 4.6 The electromagnetic spectrum of radiation, with the main bands indicated at top along with the associated wavelengths in meters. Most of the Sun's rays are in the visible (in yellow up top) which is broken down below into colors, that are readily seen in a rainbow. The scale there is in micrometers (10^{-6} m).

References and Further Reading

Coddington, O., J. L. Lean, P. Pilewskie, M. Snow, and D. Lindholm, 2016: A solar irradiance climate data record. *Bulletin of the American Meteorological Society*, **97**, 1265–1282. doi: 10.1175/BAMS-D-14-00265.1.

Kopp, G., 2016: Magnitudes and timescales of Total Solar Irradiance variability. *Journal of Space Weather and Space Climate*, **6**, A30. doi: 10.1051/swsc/2016025.

Loeb N.G., B.A. Wielicki, D.R. Doelling, et al., 2009: Toward optimal closure of the earth's top-of-atmosphere radiation budget. *Journal of Climate*, **22**, 748–766. doi: 10.1175/2008jcli2637.1.

Shindell, D. T., G. Faluvegi, and G. A. Schmidt, 2020: Influences of solar forcing at ultraviolet and longer wavelengths on climate. *Journal of Geophysical Research: Atmospheres*, **124**. doi: 10.1029/2019JD031640.

SolarCycleScience.com: http://solarcyclescience.com/solarcycle.html

Trenberth, K. E., and B. L. Otto-Bliesner, 2003: Toward integrated reconstructions of past climates. *Science*, **300**, 589–591. doi:10.1126/science.1083122

5 Observations of Temperature, Moisture, Precipitation, and Radiation

The patterns and variables of most importance in describing and moving energy around in the climate system are described. Timescales from hourly, to seasons, to the mean climate are included.

5.1 State Variables

Radiation is a fundamental variable for the whole flow of energy, and variations are perceived mainly by changes in cloudiness. Accordingly, the incoming and outgoing radiation are included here.

However, temperature and rainfall, or more generally precipitation, are probably the most vital variables of interest to humans. They determine whether life is possible and food and water are viable. Historically, this was especially the case, and it is only relatively recently that large volumes of food and material goods have been shipped around the globe to enable more optimal use of natural resources.

Outside of the tropics, there is a very large annual cycle of seasonal effects on both food and water, and traditionally the seasons have been defined in terms of temperature: summer is the warm season and winter is the cold season, with spring and autumn (fall) as transition seasons. In the tropics, on the other hand, temperature variations are smaller, and the seasons are more commonly and importantly associated with the wet and dry seasons, linked with monsoons in particular. The summer monsoon arises from the land warming more rapidly than the ocean, creating continental-scale sea breezes that are strongly influenced by Earth's

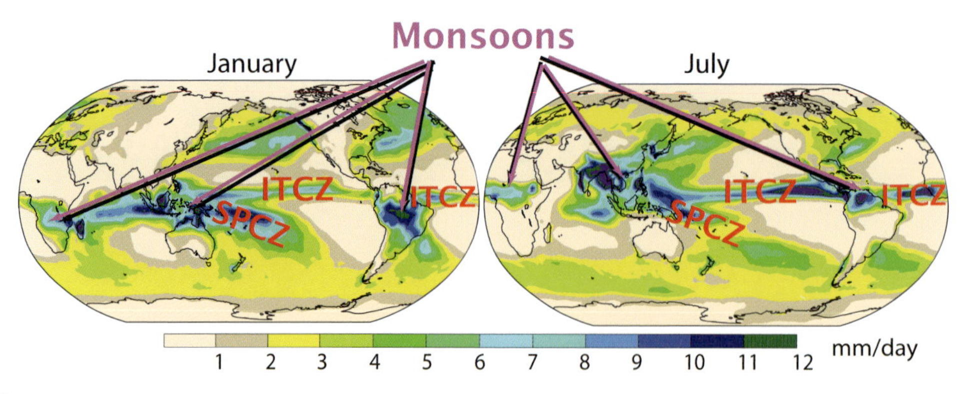

Fig. 5.1 Superposed on the long-term average precipitation patterns for January (left) and July (right) (see also Figs. 5.4 and 5.5) are pointers to the main monsoon rains and the ITCZs and SPCZ.

rotation. The summer monsoon is therefore wet, in general, as warm air rises, but this means that the denser air subsides elsewhere, in the winter monsoon. Hence the monsoon circulation consists of a large-scale overturning, and the wet season in one region is linked to the dry season in another region.

Figure 5.1 gives the average precipitation for January and July (see also below for other times) with arrows pointing to the main monsoon rains over Africa, Asia-Australia, and the Americas. In January these rains are in the southern hemisphere, while southern Asia, for instance, is dry. In July the monsoon rains are in the northern hemisphere and most of Australia is dry. Also indicated are the main convergence zones over the oceans, the Inter-Tropical Convergence Zone (ITCZ) and South Pacific Convergence Zone (SPCZ). The ITCZs over both the Pacific and Atlantic remain north of the equator, in general, although in the Pacific the SPCZ strengthens while the ITCZ weakens somewhat in southern summer (January). A weaker South American Convergence Zone, extending from South America southeast over the Atlantic, is also apparent, mainly in January. Exceptions to these regular migrations occur in times of El Niño events.

5.2 Earth's Temperatures

The Sun–Earth geometry and radiation received at Earth (Section 5.6) drive the seasons (Chapter 4). In some countries, the radiation is used to define seasons, but not in a logical way. In particular, the mid-latitude oceans warm from spring until the autumnal equinox, accumulating heat at and below the surface. Therefore, the annual cycle of sea surface temperatures (SSTs) does not follow the Sun's incoming radiation, and SSTs peak about March in the southern hemisphere and September in the northern hemisphere, which is, not coincidentally, the most active time for tropical storms and hurricanes in each hemisphere. However, in the United

States, an astronomical definition is commonly used for the seasons, although with no sensible rationale, because in that framework, summer is defined to begin on the longest day, rather than the longest day being in the middle of the season.

Instead, when one examines the definition of summer as the warmest season, and winter the coldest, surface temperatures lag the Sun by 32½ days in mid-latitudes of the northern hemisphere, 27½ days over the United States, and 44 days in the southern hemisphere. This reflects the quicker response and lower heat capacity of land, namely a "*continental climate*" versus the strong ocean influences in the southern hemisphere. The astronomical definition makes sense only in the southern hemisphere over oceans, while in the northern hemisphere over land, the warmest and coldest seasons begin about 3–4 days after the beginning of June and December, respectively, and hence the traditional meteorological seasons of summer as June–July–August, and winter as December–January–February are standard in climate.

The term "surface temperature" refers to surface air temperature which is taken in standard ways 2 m above the actual surface, preferably over a grass surface and in a special screen that allows ventilation but shades out direct sunshine. The surface ground temperature is much more quixotic and variable; it varies with surface type and whether it is daytime and cloudy or not. It is important to ensure that the temperature being taken is of the free air, and not an artificial temperature of a rooftop, nearby wall, or the thermometer material.

It is difficult to determine the actual surface temperature of the whole planet owing to a lack of observations in some areas, and until fairly recently, not even very good determination of where the surface was in terms of altitude. This was mainly an issue in Antarctica, and mountain areas have generally been poorly observed. Accordingly, the best historical estimate gave a global mean value of about 14°C for 1961–90. Since then, climate change has led to a warming of about 1°C, which would now put the value at 15°C. That value was chosen as a standard for some atmospheric conditions (International Standard Metric Conditions).

With satellite observations and improved atmospheric modeling at high resolution that properly define mountain heights, the latest values for the surface 2 m air temperature, from ERA-Interim *reanalyses* for 1979 through 2018 are given in Table 5.1 and come to: Annual: 12.84°C; January: 10.92°C; April: 12.81°C; July: 14.70°C; and October: 12.92°C. The lower values compared with earlier estimates arise from properly accounting for Antarctica.

Table 5.1 Global mean temperatures, in K, for 1979–2018 from ERA-Interim reanalyses

	Annual	Jan.	Apr.	July	Oct.
Global	286.0	284.1	286.0	287.9	286.1
N. hemisphere	286.9	280.2	286.2	293.3	288.3
S. hemisphere	285.2	288.1	285.9	282.5	284.2

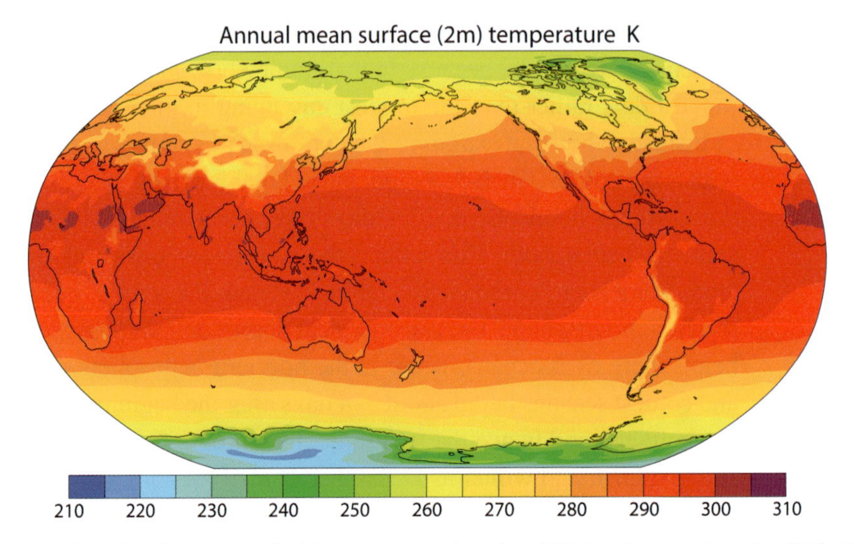

Fig. 5.2 Mean annual surface (2 m atmospheric) temperatures based on ERA-interim reanalyses for 1979–2018 in K.

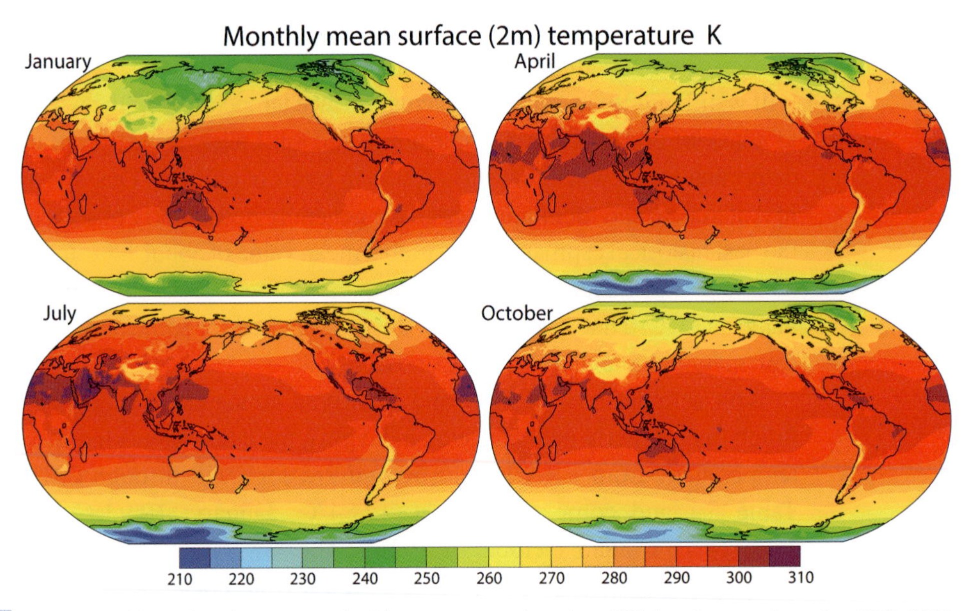

Fig. 5.3 Mean monthly surface (2 m atmospheric) temperatures based on ERA-interim reanalyses for 1979–2018 in K for January, April, July, and October.

The average annual mean surface temperatures (Fig. 5.2) show the highest temperatures in the tropics usually over land in the dry season. For instance, it is much hotter over India in the pre-monsoon season (April) than in peak summer (July) (Fig. 5.3). Australia in the southern hemisphere is hottest in January. Note how the high topography and icy or mountainous regions stand out as cold spots (Himalayas, Andes, Greenland, and Antarctica). In part because of Antarctica, the northern hemisphere is over 1°C warmer on average than the southern hemisphere. Because of the larger amplitude over land, the global mean annual cycle follows the northern hemisphere, with peak values in July (14.7°C) and a minimum in January (11.0°C), showing the dominance of land effects.

5.3 Links between Temperatures and Precipitation: Humidity

There is a strong relationship between precipitation patterns and surface temperatures, especially in the tropics, but the gradients in temperatures matter as much as the absolute values. A key link between these fields is the humidity which relates to moisture availability that is strongly controlled by temperature but in a nonlinear way.

Moisture in the atmosphere is measured by the humidity. The *specific humidity* and *mixing ratio* are measures of the moisture amount in the form of the mass of water vapor relative to the mass of air, for instance in grams per kilogram, and the difference is whether the total mass (specific humidity) or the mass of dry air (mixing ratio) is used as a reference. The *relative humidity* measures how much moisture there is as a percentage of the total moisture-holding capacity for that air. Air with 100% humidity is saturated, and may easily form clouds or rain, while very dry air, such as from a desert region, may have relative humidities of less than 20%. Also of interest is the total moisture in a column, called the total column water vapor (TCWV) and also called the *precipitable water* since that is the maximum amount of rain that could be extracted if all moisture were condensed from the column of air.

Moisture availability and flows matter, and form a key link between temperatures and precipitation. Warmer air can hold more moisture per degree Celsius at a rate of about 7% of the total amount (not to be confused with relative humidity). This comes from the Clausius–Clapeyron (C-C) equation which involves an exponential function, and this is why the relationship is best expressed as a percentage rather than an absolute value. It means, however, that warmer environments have the potential for much heavier rainfall or snowfall amounts. Another key factor for precipitation is whether moisture is available in the atmosphere from a surface source. It always is over the ocean areas, but not over deserts or arid regions, and over land it can depend a lot on season and proximity to surface water (seas, lakes) and whether mountains or other topography get in the way. Hence, over the oceans, there is a very strong relationship between TCWV and sea surface temperatures (SSTs) in both their overall patterns and their variations over time (e.g., Fig. 5.4 for January and July). For a given SST change, atmospheric

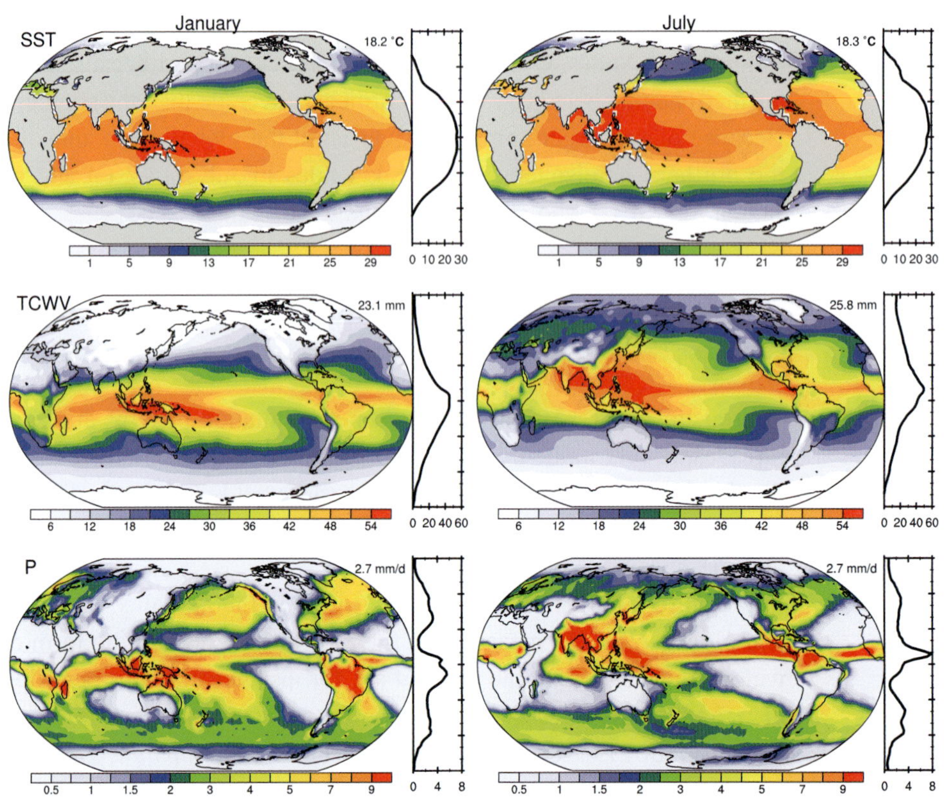

Fig. 5.4 Mean sea surface temperature (SST) in °C, total column water vapor (TCWV) in mm, and precipitation amount (P) in mm day^{-1}, for January (left panels) and July (right panels). Graphs to the right of each panel: zonal means; values at top right of each panel: global mean (for oceans in top plot). From Trenberth (2011), published with permission

temperature perturbations tend to be larger and increase with height, especially in the tropics, which means that the TCWV change can often be somewhat larger than given by the SST alone.

The above principles mean that the largest amount of water vapor occurs where SSTs are highest, while interiors of continents are prone to be dry. Accordingly, the largest average TCWV values occur over the Warm Pool in the tropical western Pacific, where highest large-scale values of SSTs typically reside (see Fig. 5.4). Global surface air temperatures are higher in northern hemisphere summers, and this is reflected in much higher TCWV values over land versus the ocean and southern hemisphere. The observed TCWV and precipitation annual cycles also show very strong relationships in the tropics and subtropics, but not over the extratropical oceans (Fig. 5.4).

The atmospheric winds near the surface play a major role. Warmer air expands, becomes less dense and tends to rise, creating lower surface pressures, while denser air tends to subside over cooler areas, where the air also shrinks and becomes compacted, creating higher surface pressures (see Fig. 11.1). The resulting surface pressure gradients are what drives the surface winds that then tend to flow from cooler to warmer SST regions, or even to land regions in monsoons. Mass conservation guarantees that when air rises in one place it must come down someplace else. The winds carry moisture as water vapor, which then condenses as the warm air rises, while in subsidence regions, the sinking air is dry because the moisture was rained out as it ascended. Accordingly, in the tropics, the precipitation patterns tend to mimic the patterns of TCWV and thus SSTs, but with much more structure and sharper edges. The reason is because overturning circulations, such as those associated with the Hadley Circulation and monsoons (Chapter 7), develop and create strong areas of subsidence that dry out the troposphere above the near-surface atmospheric boundary layer.

The patterns of SST are themselves strongly influenced by the surface winds and their associated stress on the surface, as the winds blow the waters along, and upwelling occurs as the winds pull the surface waters in different directions, especially near the coasts. The Pacific Warm Pool is a consequence of the warm surface waters piling up as the trade winds blow from east to west (see Chapter 12 for details). It is not sufficient for SSTs to be high to attract converging winds and moisture, they have to be the highest in that neighborhood. Meanwhile Earth's rotation plays a major role as the scales of motion become large, larger than 1000 km or so, affecting both atmospheric and ocean currents.

The mean TCWV falls off at higher latitudes in both hemispheres, along with the SSTs, but precipitation has a secondary maximum over the oceans associated with mid-latitude storm tracks, described in Section 5.4. In the troposphere the largest fluctuations in TCWV occur with El Niño–Southern Oscillation (ENSO) events, with increases associated with the high SSTs in the central Pacific (Chapter 12).

As the climate warms, water-holding capacity increases with higher temperatures according to the C-C relationship, and hence it is natural to expect increases in water vapor amounts because relative humidity is more likely to remain about the same, although this is complicated by movement of air between oceans and continents. Changes in TCWV and humidity are evident in observations since 1987 (when good global observations began from satellite). At the surface, as temperatures have increased, there are clear indications of increases in specific humidity and TCWV. Over oceans, the increases are consistent with C-C expectations with a constant relative humidity, but increases are somewhat lower over land, especially where water availability is limited. This is due to land warming faster than the oceans while the source of most moisture is from the oceans. Hence while the atmospheric water vapor increases over oceans in line with C-C expectations, once the air is advected onto land, where it is warmer, the relative humidity drops even though the specific humidity has increased.

The observed increases in water vapor affect the greenhouse effect, thus providing a positive feedback to climate change (Section 13.2.3), and the hydrological cycle (Chapter 10), by providing more atmospheric moisture for all storms to feed upon. Accordingly, it has ramifications for precipitation.

5.4 Earth's Precipitation

Precipitation is a difficult quantity to deal with because it is inherently intermittent. Often, as above, the term precipitation is synonymous with precipitation amount, but the intensity (how hard it rains or snows), frequency (how often), duration (how long), and type (whether rain or snow) matter every bit as much as amount.

Difficulties in the measurement of precipitation remain an area of concern in quantifying the extent to which global and regional-scale precipitation have changed. In situ measurements are especially impacted by wind effects on the gauge catch, especially for snow and light rain, and few measurements exist in areas of steep and complex topography. For remotely sensed measurements (radar and space-based), the problems are that only measurements of instantaneous rates can be made and there are uncertainties in algorithms for converting radiometric measurements (radar, microwave, infrared) of cloud droplets into precipitation rates at the surface. Moreover, measurements of snowfall are difficult from space. Only recently have global fields of precipitation become available every hour, but they contain considerable uncertainty.

To gain confidence in observations, it is useful to examine the consistency of changes in a variety of complementary moisture variables in the hydrological cycle, including both remotely sensed and gauge-measured precipitation, drought, evaporation, atmospheric moisture, soil moisture, and streamflow, although uncertainties exist with all of these variables as well. In other words, by insisting on conservation of water substance, and tracking the flow of moisture through the climate system, one can gain confidence in the results and expose where the greatest limitations occur (see also Chapter 10).

It was already noted above that seasons in the tropics are more usefully defined in terms of rainfall. Unlike temperature fields, the precipitation patterns have very strong gradients, and wet and dry regions can coexist not that far apart. Moreover, as described in Section 5.3, there is a strong relationship between precipitation patterns and surface temperatures.

In the tropics and subtropics, the temperatures are nearly always warm enough to make precipitation occur as rainfall. However, when surface temperatures are less than about 2°C the precipitation is apt to be snow (because it is below freezing above the surface where snowflakes form). With a surface temperature close to 0°C the precipitation is apt to be wet heavy snow, and largest snowfall amounts occur when temperatures are about −2°C. When temperatures are less than about −10°C snowflakes are small and light, and fail to bond together well, so that making snowballs, or a snowperson, is not possible.

The atmosphere is relatively transparent to incoming solar radiation, which therefore heats the surface. Land gets hot by day and SSTs rise. Winds greatly increase evaporation and carry the evaporated moisture away, allowing more evaporation to occur, and associated evaporative cooling of the ocean and land (if wet) take place. When precipitating clouds form, heat from the evaporation is given back to the atmosphere as the moisture condenses in the form of latent heat. This warms the buoyant rising air even more and intensifies the atmospheric circulation. Latent heating from rainfall is the dominant heating of the atmosphere, especially in terms of the contrasts from one region to another. This kind of feedback is what concentrates the precipitation into zones in the tropics, leaving the subtropics as dry deserts with sunny cloudless skies but also as major sources of evaporated water over oceans.

Accordingly, the precipitation fields (Figs. 5.5 and 5.6; derived from the Global Precipitation Climatology Project [GPCP] data) exhibit strong character and large annual cycles, as the Sun migrates (relative to Earth) from one hemisphere to the other (see Chapter 4). These figures elaborate on Fig. 5.1 with the ITCZs over the Pacific and Atlantic Oceans, and monsoons over Asia versus Australia, the Americas, and Africa; compare January versus July, in particular. The main arid regions are in

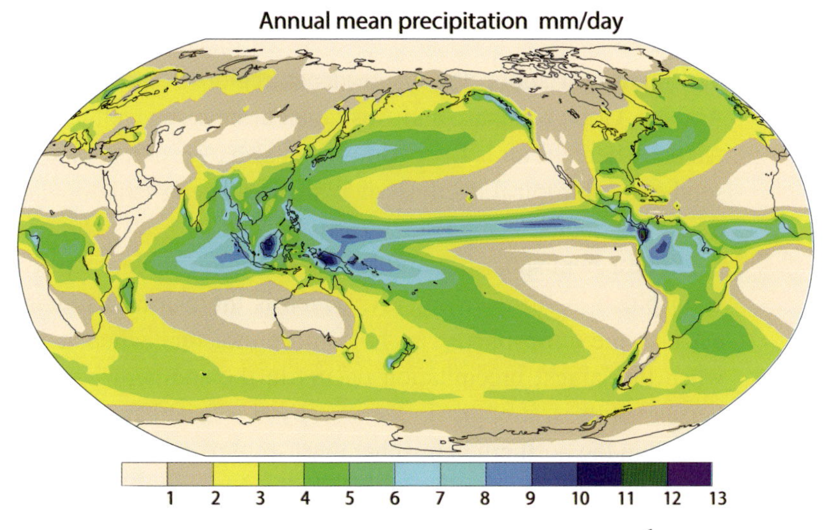

Fig. 5.5 Annual mean precipitation amount 1979–2018 from GPCP data in mm day^{-1}.

Table 5.2 Mean precipitation for 1979–2018 for GPCP in mm day^{-1}. The annual value is the mean of all daily values

	Annual	Jan.	Apr.	July	Oct.
Global	2.69	2.72	2.63	2.72	2.66
N. hemisphere	2.74	2.26	2.17	3.28	3.13
S. hemisphere	2.63	3.17	3.10	2.15	2.19

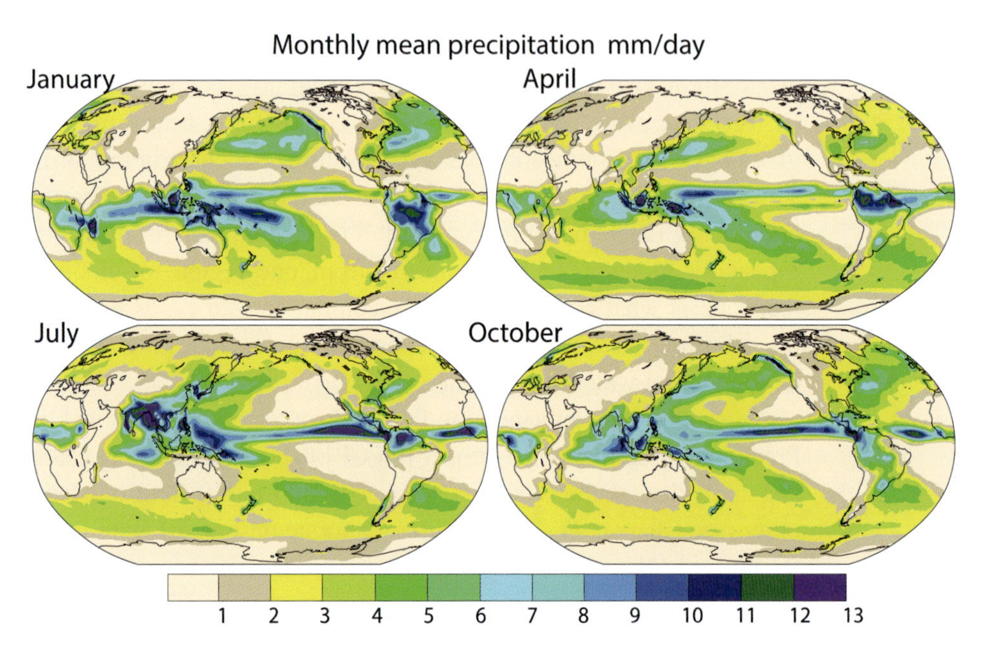

Fig. 5.6 Monthly mean precipitation amount 1979–2018 from GPCP data in mm day^{-1}.

the subtropical anticyclonic subsidence regions, especially over land (the Sahara, Arabian, Gobi, Kalahari, and Australian deserts), as well as over Antarctica and the Arctic (where the air is freeze dried), but also over parts of the oceans (especially the southeast subtropical Pacific, where the most persistent circulation feature on Earth exists: the South Pacific High pressure system).

Highest hemispheric monthly precipitation (Table 5.2) occurs in the northern hemisphere in summer: in July, which is the warmest month (Table 5.1) largely because the atmosphere can hold more moisture when it is warmer. January is when the largest precipitation occurs in the southern hemisphere, and the north versus south is strongly influenced by the fact that the ITCZs over the oceans remain in the northern hemisphere year-round, in association with a cold tongue of water evident in the SST field, linked with upwelling from the southeast trade winds along the equator (Fig. 5.4). The warmth of the northern hemisphere is also related to the land and its distribution, while precipitation in Antarctica (snow) is very low year-round.

In January, mid-latitude precipitation maxima are strongly evident over the North Pacific and Atlantic Oceans and, to a lesser extent, throughout the southern oceans (Fig. 5.6). The Southern Ocean storm tracks are strongest in the transition seasons and broadest in the southern winter, while the northern ocean storm tracks become weaker in the northern summer. The absence of relationships between average TCWV and precipitation in the extratropics highlights the very transient nature of precipitation events that are associated with extratropical cyclones, while in the tropics there is a much stronger mean-flow component associated with monsoons and the Hadley Circulation.

5.5 How Often Does It Rain?

The perception about whether somewhere is a nice place to live often depends on how frequently it rains (or snows). This relates to how dreary the weather appears, and it is often the duration much more than the amount that influences perceptions. Yet information about the frequency of rainfall, or precipitation in general, is spotty. The intermittent nature of precipitation means that it is important to quantify not just the amount, but also the frequency and duration of events, and the intensity of precipitation when it occurs. The intensity likely relates most strongly to risk of flooding, especially when precipitation is sustained over many hours.

It has been shown that it is important to get down to hourly values to gain a proper appreciation of the true frequency. Here some values are given for near global averages, excluding only the polar regions. At 3-hourly resolution the frequency of precipitation is about 25% higher than hourly, and at daily resolution it is about 250% higher on average, because it counts as a rainy day even if it lasts for only a few minutes. Use is made of a threshold of 0.02 mm h^{-1} for when precipitation begins. Results also depend on spatial resolution of measurements because of the fine structure of many precipitation events. Across a 1° by 1° cell compared with ¼° values the frequency is 70% higher because precipitation counts no matter where or how small within that grid cell. Overall for ¼° and hourly data, precipitation occurs $11.0 \pm 1.1\%$ (the uncertainty here is plus or minus one standard deviation) of the time. Alternatively, 89.0% of the time it is not precipitating. But outside of the ITCZ and SPCZ where values exceed 30%, and the arid and desert regions where values are below 4%, the rates are more like 10% or so, and over land where most people live, values are closer to about 8% (Fig. 5.7).

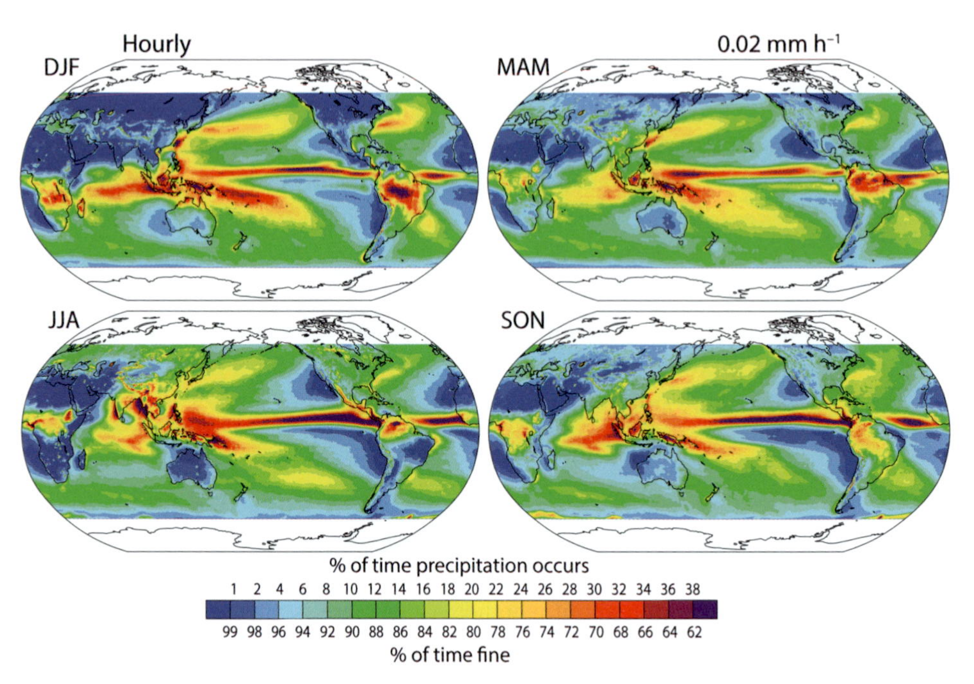

Fig. 5.7 Seasonal mean of the percentage time precipitation occurs above a given threshold of 0.02 mm h^{-1} for hourly data, or "fine" refers to precipitation below that value or not at all. From Trenberth and Zhang (2018). © American Meteorological Society. Used with permission

One of the world's great myths is Noah's flood. The tale of this flood occupies chapters 6–8 of the Book of Genesis in the Bible (Silverman 2013). Noah built an ark to save certain animals and his family. Is it possible to have rain for 40 days and 40 nights? Can it rain everywhere at the same time? Certainly, the answer to these two questions together is no. Where does the moisture come from? The hydrological cycle comes in from a science perspective. The evaporation of moisture from especially the oceans provides the main source of moisture for storms that produce precipitation. Typical evaporation rates are 1–5 mm day^{-1} and hence it might be possible to have continuous rain at those same rates. Except that what goes up must come down and the phenomena that produce precipitation do not work that way; rainy areas are always accompanied by fair or fine areas elsewhere. Instead, storms reach out and gather in moisture from surrounding areas, concentrate it, and dump it down typically at much higher rates, which can easily exceed 3 inches (75 mm) per day. The latter is our rule of thumb for when local flooding occurs.

For the world's continents except Antarctica, precipitation occurs most frequently at 25% of the time annually over the Maritime Continent (the ocean area in and around Indonesia and all of the islands there), followed by South America at 14.7% and South Asia at 9.7%, but least frequently over North Asia and Australia at around 5.3–5.4%. North America, Europe, and Africa lie in between. Strong seasonality is noticeable for all the continents except Africa (as a whole) and the Maritime Continent, and is clearly related to monsoon rains: precipitation occurs much more often in the summer than winter months over both hemispheres (Fig. 5.7). "Summer" vs "winter" differences are small for Africa and the Maritime Continent that both straddle the equator, and consequently the terms summer and winter are ill-determined. Moreover, for Africa as a whole, there is always a monsoon system present, with monsoon rains in one hemisphere or the other, and hence the continental value does not apply locally.

The average precipitation rate, during times when precipitation does occur, is about nine times the average evaporation rate. In other words, in order to supply the moisture to a weather system, it has to reach out to embrace an area nine times the area of the precipitation in order to sustain the precipitation. This simple calculation assumes precipitation at a low threshold of 0.02 mm h^{-1}, whereas if a more typical value for moderate rain of 0.2 mm h^{-1} is used, the factor is about fifteen. Therefore, if the environmental moisture goes up, as expected and observed with a warming climate, it must rain harder and increase the risk of flooding. Although these arguments are heuristic, because the frequencies obviously differ considerably with location and event, they provide an excellent rule of thumb and conceptual basis for understanding how often rain falls and how it may change in the future.

5.6 Earth's Radiation

The biggest driver of the climate system is the Sun and the Sun–Earth geometry. However, the changes from radiative energy to other forms lends an incredible richness to the resulting phenomena that store and move energy around. In this section, the regional radiation patterns are explored to see what lessons can be learned that need to be delved into further elsewhere. Section 14.3 presents the time series of global values and trends in the EEI.

The annual mean *absorbed solar radiation* (ASR) and *outgoing longwave radiation* (OLR) are shown in Fig. 5.8. The net radiation R_T = ASR − OLR. Most of the atmosphere is relatively transparent to solar radiation, with the most notable exception being clouds. At the surface, snow and ice have a high albedo and consequently reflect incoming radiation. Therefore, the main departures in the ASR from what would be expected simply from the Sun–Earth geometry are the signatures of persistent clouds. Bright clouds occur in the ITCZs and especially

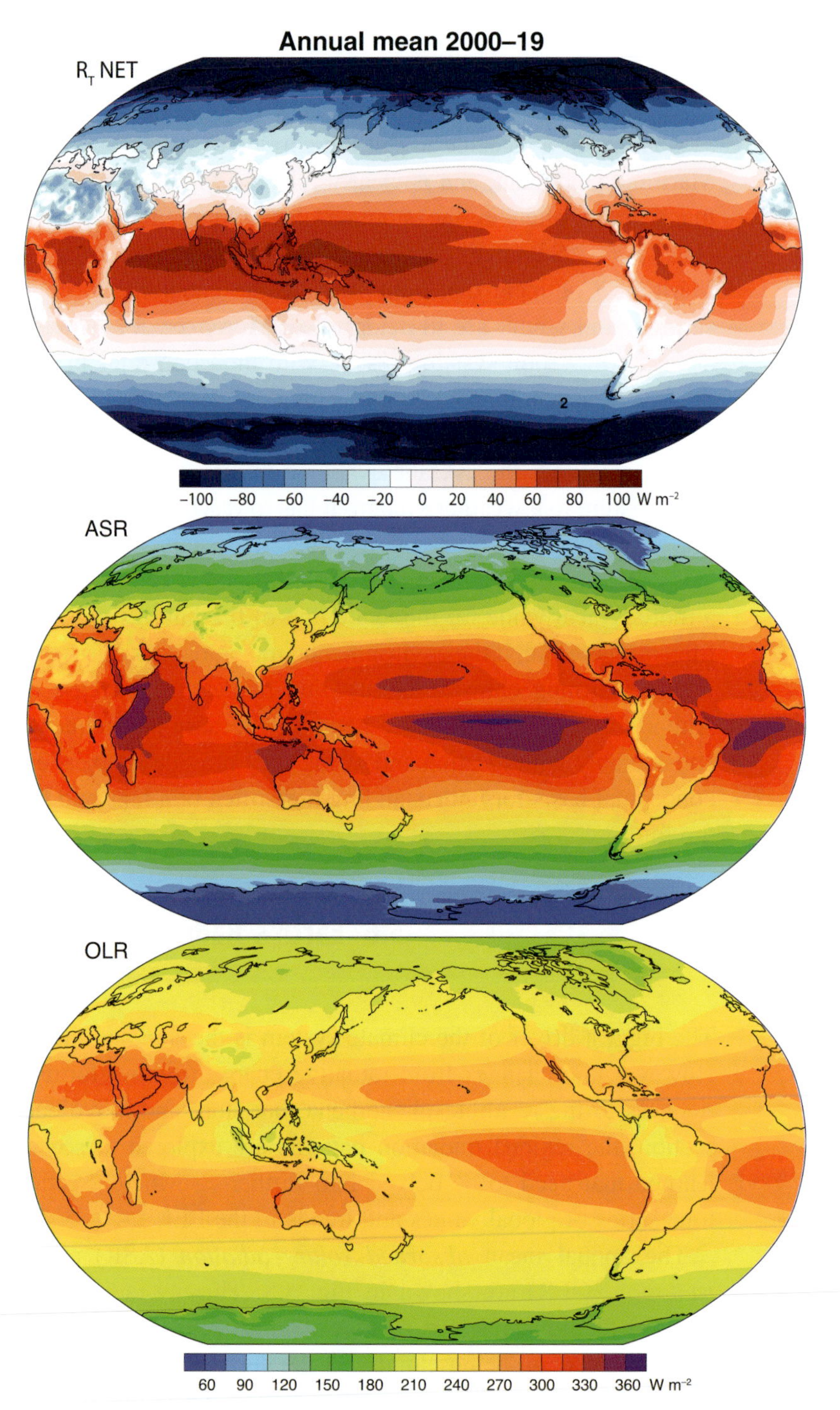

Fig. 5.8 Maps of the annual mean net radiation, ASR, and OLR at the top of the atmosphere based on the period March 2000 to December 2019 from CERES EBAF4. The units are W m^{-2}. Updated from Trenberth et al. (2019). (CERES is the NASA Clouds and the Earth's Radiant Energy System project and EBAF4 is the fourth generation "Energy Balanced and Filled" product.)

over Indonesia and Malaysia, across the Pacific near 10° N, and over the Amazon in southern summer, contributing to the relatively low values in these locations, while dark oceanic cloud-free regions along and south of the equator in the Pacific and Atlantic and in the subtropical anticyclones absorb most solar radiation. Large deserts, such as the Sahara, tend to be cloud-free but also bright with a high albedo, and hence reflect solar radiation.

The OLR is greatly influenced by water vapor and clouds but is generally more uniform with latitude than ASR (Fig. 5.8) because the temperature dependence is for absolute temperature to the fourth power (in kelvin). Nevertheless, the signature of high-top, and therefore cold clouds is strongly evident in the OLR. Similarly, the dry cloud-free regions are where most surface radiation escapes to space. There is a remarkable cancellation between much of the effects of clouds on the net radiation (Fig. 5.8). In particular, the high convective clouds are bright and reflect solar radiation but are also cold and hence reduce OLR. The main remaining signature of clouds in the net radiation from Earth is seen from the low stratocumulus cloud decks that persist above cold ocean waters most notably off the west coasts of California and Peru. Such clouds are also bright but, as they have low tops, they radiate at temperatures close to those at the surface, resulting in a net cooling of those regions. Note that the Sahara Desert has a high OLR, consistent with dry cloud-free and warm conditions, but as it is also bright and reflects solar radiation, it stands out as a region of net radiation deficit.

Immediately, these observations indicate that much more is going on. How can the desert be so hot by day and yet reside as a deficit of radiation? Of course, the answer has to be the role of the atmospheric dynamics: the movement of heat from other areas into the desert regions through the atmospheric winds. In this case, strong anticyclonic conditions lead to systematically subsiding air that warms adiabatically (simply because the air is compressed, not by addition of heat) which also suppresses clouds and rain. The clear skies allow ample sunshine in by day, making for very hot surface conditions and promoting the desert, but by the same token, the clear skies and absence of water vapor means strong radiative cooling to space by night and hence dramatic changes in temperature throughout the daily cycle. The general tendency for subtropical anticyclones to form as part of the downward branch of the Hadley Circulation also provides the reason that most deserts are in the subtropics.

For Earth, on an annual mean basis, there is an excess of solar over outgoing longwave radiation in the tropics and a deficit at mid- to high latitudes (Fig. 5.8) that sets up an equator-to-pole temperature gradient. These combine with Earth's rotation to produce a broad band of westerlies and a jet stream in each hemisphere in the troposphere (Fig. 1.2; and see Sidebar 7.2). Embedded within the mid-latitude westerlies are large-scale weather systems which, along with the ocean, act

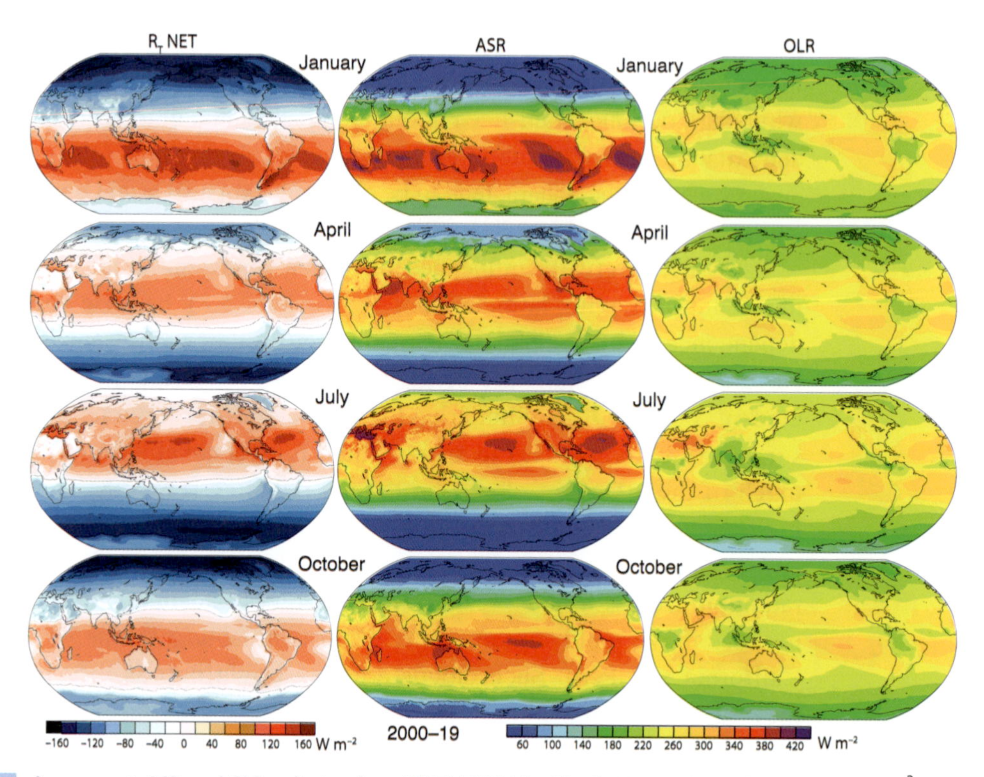

Fig. 5.9 Average net, ASR, and OLR radiation from CERES EBAF4 for March 2000 to December 2019 in W m^{-2}. The scale below is the same for each month but differs for net versus ASR and OLR.

to transport heat polewards to achieve an overall energy balance, as described below (see Chapter 9 for details).

The annual mean values of radiation (Fig. 5.8) hide a lot of important details, however. As well as the diurnal cycle, discussed more in Section 7.4, there is a strong annual cycle of the seasons (Fig. 5.9) as represented by long-term averages for the central months of each season: January, April, July, and October. In January, the southern summer and northern winter, a huge excess of radiation into the southern hemisphere is mostly taken up by the oceans (Section 8.4). In July it is the reverse: winter reigns in the south and excess radiation occurs in the northern hemisphere. The excess heat taken up by the oceans is then largely sent back into the atmosphere in winter. This is what gives the maritime climate for the oceans and adjacent areas, where the changes in surface temperature are greatly moderated by the ocean. In contrast, on land, where there is far less heat capacity (see Section 6.5) and hence the ability to store energy is much reduced compared with the oceans, there is instead a greater annual cycle of surface temperature.

Hence from a radiation perspective, in the transition seasons spring and autumn, the view from the standpoint of radiation is skewed. October looks a bit more like January, while April looks a bit more like July.

References and Further Reading

Jones, P.D., M. New, D.E. Parker, S. Martin, and I.G. Rigor, 1999: Surface air temperature and its changes over the past 150 years, *Reviews in Geophysics*, **37**, 173–199.

Silverman, J., 2013: *Opening Heaven's Floodgates: The Genesis Flood Narrative, Its Context, and Reception*. Piscataway, NJ: Gorgias Press, 548pp.

Trenberth, K. E., 1983: What are the seasons? *Bulletin of the American Meteorological Society*, **64**, 1276–1282.

Trenberth, K. E., 2011: Changes in precipitation with climate change. *Climate Research*, **47**, 123–138, doi: 10.3354/cr00953.

Trenberth, K. E., and Y. Zhang, 2018: How often does it really rain? *Bulletin of the American Meteorological Society*, **99**, 289–298. doi: 10.1175/BAMS-D-17-0107.1.

Trenberth, K. E., A. Dai, R. M. Rasmussen, and D. B. Parsons, 2003: The changing character of precipitation. *Bulletin of the American Meteorological Society*, **84**, 1205–1217. doi: 10.1175/bams-84-9-1205.

6 The Climate System

The basic components of the climate system, the atmosphere, oceans, land and ice are described along with their structure, and their role in heat storage.

6.1 Atmosphere

6.1.1 Composition and Structure

The atmosphere is the layer of air surrounding Earth. The atmosphere consists of four layers: the troposphere, stratosphere, mesosphere, and thermosphere. Figure 6.1 shows the different layers and how the temperature changes with height from the ground to the very thin top of the atmosphere. The troposphere, the lowest layer of the atmosphere, is where we live and where weather occurs. Temperatures generally decrease with height. The boundary between the troposphere and the stratosphere is called the tropopause, which is higher in the tropics and lower at mid to high latitudes. The temperature in the stratosphere generally increases with height because it encompasses the ozone layer which absorbs ultraviolet (UV) rays from the Sun. In the mesosphere the temperature again decreases with height and it is well mixed with a similar composition to the troposphere, except there is little water vapor. The thermosphere is the uppermost layer of the atmosphere where the

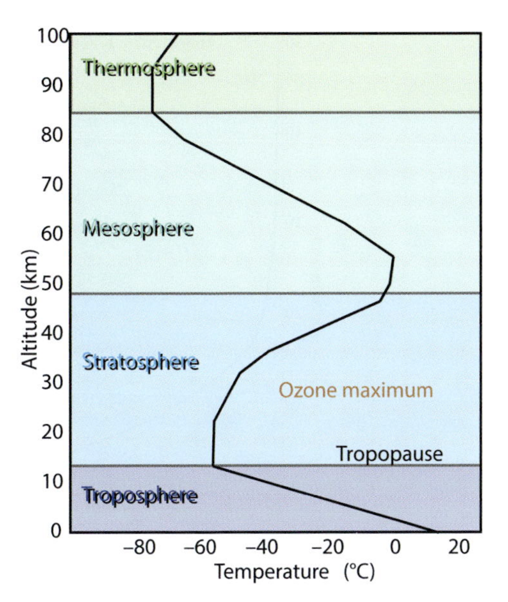

Fig. 6.1 The vertical structure and layers of the atmosphere based upon the temperature profile shown. The tropopause is the top of the troposphere; the lower stratosphere is where the predominant ozone layer begins and, as ozone absorbs radiation, temperatures increase up to the stratopause that divides the stratosphere from the mesosphere, where temperatures fall off with altitude again. The thermosphere, above 80–90 km altitude, is where the molecules become broken up into ionized particles.

temperature increases with height because it is directly heated by the Sun. Above about the mesopause, which divides the mesosphere from the thermosphere, the molecules of the atmosphere become ionized by the active very shortwave radiation from the Sun (X-rays, UV) or cosmic rays (which may come from elsewhere) that do not make it to the lower levels, and their abundance is greatly affected by solar activity. The ionosphere, which overlaps with the other layers defined by temperature, plays a major role in transmission of radio waves over great distances around the world. At these very high levels of the atmosphere, above about 80 km altitude, the composition of the atmosphere is very different than at lower levels.

The mass of the atmosphere averages 5.1480×10^{18} kg and is equivalent to a surface pressure globally of 985.5 hPa (also called millibars); see Fig. 6.2. Although the mass of the dry atmosphere is virtually constant, the water vapor component varies considerably and depends a lot on temperature. On average it has a mass of 1.27×10^{16} kg, but this varies with season. The equivalent surface pressure is highest in northern summer in July (2.62 hPa) and lowest in December and January in northern winter (2.33 hPa), and hence the total mass varies similarly. Water vapor also varies enormously with latitude and height. The atmospheric mass falls off exponentially with height, and the pressure is about

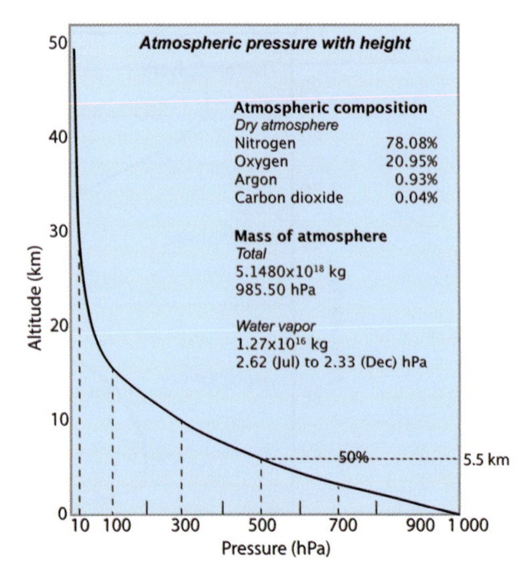

Fig. 6.2 The vertical structure of the atmosphere in terms of pressure levels and thus the mass of the atmosphere above each level. The halfway point is indicated. The composition of the atmosphere is also given.

500 hPa at 5.5 km altitude. The tropopause may be just above 300 hPa in the extratropics, near 10 km, but close to 100 hPa in the tropics. Hence only 10% of the atmosphere is above that level. The value 10 hPa (1% of the surface value) occurs at close to 30 km altitude in the middle of the stratosphere. At 80 km altitude the pressure is down to about 0.01 hPa and thus 99.999% of the atmosphere is below that level.

Note that the standard surface pressure value used is often 1013.25 hPa, equivalent to 760 mm of mercury (in a mercury barometer), and this was defined as 1 atmosphere until 1982, but the value was changed to 1000 hPa in 1982. These values are common at sea level, but mountains occupy a lot of the lower atmosphere and make for the difference with the observed global mean surface value of 985.5 hPa. Sea-level pressure (SLP) is much higher, but commonly used to compare neighboring areas in meteorology in mean SLP maps to determine the locations of anticyclones (high-pressure systems) and depressions or cyclones (low-pressure systems).

Most of the atmosphere is made up of nitrogen N_2 (78% of dry atmosphere) and oxygen O_2 (21%). Nitrogen is odorless, colorless, and tasteless, and mostly unreactive at room temperature. Oxygen is used by all living creatures and is essential for respiration. It is also essential for combustion or burning. Argon (1%) is an inert gas (not reactive) and hence is used in lightbulbs (to prevent corrosion), double-paned windows (for insulation), and welding. Other gases, in smaller amounts, are known collectively as trace gases.

When a molecule absorbs electromagnetic radiation, it can undergo various types of excitation: principally translational energy through motion of the molecule as a whole; rotational energy due to bodily rotation of the molecule about an

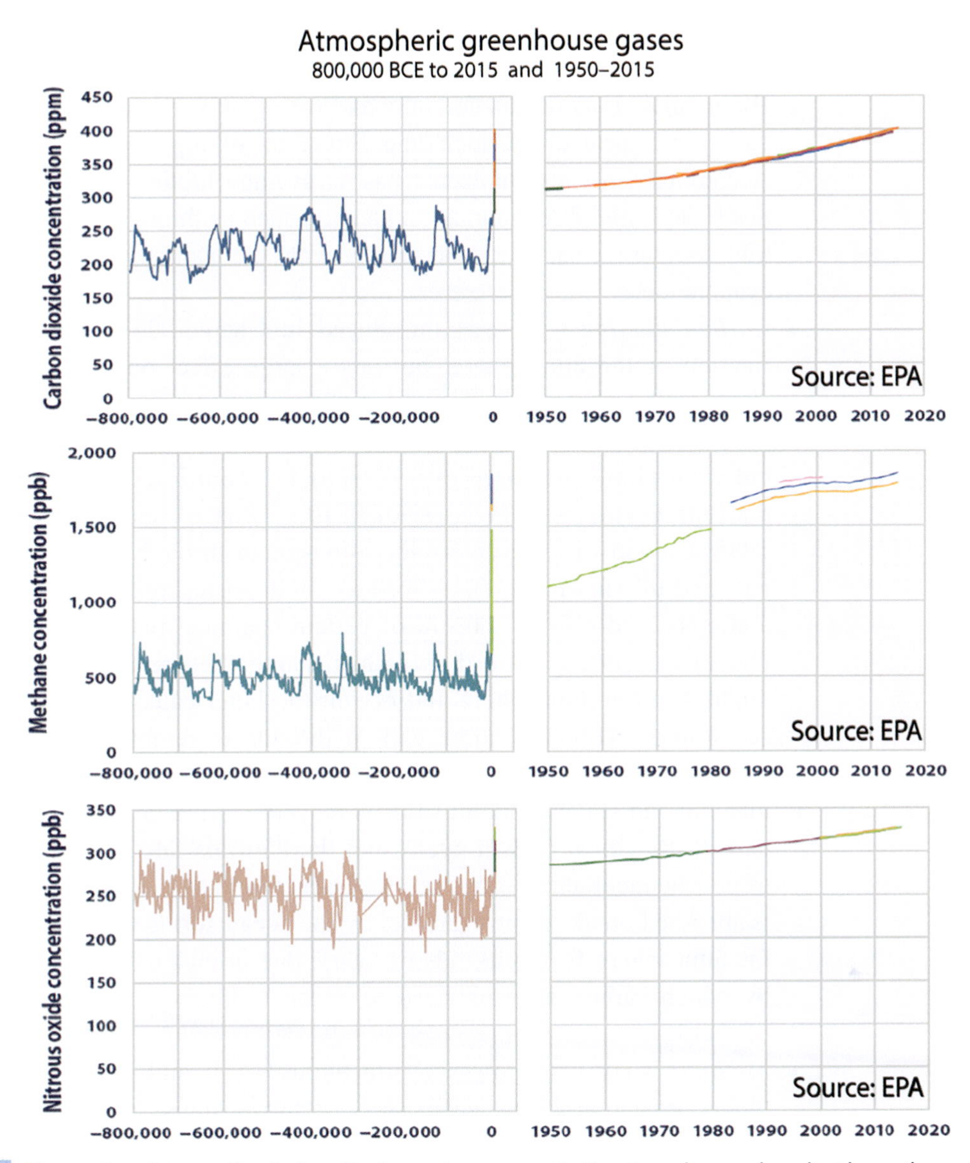

Fig. 6.3 Time series of the total main long-lived greenhouse gases in the atmosphere: carbon dioxide, methane and nitrous oxide for the past 800 000 years (left) and after 1950 (right), from EPA (2020).

axis passing through the center of gravity of the molecule; and vibrational energy due to the periodic displacement of its atoms about their equilibrium positions. Molecules with three or more atoms can vibrate when they absorb heat, and as a rule, gas molecules with three or more atoms are greenhouse gases. However, because N_2 and O_2 consist of only two atoms per molecule, they are restricted in how they are excited, and they are not greenhouse gases. Hence water vapor H_2O, ozone O_3, nitrous oxide N_2O, and methane CH_4 are all powerful greenhouse gases, and most are increasing because of human activities (Fig. 6.3).

6.1.2 Greenhouse Gases

- ***Water vapor*** H_2O occurs naturally and varies with weather variations as part of the hydrological cycle, including formation of fog, clouds, rain, and snow. Accordingly, a water molecule has an average lifetime in the atmosphere of just 9 days, and direct human influences, such as through irrigation and industrial use, are small by comparison with the natural sources from surface evaporation.
- ***Carbon dioxide*** CO_2 was introduced in Chapter 2. It makes up a small fraction of the atmosphere, but has a large effect on climate. There were about 280 parts per million by volume (ppm) of carbon dioxide in the atmosphere in the mid-19th century at the start of the industrial revolution, but values have increased over 45% to 410 ppm in 2020 mainly because of the burning of fossil fuels. About half of this increase has occurred since 1985. Carbon dioxide enters the atmosphere through burning coal, natural gas, and oil (fossil fuels), as well as solid waste, trees and other biological materials, and also as a result of certain chemical reactions (e.g., manufacture of cement). Carbon dioxide has a long lifetime and accordingly accumulates in the atmosphere and ocean; the latter causing acidification of the sea waters. Although some carbon dioxide is removed quickly by being involved in photosynthesis and taken up by trees and plants to make woody material and leaves, much of this is recycled in the fall. Atmospheric CO_2 is part of the global carbon cycle, and therefore its fate is a complex function of geochemical and biological processes. Some of the excess carbon dioxide is absorbed quickly (for example, by the ocean surface), but some remains in the atmosphere for thousands of years, due in part to the very slow process by which carbon is transferred to ocean sediments or through weathering of rocks.
- ***Methane*** CH_4 is a powerful greenhouse gas, and accounts for 16% of global greenhouse gas emissions. Methane stays in the atmosphere for about 9 years (note that this is an exponential lifetime; see Box 10.1), much shorter than carbon dioxide. However, its global warming potential is 86 times higher than carbon dioxide when averaged over 20 years, and 28 times higher over 100 years. Global concentrations rose from 722 parts per billion by volume (ppbv) in pre-industrial times to 1875 ppbv by the end of 2019, an increase by a factor of more than 2.5, and the highest value in at least 800 000 years. Methane is used as a fuel as natural gas, and forms naturally in wet boggy areas where it is known as marsh gas. About 60% of methane emissions are man-made. Methane is emitted during the production and transport of coal, gas, and oil. Recent growth

in emissions has arisen through the fracking boom, through leaks and fugitive emissions. Methane emissions also result from livestock and other agricultural practices and from the decay of organic waste in municipal solid waste landfills. When burned, methane releases carbon dioxide into the atmosphere.

- ***Nitrous oxide*** N_2O is emitted during agricultural (especially synthetic nitrogen fertilizer) and industrial activities, combustion of fossil fuels and solid waste, as well as during treatment of wastewater. It has a lifetime of about 116 years before being removed or destroyed through chemical reactions. The impact of N_2O on warming the atmosphere is almost 300 times that of the same mass of carbon dioxide. Globally, about 40% of total N_2O emissions come from human activities.
- ***Ozone*** O_3 mainly resides in the lower stratosphere, but overall amounts decreased by about 3% between 1979 and 2014. All of the decrease happened in the stratosphere, mainly between 1979 and 1994 because of human-made ozone-depleting substances, such as chlorofluorocarbons (CFCs), halocarbons, and halons. Biggest decreases occurred in the southern spring from September to November over and around Antarctica and formed the "ozone hole." International efforts since 1987 have reduced emissions and phased out the use of CFCs which were formerly used extensively as refrigerants and in spray cans. Meanwhile the amount of ozone in the *troposphere* increased by about 3% between 1979 and 2014. Shortwave (289–315 nanometer wavelength) UV radiation (see Sidebar 4.1), called UV-B, is usually absorbed in the ozone layer, and hence reductions in ozone increase the UV-B reaching Earth's surface. UV-A (315–400 nanometer wavelength) and other solar radiation are not strongly absorbed by the ozone layer. Human exposure to UV-B increases the risk of skin cancer, cataracts, and a suppressed immune system. UV-B exposure can also damage terrestrial plant life, single cell organisms, and aquatic ecosystems.

6.1.3 Aerosols

Aerosols are tiny micron-sized particulates in the atmosphere. They occur in the atmosphere from natural causes; for instance, they are blown off the surface of deserts or dry regions. Desert dust is quite common and becomes fairly widely dispersed by dust storms. Over the ocean, sea salt from ocean spray can form particles that very effectively provide cloud condensation nuclei. Dimethyl sulfide (DMS), produced by many marine organisms, is the major natural source of natural marine sulfate aerosols.

Volcanic aerosol is often big and heavy enough to quickly fall out of the atmosphere, but much longer-lasting volcanic aerosol may form in the stratosphere predominantly through oxidation of sulfur dioxide gas which is converted to droplets of sulfuric acid over the course of a week to several months after the eruption and thus to sulfate particles. The eruption of Mt. Pinatubo in the Philippines in June 1991 added considerable amounts of aerosol to the stratosphere which, for about 2 years, scattered solar radiation leading to a loss of radiation at the surface and a cooling there.

Human activities contribute significantly to aerosol particle formation through injection of sulfur dioxide into the atmosphere (which contributes to acid rain) particularly from power stations, and through biomass burning. Sulfur dioxide initially disperses as a gas which is not washed out of the atmosphere, and sulfate aerosols can therefore become widely distributed. They can be seen from airplane windows across the northern hemisphere as a pervasive whitish haze.

The main direct effect of aerosols, such as sulfate particles, is the scattering of some solar radiation back to space; which tends to cool Earth's surface. Carbonaceous aerosols influence the radiation budget by directly absorbing solar radiation, leading to local heating of the atmosphere and, to a lesser extent, by absorbing and emitting thermal radiation.

A further influence of aerosols such as DMS is indirect, as many act as nuclei on which cloud droplets condense. A changed concentration therefore affects the number and size of droplets in a cloud and hence alters the reflection and the absorption of solar radiation by the cloud. Particulate pollution tends to make more, smaller cloud droplets, which makes clouds brighter and more reflective of solar radiation. However, the effects on onset of precipitation, and thickness, size and lifetime of clouds can vary enormously. There is increasing understanding of aerosols, especially from observations and research in microphysics to better appreciate the complex indirect effects of aerosols on clouds.

In the troposphere, aerosols typically fall out or are washed out of the atmosphere as they get caught up in weather systems. Hence, human-made aerosols typically remain in the atmosphere for only a few days and they tend to be concentrated near their sources such as industrial regions. The radiative forcing therefore possesses a very strong regional pattern, and the presence of aerosols adds complexity to possible climate change as it can help mask, at least temporarily, any global warming arising from increased greenhouse gases. However, the aerosol effects do not cancel the global-scale effects of the much longer-lived greenhouse gases, and significant climate changes still result. The variety of aerosols and their sources adds to the complexity, as does the environment in which they are inserted, such as whether there is water, ice, or mixed phase droplets.

Sidebar 6.1: Hydrostatic Approximation

The atmosphere and the oceans for the most part can be considered to be in what is called *hydrostatic balance*, which is to say that at any point the pressure experienced is simply the weight of all the air (and water) above that point due to gravity. The main exception is for small scales, such as in thunderstorms, where vertical motions are substantial, and then the weight can be offset by the air movement.

For the compressible atmosphere, this means integrating the mass in an infinitesimal layer of thickness δz, which is $\rho \delta z$, where ρ is the density. In turn, ρ depends on temperature because the atmosphere expands when warm. The equation of state for the atmosphere (for an ideal gas) is $p = \rho RT$, where R is the gas constant and T is the temperature. The weight is then $\rho g \delta z$, where g is gravity and this applies to the difference in pressure experienced across that layer δp. If z is height going upwards, then pressure increases downwards (see Fig. 6.2). Hence the hydrostatic equation is $\delta p/\delta z = -\rho g$. For the ocean, where the density is approximately constant, this means that at the bottom of a layer of thickness dz, the incremental pressure is $\rho g dz$.

For the compressible atmosphere, however, if it is assumed that the temperature is constant, then $dp/p = -g/RT\ dz$, which can be integrated to give $p(z) = p_0 \exp(-z/H)$ and the quantity $H = g/RT$ is called the scale height, as it is the distance to go upwards or downwards to change values of pressure by a factor of the exponential value $e \approx 2.718$, and p_0 is a reference level p.

Strictly speaking, because gravity varies slightly with location and altitude this involves a model of the gravity on Earth, called geopotential, that enables a constant standardized value of g to be used. The height z is then called the geopotential height. Then the change in geopotential height in going from a level of pressure p_1 to one of pressure p_2 is $\delta z = RT/g \ln p_1/p_2$, where ln is the natural logarithm and it is again assumed that T is a constant. Because T varies in practice, these expressions have to be integrated.

6.2 Oceans

The relationships among temperatures and density of the ocean are quite complex, in part because of the peculiar characteristics of pure water and also because of various salts and impurities. Hence, the equation of state of the ocean is quite complex. The first clue is that fresh water has a maximum density at about 4°C with a value of 1000 kg per cubic meter (Chapter 8). Density decreases as waters warm above this value, by as much as 0.4% by about 30°C. But density also decreases at colder temperatures and, at freezing point (0°C), is 0.01% less. Although this is not much, it has enormous implications for waters in a lake, for instance, where the coldest water rises to the surface as temperatures drop, and ice then forms at the surface. Ice itself has a density of about 91.7% of water, and hence it floats. Saline water is heavier than fresh water. The density of surface ocean waters varies from about 1020 to

1029 kg m^{-3}. With salinity included there is no maximum density except at the coldest values, and sea water tends to freeze about −1.8°C. In the process, salt (brine) tends to get rejected, making the icy waters even saltier and denser still. Hence ice also floats in the ocean, as seen from icebergs.

For typical salinities between about 34 and 35 parts per thousand, the density ranges from about 1027.8 kg m^{-3} at 0°C to 1027.3 kg m^{-3} at 5°C, to 1025.7 kg m^{-3} at 15°C. Hence, the changes of density with temperature are less if the temperature is below about 5°C than for higher temperatures. Indeed, at high latitudes, polewards of about 50° in both hemispheres, the annual mean values do not change much with depth (Fig. 6.5, further below).

The ocean is the principal time-varying reservoir of heat for the climate system. Its density is 1000-fold more than the atmosphere and the total heat capacity of the atmosphere is equivalent to that of only 3.2 m of the ocean. The role of land is much less, mainly because the thermal conductivity of land is 1000-fold less than the turbulent conductivity of the ocean arising from mixing and convection. The average depth of the ocean is about 3800 m, although the deepest spot in the Mariana Trench is almost 11 km deep. Accordingly, if the atmospheric temperatures depart significantly from the sea-surface temperatures, then large exchanges of heat and moisture occur between the two fluids, resulting in the atmosphere mostly adjusting to the ocean temperatures. Mean SSTs (Fig. 5.4) seldom exceed 30°C, and the warmest waters are found in the tropical western Pacific, called the Warm Pool. In high latitudes, sea water freezes at −1.8°C, and sea ice forms in winter but melts in the summer, unless it is permanent.

The oceans are stratified opposite to the atmosphere, with warmest waters near the surface. The vertical average temperature structure of the oceans (Fig. 6.5, further below) is for a shallow so-called mixed layer above a much deeper cool layer, with the divide referred to as the *thermocline*. The surface is in contact with the atmosphere, and radiation may penetrate tens of meters into the ocean. This surface layer is also called the sunlight zone or epipelagic zone. The upper layers are mechanically mixed strongly by surface winds and may also be mixed by convection.

In the ocean, convection arises from cooling at the surface and transport of heat upwards occurs through colder, denser waters sinking and being replaced by lighter, more buoyant waters. Another vital factor in convection is the salinity of the waters, because this also affects density. Consequently, the densest waters are those that are cold and salty, and these are found at high latitudes where there is a deficit in radiation and atmospheric temperatures are low, but also the formation of sea ice leads to a rejection of brine and increases the salinity of the surrounding waters. The cold deep abyssal ocean turns over only very slowly on timescales of hundreds to thousands of years.

On average, sea water in the world's oceans has a salinity of approximately 3.5%, or 35 parts per thousand (ppt) (also expressed as 35‰). Salinity is an important factor in determining many aspects of the biogeochemistry of sea waters; that is to say it plays a role in both the chemical and biological processes. It varies from low values where rivers enter the ocean, while higher values occur where evaporation exceeds precipitation, especially in the subtropics. Values as high as 200 ppt occur in the Dead Sea. The main salt is common salt or sodium chloride, but other salts are also typically present, such as magnesium chloride, potassium nitrate, and sodium bicarbonate.

Significant exchanges with the atmosphere occur for carbon dioxide and nitrous oxide, both greenhouse gases. The former leads to acidification of the oceans by forming carbonic acid. Solubility of carbon dioxide is greater at cold temperatures and less in more saline waters.

The oceans cover 70.8% of the surface, although with a much greater fraction in the southern hemisphere (80.9% of the area) than the northern hemisphere (60.7%), and through their fluid motions and high heat capacity, have a central role in shaping Earth's climate and its variability. Ocean waters are typically dark with an albedo of 6–7%. The fraction of land versus ocean, or more completely water (to account for lakes, etc.) (Fig. 6.4) shows the almost opposite character of the two hemispheres, with the Antarctic entirely land, while the Arctic is open water, switching to all ocean at 60° S versus 70% land in the northern hemisphere at 65° N. The southern hemisphere is the ocean hemisphere. Only in the tropics do the two hemispheres have comparable amounts of land versus water.

The continents divide the ocean into the Atlantic, Pacific, Indian, Arctic, and Southern basins and further segregation occurs through mid-ocean ridges that can

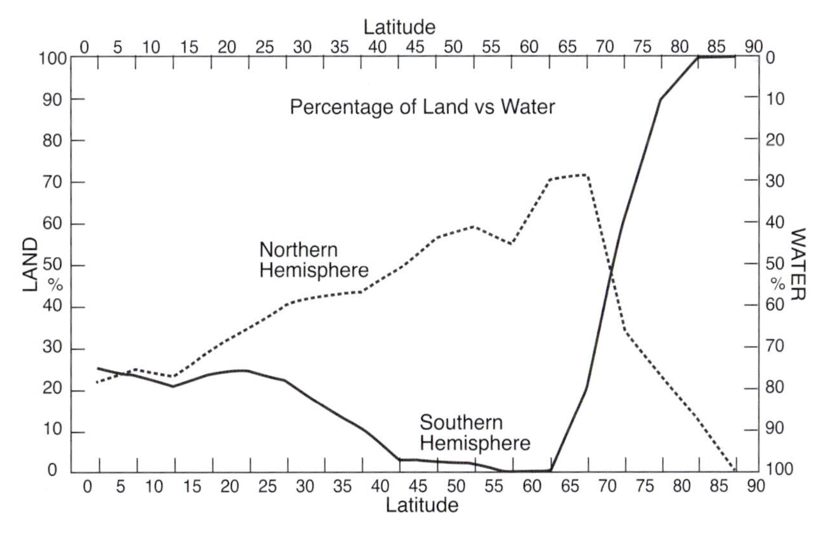

Fig. 6.4 Fraction of the zonal area around each latitude band in 5° increments for the two hemispheres. Adapted from Trenberth (1983). © American Meteorological Society. Used with permission

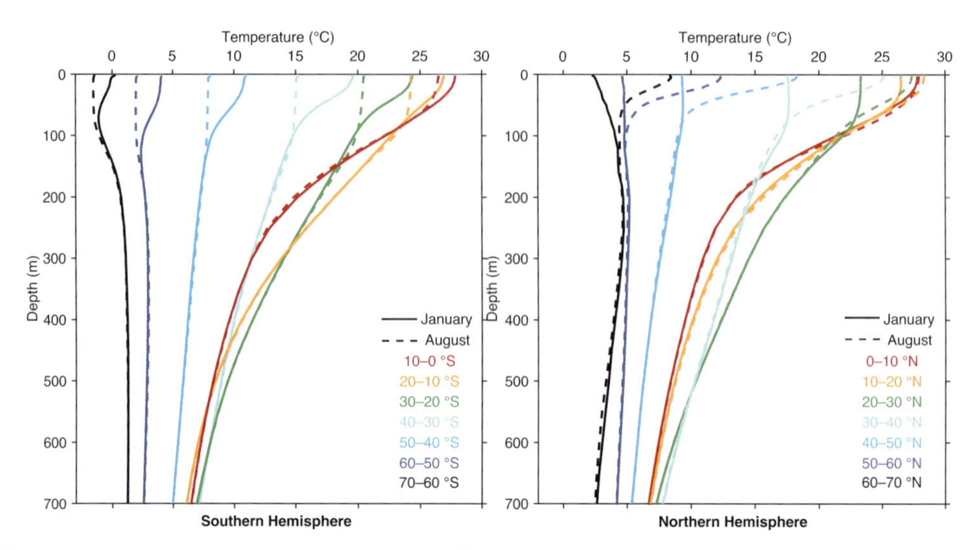

Fig. 6.5 Vertical average temperature profiles for each 10° latitude for January (solid curve) and August (dashed curve) for the southern hemisphere (left) and northern hemisphere (right) in °C. Data kindly provided by Lijing Cheng and based upon Cheng and Zhu (2014)

rise a few thousand meters from the sea floor. The largest ocean is the Pacific with a surface area of 168.7 million square kilometers (45%), followed by the Atlantic (85.1; 24%), Indian (70.6; 20%), Southern Ocean (22.0; 6.1%) and Arctic Ocean (15.6; 4.3%). The pattern of surface winds plus heat and moisture exchanges with the atmosphere causes the water to circulate (Chapter 8).

In Fig. 6.5, the vertical average temperature profiles from the far north to Antarctica are given every 10° latitude for the warm and cold seasons; Chapter 8 contains a lot more detail. Stratification is strongest in lower latitudes and summer, and the differences between January and August at each latitude show the main mixed layer and thermocline depth of about 90 m in many locations. At very high latitudes where sea ice formation plays a role, surface temperatures may be lower than those below 100 m depth in winter owing to relatively fresh waters. At 700 m depth, ocean temperatures are typically 3–7°C, much less than values near the surface. The somewhat idealized temperature changes throughout the year are characterized in the mid latitudes of the northern hemisphere oceans as being approximately isothermal for the top 100 m in March, warming and mixing through the upper 50 m or so into June by perhaps 3°C, warming by 10°C from the March value through an increasingly shallow layer, perhaps 20 m depth, until August, then declining but mixing through deeper layers for the rest of the annual cycle. Hence, in September the surface temperature starts to decline but the warm mixed layer becomes deeper by 10 m or so. By November the mixed layer is down to 70 m depth, but temperatures are less than 3°C above the March values.

The vertical distribution of temperature cannot be maintained for centuries in a fluid that conducts heat unless a volume of cold water continually replenishes the deep waters while net heating restores the warmth of the upper layers. Because the bottom of the ocean is insulating, for the most part (with exceptions for hot spots where tectonic activity occurs), the main exchanges of heat occur with the atmosphere. Therefore, the uptake of heat in the tropics and low latitudes by the ocean has to be balanced by the loss of heat at high latitudes, and hence there must be a transport of heat from the places where it is absorbed in the ocean to places where it is lost. These aspects are amplified in Chapters 8 and 9, where recent observations and analyses have allowed them to be determined in new ways.

The most important characteristic of the oceans is that they are wet. Water vapor, evaporated from the ocean surface, provides latent heat energy to the atmosphere when precipitated out. Wind blowing on the sea surface drives the large-scale ocean circulation in its upper layers (Fig. 8.7). The ocean currents carry heat and salt along with the fresh water around the globe (Fig. 1.2). The oceans therefore store heat, absorbed at the surface, for varying durations and release it in different places, thereby ameliorating temperature changes over nearby land and contributing substantially to variability of climate on many timescales. Additionally, the ocean thermohaline circulation (Section 8.3), which is the circulation driven by changes in sea water density arising from temperature (thermal) or salt (haline) effects, allows water from the surface to be carried into the deep ocean where it is isolated from atmospheric influence and hence it may sequester heat for periods of 1000 years or more.

The main energy transports in the ocean are those of heat associated with overturning circulations as cold waters flow equatorwards at some depth while the return flow is warmer near the surface. There is a small poleward heat transport by gyre circulations whereby western boundary currents, such as the Gulf Stream in the North Atlantic and the Kuroshio in the North Pacific, move warm waters polewards while part of the return flow is of colder currents in the eastern oceans. However, a major part of the Gulf Stream return flow is at depth, and the most pronounced thermohaline circulation is in the Atlantic Ocean. In contrast, the Pacific Ocean is fresher and features shallower circulations. A key reason for the differences lies in salinity (Section 10.2). Because of the net atmospheric water vapor transport by tropical easterly trade winds from the Atlantic to the Pacific across the Central American isthmus, the North Atlantic is much saltier than the North Pacific.

6.3 Land

The land surface encompasses an enormous variety of topographical features and soils, differing slopes (which influence runoff and radiation received) and water

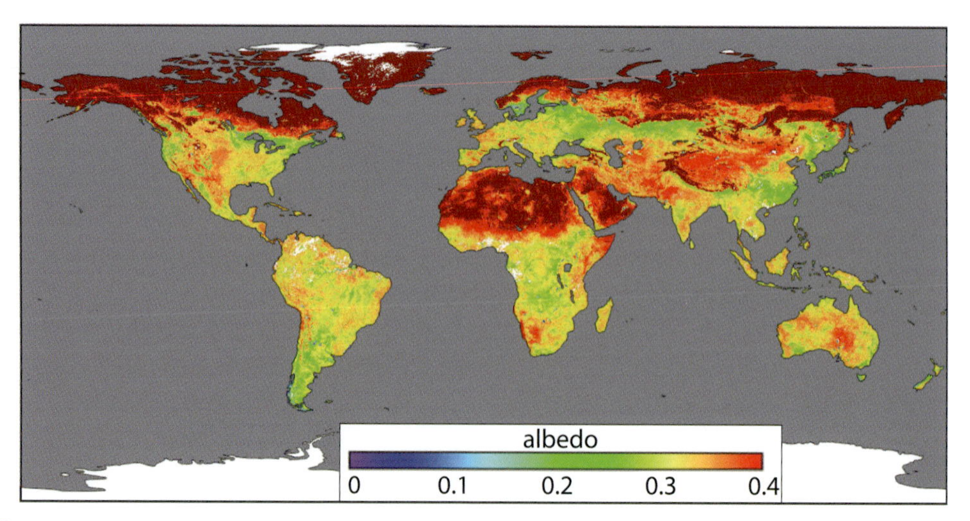

Fig. 6.6 Surface albedo as a fraction. Areas colored red show the brightest, most reflective regions; yellows and greens are intermediate values; and blues and violets show relatively dark surfaces. White indicates where no data were available, and no albedo data are provided over the oceans. This image was produced using data composited over a 16-day period, from April 7 to 22, 2002, from MODIS (Moderate Resolution Imaging Spectroradiometer), courtesy NASA Earth Observatory.

capacity. The highly heterogeneous vegetative cover is a mixture of natural and managed ecosystems that vary on very small spatial scales. About 71% of land is regarded as habitable. Of that, half is for agriculture (77% livestock including grazing land, 23% crops), 37% is forests, 11% shrubs, 1% is urban built-up areas, and another 1% is fresh water. The surface albedo can vary enormously, from dark moist soil 10–20%, grass 35%, sand 40%, concrete pavement 15–55%, forest 8–15%, and fresh snow 80%. An example for April (Fig. 6.6) illustrates the rich variety and shows up snow-covered regions in the north and northern mountains. Desert regions also stand out as bright surfaces.

The very different distribution of land with latitude in the two hemispheres is apparent (Fig. 6.4). Most land is in the northern hemisphere: about 40% of the surface, versus 20% in the southern hemisphere. One quarter of the northern hemisphere and 17% of Earth's exposed land surface is underlain by permafrost; that is ground with a temperature remaining at or below 0°C for at least two consecutive years. The thermal state of permafrost is sensitive to changing climatic conditions and in particular to rising air temperatures and changing snow regimes.

The heat penetration into land occurs mainly through conduction and diffusion except where water plays a role, so that heat penetration is limited and slow. The subsurface is far from uniform and features fissures and cracks that allow water seepage. Changes in soil moisture affect the disposition of heat at the surface and whether it results in increases in air temperature or increased evaporation of moisture. The latter is complicated by the presence of plants which can act to pump moisture out of the root zone into the leaves, where it can be released into the atmosphere as the plant participates in *photosynthesis*; a process called *transpiration*. The behavior of land ecosystems can be greatly influenced by changes in atmospheric composition and climate. The availability of surface water and the use of the Sun's energy in photosynthesis and transpiration in plants influence the uptake of carbon dioxide from the atmosphere as plants transform the carbon and water into usable food for the plant. Changes in vegetation alter how much sunlight is reflected and how rough the surface is in creating drag on the winds, and the land surface and its ecosystems play an important role in the carbon cycle and fluxes of water vapor and other trace gases.

Temperature profiles taken from boreholes into land, permafrost or ice caps provide a blurry coarse estimate of temperatures in years long past. Increases in land energy cause increased evapotranspiration and increased temperatures, so that a consequence is decreases in soil moisture that may enhance the occurrence of extreme drought and heatwaves. Consequently, surface air-temperature changes over land occur much faster and are much larger than over the oceans for the same heating and, because we live on land, this directly affects human activities. Such extreme events often have negative health effects in vulnerable sectors of the human and animal population.

Soil temperature varies from month to month as a function of incident solar radiation, rainfall, seasonal swings in overlying air temperature and cloudiness, local vegetation cover, type of soil, and depth in the earth. Due to the much higher heat capacity of soil relative to air and the thermal insulation provided by vegetation and surface soil layers, seasonal changes in soil temperature deep in the ground are much less than and lag significantly behind seasonal changes in overlying air temperature. Thus, in spring the soil naturally warms more slowly and to a lesser extent than the air, and by summer it has become cooler than the overlying air. Likewise, in autumn the soil cools more slowly and to a lesser extent than the air, and by winter it is warmer than the overlying air. At soil depths greater than 5–8 m below the surface (6 m for light dry soil), the soil temperature is relatively constant, and corresponds roughly to the water temperature measured in groundwater wells 6–15 m deep, referred to as the "mean earth temperature." These numbers are important for geothermal heat pumps.

Although the land surface is naturally highly heterogeneous, human civilization has undertaken management of huge tracts of land for farming, forestry, and other purposes. It has been estimated that more than half of the world's habitable land is used for agriculture. More than 50% of the population live in urban areas; however, cities in total occupy less than 0.5% of Earth's land area. This depends on the definition of "urban," which here refers to the built environment that is human-constructed (such as roads and buildings). The urban land area varies from only 0.17% of total continental land area in Africa to 0.67% in North America, with most regions near the continental average of 0.5% urbanized (e.g., South America, 0.47%; Asia, 0.53%). The exception is the European land mass (1.78%), where there is extensive urban morphology. Changes in land use and land cover are considered in Chapter 14.

6.4 Ice

Terrestrial ice accounts for about 1.5% of all water, compared with about 94% stored in the world's oceans. However, it contains about 99% of the fresh water. Nevertheless, small changes in land ice can greatly influence sea level, with corresponding coastal effects. The mass of ice on Earth is changing and has only been somewhat reliably determined recently. East Antarctica is the largest ice sheet on the planet, with thicknesses greater than 4600 m. West Antarctica is classified as a marine ice sheet because the topography beneath the ice sheet is largely below sea level. Antarctica has about 30 000 000 km^3 of grounded ice, about 10 times that of Greenland, which in turn is about 30 times that of small ice caps and glaciers. The equivalent sea-level rise if all the land ice were to melt is estimated at 60 m for Antarctica, 7 m for Greenland and 0.4 m for glaciers and small ice caps. The density of ice is 917 kg m^{-3} (or 0.917 times that of fresh water) and hence ice floats on water.

A glacier or ice sheet is nourished by accumulation from precipitation, especially snow, while losses occur from surface melting and ablation, and calving of icebergs at the margin. Redistribution is achieved through flows of ice. Some 11 000 years ago, toward the end of the Last Glacial Maximum, the western edge of the Greenland ice sheet was about 200 m farther advanced than today. However, glaciers of ice rarely achieve steady state, even when climate is not changing, because accumulation and ablation are always changing as the atmospheric weather evolves, and the responses in ice changes are ponderous. In addition, internal instabilities can arise in the form of ice surges.

Major ice sheets, like those over Antarctica and Greenland, have a large heat capacity but, like land, the penetration of heat occurs primarily through conduction or diffusion so that the mass involved in temperature changes from year to year is small. Temperature profiles can be taken directly from boreholes into ice and the estimate is that terrestrial heat flow is 51 milliwatts per square meter. On century timescales, however, the ice sheet heat capacity becomes important. Unlike land, ice can melt which has major consequences through changes in sea level on longer timescales.

On land, permafrost is frozen soil, rocks and sand held together by ice. There may be a shallow "active layer" near the surface that thaws in summer and refreezes in winter. Ordinarily, the heat from the center of Earth means that permafrost exhibits the warmest temperatures at the bottom of a borehole. In recent decades, however, there is a secondary peak at the surface and a minimum in between, clearly signifying the recent warming from the surface. Nevertheless, the year-to-year variability in surface temperatures is huge, and while the land and permafrost filter out a lot of the variability, the signal of warming can be quite noisy. Moreover, it leads to thawing of organic material that may have been frozen for centuries, leading to methane or carbon dioxide emissions that amplify warming through their greenhouse effect.

Sea ice is an active component of the climate system that is important because it has a high albedo. A warming that reduces sea ice also reduces the albedo, and hence enhances the absorption of solar radiation, amplifying the original warming. This is known as the *ice–albedo feedback* effect (see Section 13.2.2). Sea ice varies greatly in areal extent with the seasons, but only at higher latitudes. In the Arctic, where sea ice is confined by the surrounding continents, mean sea ice thickness has been 3–4 m deep and multi-year ice can be present, but is diminishing rapidly with global warming. Around Antarctica the sea ice is unimpeded and spreads out extensively but, as a result, the mean thickness is typically 1–2 m, and it is greatly influenced by winds.

6.5 The Role of Heat Storage

The different components of the climate system contribute on different timescales to climate variations and change. The atmosphere and oceans are fluid systems and can move heat around through convection and advection in which the heat is carried by the currents, whether small-scale short-lived eddies or large-scale atmospheric jet streams or ocean currents. Changes in phase of water, from ice to liquid to water vapor, affect the storage of heat. However, even ignoring these complexities, many facets of the climate are determined simply by the heat capacity of the different components of the climate system. The total heat capacity considers the mass involved as well as its capacity for holding heat, as measured by the specific heat of each substance.

The atmosphere does not have much capability to store heat. The heat capacity of the global atmosphere corresponds to that of only 3.2 m of the ocean. However, the depth of ocean actively involved in climate is much greater than that. The specific heat of pure water is 4182 J kg^{-1} K^{-1}, while for inorganic soil it is 700–1000 J kg^{-1} K^{-1}, and perhaps double that for soil with organic materials. Thus, the specific heat of dry land is roughly a factor of 4½ less than that of sea water (for moist land the factor is probably closer to 2). The mass is linked also to density, and the density of soil is 150–260% of that of water. Moreover, heat penetration into land is limited (see also Section 14.7) and only the top 3 m or so typically plays an active role (as an e-folding[1] depth for the variations on annual timescales, say), with a delay of 2.5 months or so at 3 m depth. Accordingly, land plays a much smaller role in the storage of heat and in providing a memory for the climate system, although it can be useful for geothermal heating applications. Similarly, the ice sheets and glaciers do not play a strong role, while sea ice is important where it forms.

The seasonal variations in heating penetrate into the ocean through a combination of radiation, convective overturning (in which cooled surface waters sink while warmer more buoyant waters below rise) and mechanical stirring by winds, through the mixed layer, and on average involve about 90 m of ocean. The thermal inertia of the 90 m layer would add a delay of about 6 years to the temperature response to an instantaneous change. This value corresponds to an exponential time constant in which there is a 63% response toward a new equilibrium value following an abrupt change. Hence the actual change is gradual. The total ocean, however, with a mean depth of about 3800 m, if rapidly mixed would add a delay of 230 years to the response. But mixing is not a rapid process for most of the ocean so that in reality the response depends on the rate of ventilation of water between the mixed upper layers of the ocean and the deeper more isolated layers through the thermocline. Such mixing varies greatly geographically; see Fig. 6.5. An overall estimate of the delay in surface temperature response caused by the oceans is 10–100 years. The slowest response should be in high latitudes where deep mixing and convection occur, and the fastest response is expected in the tropics. Consequently, the oceans are a great moderating effect on climate variations, especially changes such as those involved with the annual cycle of the seasons.

Generally, the observed variability of temperatures over land is a factor of 2–6 greater than that over the oceans. At high latitudes over land in winter there is often a strong surface temperature inversion. In this situation the temperature increases with altitude because of the cold land surface and it makes for a very stable layer of air below an "inversion cap" that can trap pollutants. The strength of an inversion is very sensitive to the amount of stirring in the atmosphere. Such wintertime

[1] See Box 10.1. The annual temperature variation at an e-folding depth of about 3 m in the land is about 37% of the surface value.

inversions are greatly affected by human activities; for instance, an urban heat island effect exceeding 10°C has been observed during strong surface inversion conditions in Fairbanks, Alaska. Strong surface temperature inversions over mid-latitude continents also occur in winter. In contrast, over the oceans, surface fluxes of heat into the atmosphere keep the air temperature within a narrow range. Thus, it is not surprising that over land, month-to-month persistence in surface temperature anomalies is greatest near bodies of water. Consequently, for a given heating perturbation, the response over land should be much greater than over the oceans; the atmospheric winds are the reason why the observed factor is only in the 2–6 range.

A further example is the contrast between the northern hemisphere (60.7% water) and southern hemisphere (80.9% water) mean annual cycle of surface temperature. The amplitude of the 12-month cycle between 40° and 60° latitude ranges from <3°C in the south to ~12°C in the north. Similarly, in mid-latitudes, the average lag in temperature response relative to the Sun for the annual cycle is 33 days in the northern hemisphere versus 44 days in the southern hemisphere, again reflecting the difference in thermal inertia.

6.6 Atmospheric–Ocean Interactions: El Niño

The climate system becomes more involved as the components interact. The striking example is a phenomenon that would not occur without interactions between the atmosphere and ocean, El Niño, which consists of a warming of the surface waters of the tropical Pacific Ocean. It takes place from the International Dateline to the west coast of South America and results in changes in the local and regional ecology. Historically, El Niños have occurred about every 3–7 years and alternated with the opposite phases of below-average temperatures in the tropical Pacific, dubbed La Niña. In the atmosphere, a pattern of change called the Southern Oscillation is closely linked with these ocean changes, so that scientists refer to the total phenomenon as ENSO, short for El Niño–Southern Oscillation. Then El Niño is the warm phase of ENSO and La Niña is the cold phase.

El Niño develops as a coupled ocean–atmosphere phenomenon and the amount of warm water in the tropics is redistributed and depleted during El Niño and restored during La Niña. There is often a mini global warming following an El Niño as a consequence of heat from the ocean affecting the atmospheric circulation and changing temperatures around the world. Consequently, interannual variations occur in the energy balance of the combined atmosphere–ocean system and are manifested as important changes in weather regimes and climate around the world. Further details are explored in Chapter 12.

References and Further Reading

Cheng L., and J. Zhu, 2014: Uncertainties of the Ocean Heat Content estimation induced by insufficient vertical resolution of historical ocean subsurface observations. *Journal of Atmospheric and Oceanic Technology*, 31(6), 1383–1396. doi: 10.1175/JTECH-D-13-00220.1.

EPA (Environmental Protection Agency), 2020: www.epa.gov/climate-indicators/climate-change-indicators-atmospheric-concentrations-greenhouse-gases

Geothermal heat: www.builditsolar.com/Projects/Cooling/EarthTemperatures.htm

NASA albedo: https://earthobservatory.nasa.gov/images/2599/global-albedo

Ritchie H., and M. Roser, 2020: *Land Use*. Published online at OurWorldInData.org. Retrieved from: https://ourworldindata.org/land-use, 26 April 2020.

Schneider, A. M., A. Friedl, and D. Potere, 2009: A new map of global urban extent from MODIS satellite data. *Environmental Research Letters*, **4**, 044003, doi: 10.1088/1748-9326/4/4/044003.

Trenberth, K. E., 1983: What are the seasons? *Bulletin of the American Meteorological Society*, **64**, 1276–1282.

Trenberth, K. E., and L. Smith, 2005: The mass of the atmosphere: a constraint on global analyses. *Journal of Climate*, **18**, 864–875.

Trenberth, K. E., and D. P. Stepaniak, 2004: The flow of energy through the Earth's climate system. *Quarterly Journal of the Royal Meteorological Society*, **130**, 2677–2701. doi: 10.1256/qj.04.83.

FLOWS OF ENERGY

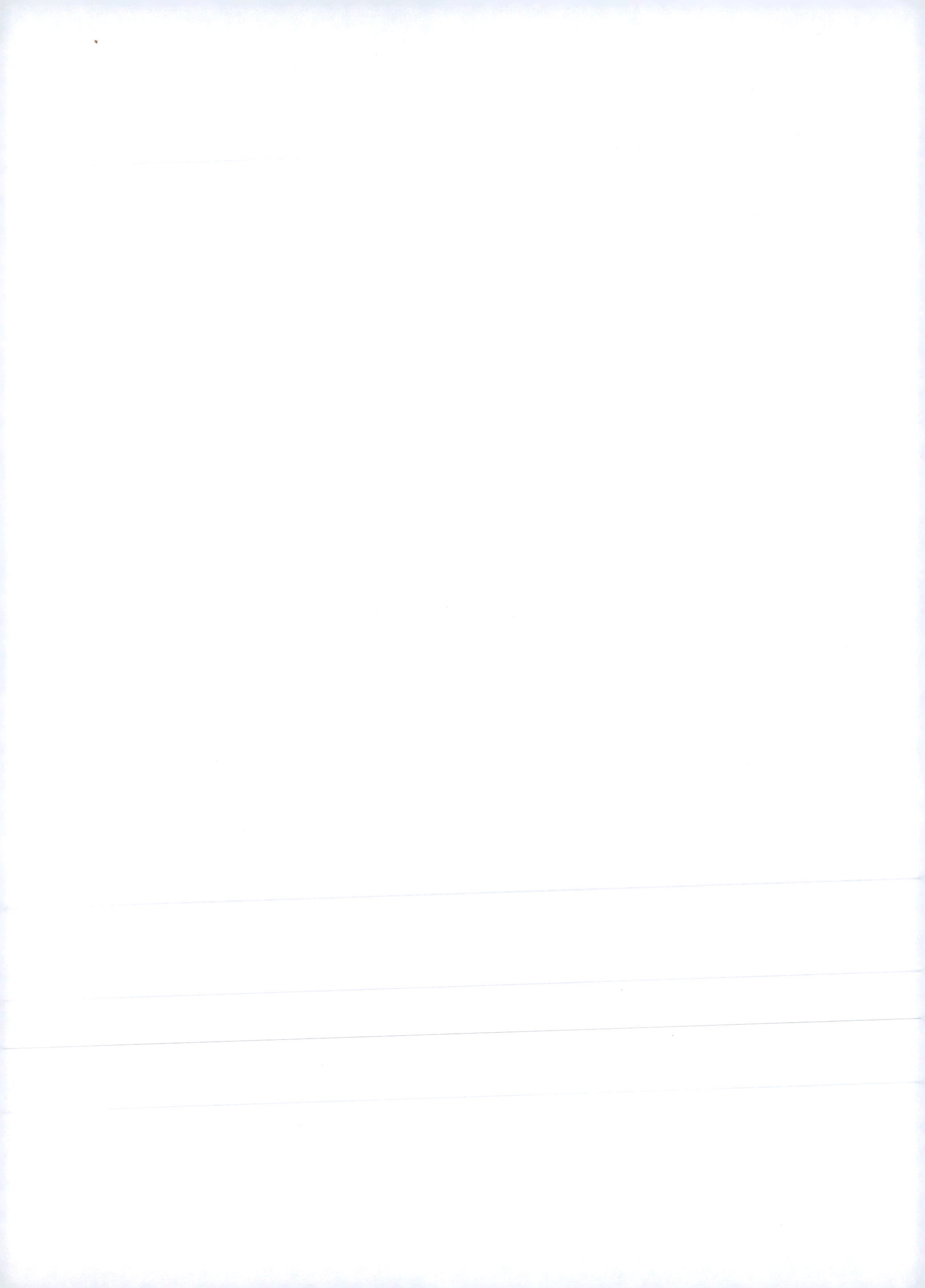

7 The Weather Machine

The atmosphere is the most dynamic component of the climate system and has many phenomena and mechanisms that move energy around, as described here.

7.1 The Atmosphere and Weather

In the atmosphere, phenomena and events are loosely divided into the realms of "weather" and "climate." Climate is usually defined to be average weather and thus is thought of as the prevailing weather, which includes not just average conditions but also the range of variations and extremes. Climate inherently involves variations in which the atmosphere is influenced by and interacts with other parts of the climate system, and the external forcings. The large fluctuations in the atmosphere from hour to hour, or day to day, constitute the weather.

Weather phenomena such as sunshine, clouds of all sorts, precipitation (ranging from light drizzle to rain, hail, and snow), fog, lightning, wind, humidity, and hot and cold conditions can all be part of much larger-scale organized weather systems that arise mainly from atmospheric instabilities driven by heating patterns from the Sun. In middle and high latitudes, the weather systems are extratropical *cyclones* (low-pressure systems, also called depressions) and *anticyclones* (high-pressure systems) and the associated warm and cold fronts. In the tropics the wet weather is dominated by convection, including thunderstorms and thunderstorm complexes that may be organized into tropical cyclones and even hurricanes. Nearly all are involved in moving energy around and, indeed, this is mainly why they occur.

Weather phenomena and weather systems mostly arise from tiny initial perturbations that grow into major events. The atmosphere, like any other system, is averse to unstable situations. This is why many triggering mechanisms exist that will push the atmosphere back toward a more stable state in which temperature contrasts are removed or at least reduced. In general, therefore, once the atmosphere has become unstable, some form of atmospheric turbulence takes place and grows to alleviate the unstable state by mixing up the atmosphere. It is not always possible to say which initial disturbance in the atmosphere will grow, only that one will grow. There is, therefore, a large component of unpredictable behavior in the atmosphere, an unpredictability that is exacerbated by and related to the underlying random component of atmospheric motions. The processes giving rise to this randomness are now referred to in mathematics as *chaos*. Because of the above factors, weather cannot be accurately forecast beyond about 10 days.

However, on average, it is known that weather systems must behave in certain ways. In spite of the infinite variety of weather, there are limits and distinct patterns manifested in the climate. So, while it may not be possible to predict the exact timing, location, and intensity of a single weather event more than 10 days or so in advance, because they are a part of the weather machine, it may be possible to predict the average statistics, which is considered to be the climate. The statistics include not only averages but also measures of variability and sequences, as well as covariability (the way several factors vary together). They also include extremes, which may do the most damage in various ways.

7.2 Energy in the Atmosphere

Energy in the atmosphere comes in several forms. The incoming radiant energy is transformed when it is absorbed into sensible energy, often called heat. At the surface, the heat may be manifested as increases in temperature or it could result in increases in evaporation, and the partitioning depends on the available moisture and nature of vegetative ground cover on land. Increased evaporation compensates the heat through evaporative cooling but adds moisture to the atmosphere and this is often referred to as latent energy, as it is realized as *latent heating* when the moisture is later condensed in cloud and subsequent precipitation. Increases in temperature increase the *internal energy* of the atmosphere, which also causes it to expand. Consequently, they change the altitude of the air and increase the *potential energy*. There is therefore a close relationship between internal and potential energy in the atmosphere, which together are combined into the concept of *enthalpy*, or *sensible heat*. Once air starts to move, usually because temperature gradients give rise to pressure gradients, some energy is converted into *kinetic energy* and manifested as winds. The sum of the internal, potential, kinetic and latent energies is a constant in the absence of a transfer of heat or loss of mass to somewhere else. The main mass changes occur from precipitation, because raindrops have their own enthalpy. Hence warm rain can transfer heat from the atmosphere to the surface, or cold rain may cause a cool pool of fresh water over the tropical oceans. Cool waters may float if they are fresh and therefore less dense than salty water. The many possibilities for conversions among all of these forms of atmospheric energy are a key part of what provides the richness of atmospheric phenomena. The main forms are briefly discussed here.

Transfers of energy may occur as radiation, as from the Sun. They may also occur through conduction, which is the transfer of heat from one substance to another by contact. While conduction works well for metals, it works very inefficiently for the atmosphere and land. For the atmosphere and oceans, an alternative way of moving heat is via convection. This is the transfer of energy within a fluid as the fluid moves around. Convection may take many forms, but all involve warm air (or water in the ocean) rising and cold air sinking. Yet another way energy is moved around is by *advection*, where the energy is transported by winds.

7.3 Convection

Much of the incoming solar radiation penetrates through the relatively transparent atmosphere to reach the surface, and hence the atmosphere is mostly heated from below (Fig. 3.1). Warm air is less dense than surrounding air and is apt to rise. The decreasing density of air with altitude also means that air expands as it rises and consequently it cools. In the troposphere, this means that temperatures generally decrease with height, and warming from below causes *convection* that transports heat upwards by warm air rising and cool air sinking, so that mass is conserved. Convection then arises as an instability in the vertical and gives rise to clouds and thunderstorms, driven by solar heating at Earth's surface which produces buoyant thermals that rise, expand and cool, and may produce rain and cloud. This process stabilizes the atmosphere by transferring heat from the lower to upper layers as the convection occurs.

Clouds that result from convection are called convective clouds. These range from small puffy cumulus clouds, to multi-celled cumulus that produce rain showers

Fig. 7.1 A spectacular example of convection building on the eastern plains of Colorado, June 3, 2008, illuminated by the setting Sun (taken by the author in Boulder, CO).

(Fig. 7.1), to large cumulonimbus clouds that may produce severe thunderstorms. In the tropics, the ocean can become quite hot in the absence of winds, and, while thunderstorms can help moderate the sea surface temperatures, in deep summer they are not enough, and instead cooperatives of thunderstorms form called tropical storms or tropical cyclones (see Section 7.9). A key difference is that while thunderstorms are accompanied by gusty surface winds, in tropical storms the winds become very large. All of these phenomena move heat upwards and stabilize the atmosphere.

7.4 Diurnal Cycle

The constant rotation of Earth creates a continual cycle of sunshine and darkness, which sets the stage for strong diurnal variations in the atmosphere (Fig. 1.2). Because of the very different heat capacities of land versus ocean, the most obvious potential for disruption arises from the contrasts in coastal regions where land warms by day and becomes warmer than the ocean, while hotter and dryer conditions over land provide the potential for greater radiative losses to space at night, and cooler conditions develop over land at night. The temperature contrasts set up pressure gradients (see Sidebar 7.1), which drive a shallow sea breeze as an overturning circulation that brings cooler sea air over land by day, often peaking in the late afternoon, but which can reverse at night, although not in such a

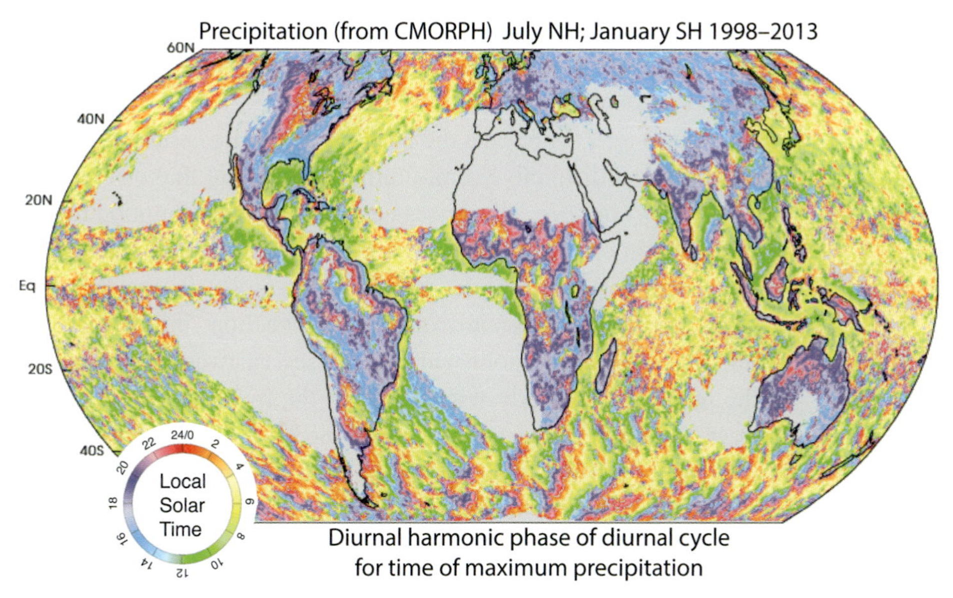

Fig. 7.2 Precipitation composites, 1998–2013, of July in the northern hemisphere and January in the southern hemisphere, for the diurnal harmonic phase, lower left, expressed as Local Solar Time of maximum precipitation. Resolution is 0.25° and hourly, and grid points where no precipitation occurs or the amplitude is low are masked gray. Data are from CMORPH: the NOAA Climate Prediction Center MORPHing technique. Updated from Covey et al. (2016)

pronounced fashion. These contrasts come to the fore when other aspects of the weather are fairly benign, and hence are present much more in summer than they are in winter when the dominance of extratropical storms and associated large-scale wind patterns prevails. However, for continents, a form of continental-scale sea breeze or monsoon circulation can develop that is more involved, and which can depend a lot on topography.

Figure 7.2 presents a map of the hour of peak precipitation in the deep summer months of July in the northern hemisphere combined with January in the southern hemisphere. In desert regions, including those over the oceans in the eastern subtropical Pacific and Atlantic, too little rainfall occurs to give reliable values. Over many ocean areas, the mottled colors indicate a lack of strong preference for when precipitation occurs and, to the extent that any patterns exist, they could be artifices of the satellite orbits. However, the sharp contrast with islands, even quite small islands such as in Indonesia or the Philippines, reveals the preferred late afternoon precipitation maximum over land. Over the immediately adjacent ocean there is often a reversal in color in the figure indicating a nocturnal or early morning maximum (orange to yellow colors). This also comes about from the destabilization aloft by radiative cooling processes. However, the heating of land by day triggers convection and rainfall. It promotes mesoscale circulations on the scales of the islands, so that the total rainfall is thought to be greater than if the islands or diurnal cycle were absent.

Note that Fig. 7.2 presents the 24-hour harmonic, whereas the real diurnal cycle often also has a distinctive semi-diurnal component that can be thought of mainly as a form of *hysteresis*. This arises because a strong afternoon maximum rainfall is not matched in opposite ways by negative rainfall at night, and so the total diurnal cycle is sharply peaked by day but flatter by night.

Over North America in summer, convection builds up over the Rocky Mountains after about noon and progresses eastward during the day embedded in the prevailing westerly winds, so that the maximum in activity is 3–4 p.m. in Denver, Colorado. However, the convection does not occur until around midnight farther east over Kansas, in conjunction with the nocturnal development of a low-level jet (within 500 m of the ground) from the Gulf of Mexico that brings moisture northwards. Disturbances and associated convection can be tracked in their eastward progression in many cases, and this accounts for some 60% of summertime rainfall between the Rockies and Appalachians. In the east and southeast United States, the rain maximum is again in the mid-afternoon. Local excitation of afternoon rainfall occurs most frequently over the Florida Peninsula and the Gulf of Mexico coastline, the Front Range of the Rockies, and the Appalachian Mountains.

Hence the continental-scale change in winds, set up in large part by the Rocky Mountains and the proximity of the Gulf of Mexico, greatly modifies the simple sea breeze picture given above. This sort of complexity also occurs in Africa, South America, and southern Asia, but in each case with their own unique developments, all driven by the warming continent as the Sun progresses from east to west relative to the land throughout the day, amidst the general westerly winds in mid-latitudes, or with prevailing easterly winds in low latitudes.

7.5 Tornadoes

Strong instability and upward energy transport occur in thunderstorms. These are a very strong form of convection. However, it is not just a matter of warm air rising and then cold air sinking, because the environmental wind conditions can create a three-dimensional circulation whereby the updraft is in one part of the storm, and the main rain shaft and downdraft in other parts. If rotation is present, then tornadoes may be spawned. The main way rotation occurs is through wind shear, both in the horizontal and the vertical. Stronger winds aloft parallel to the weak winds at low levels create horizontal rotation, because relative to the mean wind speed, the air flows in one direction above and the opposite direction below. In turn this rotation can be carried into the storm, tilted and distorted. However, more powerful rotation is readily spawned when the low-level flow is from a different direction, as well as at a lower speed, than the air aloft. The three-dimensional structure of these supercell storms increases their organization and longevity, and the threat of severe weather.

Tornadoes are more common over the eastern parts of the United States than anywhere else, in large part because of the juxtaposition of the Rocky Mountains and the Gulf of Mexico. "Tornado Alley" is a region including parts of South Dakota, Nebraska, Kansas, Oklahoma, northern Texas, and eastern Colorado, and is often home to the most powerful and destructive of these storms. US tornadoes cause about 80 deaths and more than 1500 injuries per year. The Gulf provides a source of warm, moist, potentially unstable air at low levels. In spring in the upper troposphere, the prevailing westerlies and associated jet stream flow across the country from west to east, but eastward flow is blocked at low levels by the Rockies and their extensions to the Mexican highlands. Flow from the south in the warm sector of an extratropical cycle taps into the warm moist Gulf of Mexico air, creating very unstable conditions, and strong wind shear is present. The latter can be converted into rotational energy in tornadic storms. While this can occur at almost any time, strong planetary-scale waves in winter typically set up to go around rather than over the Rockies, and similarly with the Himalayas. In mid to late summer, the monsoons dominate. Hence it is mainly in late spring or early summer that conditions conducive to supercell thunderstorms and tornadic storms are more likely.

7.6 The Hadley Circulation

In the tropics, much of the movement of energy occurs through large-scale overturning of the atmosphere. The classic example is the monsoon circulations, and also the *Hadley Circulation* (Fig. 7.3). At low levels, the moisture in the atmosphere is transported towards areas where the air is forced to rise and hence cool, forming the monsoon rains, and resulting in strong latent heating that drives the upward branch of the overturning cells. In the subtropics, the downward branch of the circulation suppresses clouds and is very dry as the "freeze-dried air" from aloft slowly descends and warms as it is compressed. Hence in these regions the surface can radiate energy to space without much interference from clouds and water vapor greenhouse effects. This is what happens, for instance, over the Sahara (Figs. 5.8 and 5.9). However, another key part in the subtropics that helps drive the monsoons is the link to mid-latitude weather systems which cool the subtropics. They transport both sensible and latent heat polewards in the cyclones and anticyclones, while warming the higher latitudes (Fig. 7.4).

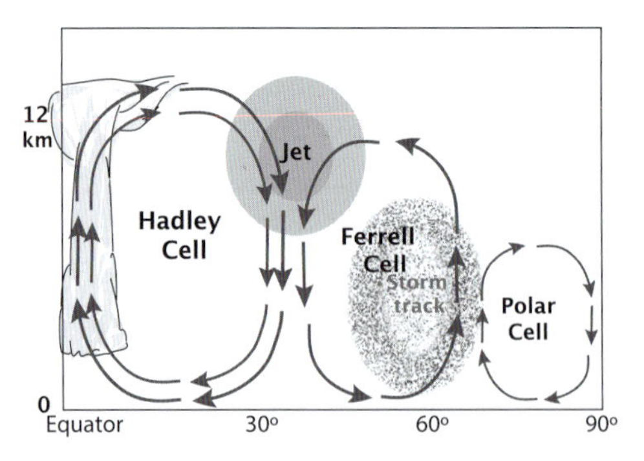

Fig. 7.3 This schematic of the zonal average annual mean troposphere illustrates the Hadley Circulation and the other two overturning cells, the Ferrell Cell and the Polar Cell. The Hadley and Polar Cells are direct in the sense that they transport energy polewards, down the equator-to-pole temperature gradient, while the reverse Ferrell Cell is indirect. The upward branch of the Hadley Cell resides in the summer hemisphere, and the winter hemisphere Hadley Cell is bigger and stronger than the summer one. However, the annual mean averages these out. The upward branch features active convection associated with the ITCZ and SPCZ over the oceans, or the monsoon trough and rains where land plays a role. The downward branch, where convection and clouds are suppressed, is in the subtropics, and strong subtropical anticyclones prevail, often leading to desert regions over continents. The subtropical jet stream forms at the poleward and upper extent of the Hadley Cell. The warmth in the downward branch feeds extratropical storms in middle to high latitudes, where the prevailing storm track of low- and high-pressure systems is active over the oceans and leads to transport of heat polewards and upwards that drives the Ferrell Cell, and which also sets up the Polar Cell.

Because the land in each hemisphere heats up or cools off according to the season, the rainy monsoons occur more or less at the same time in each hemisphere. This gives rise to the concept of a global monsoon whereby upward motion and rains are abundant in the summer hemisphere on tropical land while subsiding air and dry conditions prevail in the winter hemisphere. Indeed, there is on average such a global-scale change in the circulation accompanied by changes in the Hadley Circulation over the oceans. However, the onset of monsoons can vary considerably, and the El Niño phenomenon tends to make for substantial differences among the three main monsoons (Asian, African, and American) and a sort of competition for where the main action occurs. The Madden–Julian Oscillation (MJO), which is a 30- to 60-day variation in rain and wind in a distinctive pattern involving a cyclonic and anticyclonic couplet and is especially strong from the Indian Ocean to the western tropical Pacific (see Section 7.8), can make for large local perturbations especially in the Asia-Australia sector.

The Hadley Cell overturning is much stronger than the Ferrell Cell and varies enormously with season; it is much stronger in winter and weaker in summer,

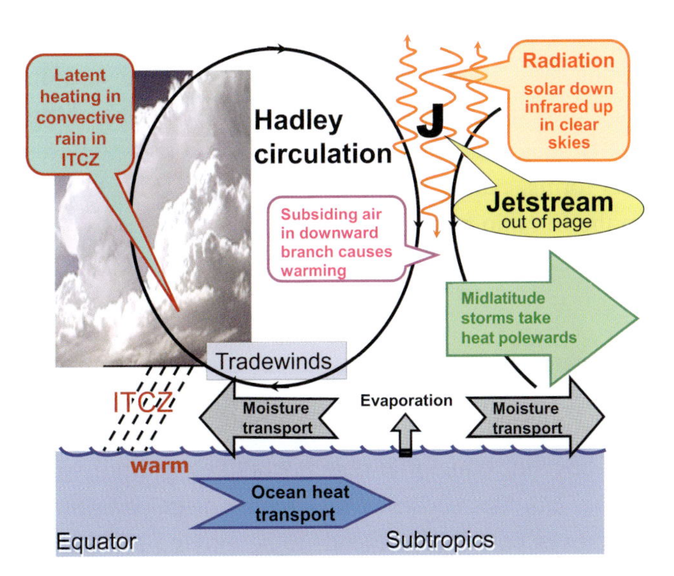

Fig. 7.4 A schematic north–south cross section in low latitudes (adapted from Trenberth and Stepaniak, 2003a) illustrates the main processes and phenomena involved. The Hadley Circulation is an overturning circulation with warm moist air rising (in the ITCZs or monsoon troughs), with plenty of rainfall. In the subtropics, downward motion prevails, and the main subtropical jet stream **J** develops in the upper troposphere. The clear skies in the subtropics facilitate a major source of heat from the Sun's rays, which warms the ocean that is in turn cooled by evaporative cooling as moisture evaporates and is transported equatorward into the rising branch of the Hadley Circulation, or poleward to higher-latitude storm tracks. Meanwhile, for the ocean, the easterly trade winds move surface water to the left south of the equator and to the right north of the equator because of the *Coriolis* effect (Sidebar 7.1), causing upwelling on the equator and an overturning within the ocean that also moves heat around (see Chapter 11 for details). © American Meteorological Society. Used with permission

especially in the northern hemisphere (Fig. 7.3). The warming on the poleward side of the mid-latitude storm track, combined with the cold polar regions, enables a smaller and weaker polar cell to occur that is also a direct cell in the sense that it transports heat down-gradient in line with the overall equator-to-pole temperature gradient. Hence, in mid-latitudes, warm air is transported polewards by storms, as detailed below, which thus warm the region near 55–65° latitude, creating upward motion as a whole and an indirect meridional overturning cell called the Ferrell Cell.

In addition to the strong north–south heating and temperature gradients, there are also significant but weaker east–west heating and temperature gradients that arise in part because of the prevailing easterly trade winds that drive ocean currents from east to west near the equator. This helps create the Pacific Warm Pool evident in SSTs (see Figs. 5.4 and 12.2 and Section 12.1). Hence there is low-level atmospheric transport of heat and moisture toward the Warm Pool where it feeds the strong upward motion there. Although much of this is linked to the overturning north–south Hadley Cell, there is also a direct east–west transverse overturning cell that forms, called the *Walker Circulation* (Fig. 12.4). This plays an important role in the El Niño phenomenon (see Chapter 12).

7.7 Winds and the Jet Streams

If the atmosphere were rotating on Earth at the same rate the planet rotates, then the air above the equatorial region would be naturally rotating eastwards faster than at higher latitudes because of its greater distance from the axis of rotation. When this air moves polewards, it tends to be moving eastwards faster than the in situ air, as angular momentum is transferred polewards. Angular momentum depends on the mass of air and its distance from the axis of rotation and it therefore has a major component associated with the rotating Earth plus an additional component associated with the winds relative to Earth's surface. Movement of air also changes the mass of air in a column and this sets up pressure gradients. The pressure gradients and Earth's rotation have to come into balance, called a geostrophic balance (see Sidebar 7.1).

The equator-to-pole temperature gradients in the troposphere are strongly associated with increasing westerly winds with height (see Sidebar 7.2). The result is a strong jet stream that forms in the subtropics near 30° in each hemisphere in winter (Fig. 7.5), characterized here by the zonal average wind. In summer, it tends to migrate polewards to 40–50° (Fig. 7.6). Because the temperature gradients are stronger in winter, the subtropical jet stream is also stronger in winter (Figs. 7.5 and 7.6) and generally exceeds 30 m s^{-1} (67 mph) in speed, but regionally at times it can easily exceed 50 m s^{-1} (112 mph). In addition, the strongest winds and major jet streams occur over the ocean, and large-scale quasi-stationary waves form especially in the northern hemisphere (Fig. 1.2) associated with the land–sea distribution and mountain ranges.

At the surface, the strongest westerlies occur at mid to high latitudes, and in winter extend upwards to a secondary polar jet stream in the southern hemisphere (Fig. 7.5). The latter is present in both hemispheres, but it is broken up in the northern hemisphere by the major continents, and the polar jet stream is highly variable and associated with storm tracks. Surface winds are detailed in Fig. 8.1 in the form of the surface wind stress.

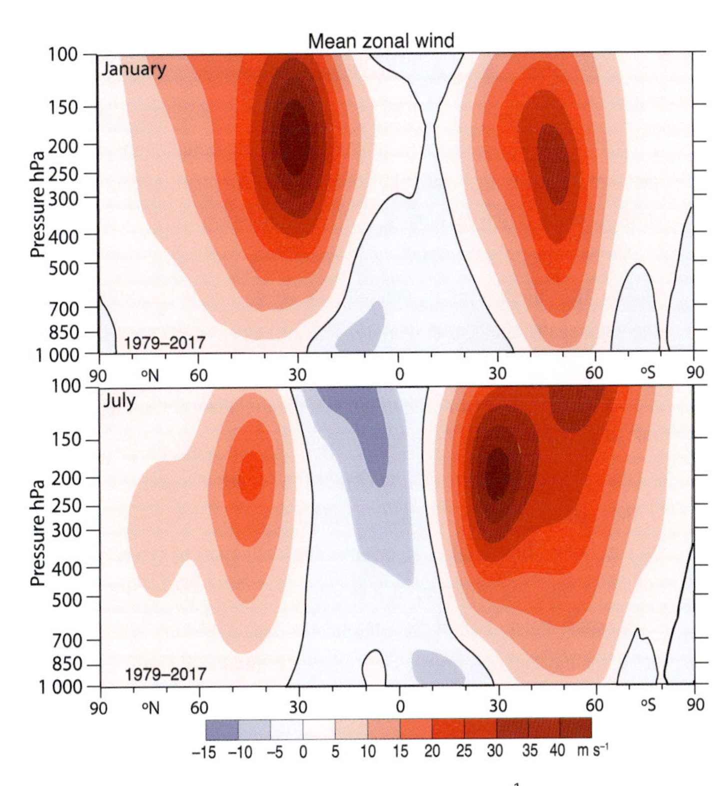

Fig. 7.5 Zonal mean winds for January and July from 1979 to 2017 in m s^{-1} based upon ERA-Interim reanalysis data. The zero line is in black. Data courtesy Yongxin Zhang and John Fasullo

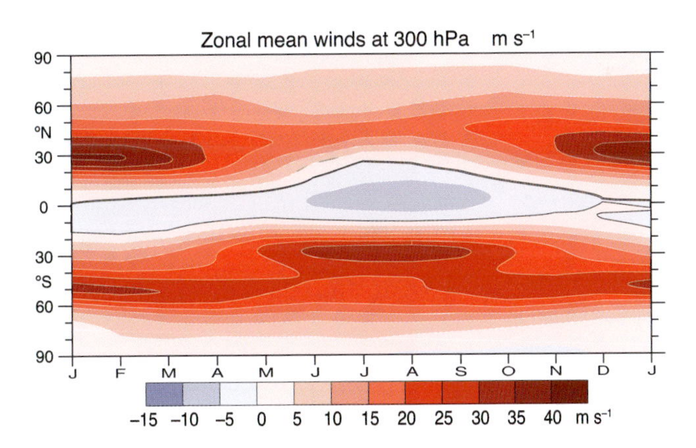

Fig. 7.6 The zonal mean westerly winds at the 300 hPa level (about 9.5 km above the surface) as a function of latitude and month for 1979–2017. The zero line is in black. Data courtesy Yongxin Zhang and John Fasullo.

Sidebar 7.1: The Coriolis Force and Geostrophic Balance

There are two dominant forces in play in the atmosphere and ocean that control the nature of the circulation and the relationship between winds and temperatures or temperature gradients. The first is the pressure gradients that naturally arise from differential heating. The second is the Coriolis force, which arises because of the rotation of Earth and movement on it.

Suppose that the atmosphere is completely uniform in temperature and mass distribution. At the surface, the surface pressure, denoting the weight of the column of air, will be uniform too. Then suppose there is a spatial gradient of heat applied, as happens because of the Sun–Earth geometry. In places where it warms, the air expands and thus moves to higher altitudes. Thus, above the surface at a given altitude, high pressure is created in the warmed region relative to the cooler regions. The pressure gradient causes the air to start to move from high pressure toward lower pressure. But as soon as it does so, there is less total air in the warmed region, so that at the surface low pressure develops compared with higher pressure in the cooler regions. This pressure gradient in turn creates a wind flow toward the warm regions, but only at low levels. The result is an overturning circulation of warm air rising, moving polewards, sinking in cooler regions, and returning at low levels (Fig. 7.7). This is the classic monsoon circulation.

On small scales, such as for a sea breeze, this circulation can remain fairly simple. But if it has any size at all, the moving air is greatly affected by the rotation of Earth. There are two main effects of rotation. The first is the centrifugal force, which is the apparent outward force experienced when a mass is rotated, such as a ball on a string. This happens on Earth, but because the force is directed away from the surface, it is experienced as a slight modification of gravity, and built into the definition of gravity.

The second effect is an apparent sideways force, called the *Coriolis force*, as experienced by a moving object on a rotating sphere, cylinder or disk, like a two-sided turntable (Fig. 7.8). If one collapses Earth

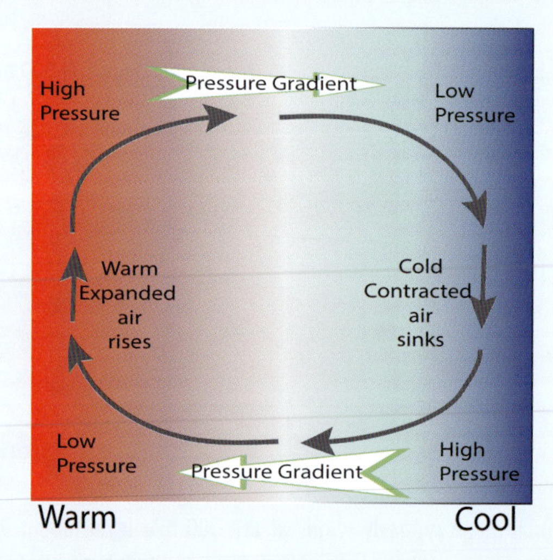

Fig. 7.7 Zonal mean cross section from equator (left) to pole (right) with relative temperatures warm in tropics and cool at high latitudes. As the warm air expands upwards, relatively high pressure is created at upper levels that pushes air polewards, down the pressure gradient. Meanwhile the surface pressure drops, creating a reverse pressure gradient at low levels, and air moves equatorward. The reverse happens at high latitudes and, together, an overturning circulation is created.

Sidebar 7.1: *(cont.)*

down to a double-sided turntable that rotates on its axis, it is readily seen that from the pole point on each side, the rotation appears to be in opposite directions. From a stationary observer in space a moving object goes straight, but the underlying surface moves by rotation as the object moves, and so it apparently deviates to the right in the northern hemisphere and to the left in the southern hemisphere on the turntable. This applies no matter which direction it starts out whether moving north–south, or east–west. The Coriolis force is not a true force in the sense that it is not there unless the object moves, and thus may be called a Coriolis acceleration.

The poles are very much like the rotating turntable, while the tropics are more like the side of the turntable or a rotating cylinder and thus there is very little rotational effect in the tropics other than the outward centrifugal force. Hence on Earth, the Coriolis effect is greatest at the poles and zero at the equator. The Coriolis parameter $f = 2\Omega \sin \varphi$, where Ω is the angular velocity of Earth with a rotation rate value of 7.2921×10^{-5} rad s^{-1}, and φ is latitude.

Since wind is a vector $\mathbf{v}$ (for velocity) with both speed and direction, it may be written with two components as $\mathbf{v} = (u,v)$ where u and v are the zonal and meridional components of velocity. Then the geostrophic winds are $fu = -1/\rho\ \partial p/\partial y$, $fv = 1/\rho\ \partial p/\partial x$, where f is the Coriolis parameter, x and y the zonal and meridional coordinates, p the pressure, and ρ the density. Alternatively, the geostrophic wind speed is given by $fV = -1/\rho\ \partial p/\partial s$, where s represents direction to the right of the wind direction with speed V.

On planet Earth, pressure gradients are ultimately balanced by the Coriolis force, allowing an equilibrium to occur (Fig. 7.8). This is called a geostrophic balance. It means that the isobars are streamlines of the actual flow. Geostrophic balance applies very generally in the atmosphere polewards of about 20° latitude or 2° latitude in the ocean. It does not apply where friction plays an important role, near Earth's surface in the atmosphere or in coastal waters in the ocean. The departures from geostrophy are small but nonetheless important in bringing about vertical motions.

In the ocean, currents can be initiated by a period of strong winds which then cease, and at that point the only force operating is the Coriolis force, so that a parcel keeps turning to the right in the northern hemisphere and may even almost complete a circle, called an inertial circle. It would be an exact circle if the Coriolis parameter did not vary. This process can create anticyclonic eddies in the ocean.

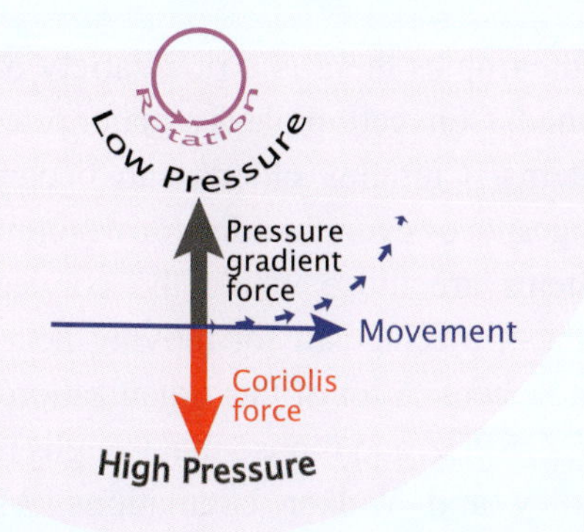

Fig. 7.8 A schematic map from the pole (top center) to the equator, where rotation of Earth is counterclockwise (as for the northern hemisphere). Depicted are the pressure gradients and Coriolis force acting on a particle moving eastwards.

Sidebar 7.2: Winds and the Jet Stream

The atmosphere prefers to be in geostrophic balance with components of the winds as

$$\mathbf{v} = (u, v) \quad \text{and} \quad fu = -1/\rho\ \partial p/\partial y, \quad fv = 1/\rho\ \partial p/\partial x,$$

where the density is given by $\rho = p/(RT)$.

It was noted that in the vertical, the large-scale atmosphere is in hydrostatic balance, so that $dp/p = -(g/RT)\ dz$. Rather than use z as the vertical coordinate, there are major advantages of a switch to p or lnp as the vertical coordinate, because this then removes the variable density from the expressions. Note that $\partial lnp = \partial p/p$. Now $\Phi = gz$ is the geopotential and z is the geopotential height. Hence the geostrophic relation can be written with components

$$v_g = 1/f(-\partial\Phi/\partial y,\ \partial\Phi/\partial x).$$

The hydrostatic equation becomes $\partial\Phi/\partial lnp = -RT$.

If the geostrophic wind components are differentiated by ∂lnp, then

$$\partial \mathbf{v}_g/\partial lnp = R/f(\partial T/dy,\ -\partial T/dx)$$

which is called the thermal wind equation, as it expresses the vertical gradients of the geostrophic wind in terms of the horizonal gradients of temperature (on a constant pressure surface).

Accordingly, if there is a strong meridional temperature gradient between equator and poles then the westerly component of the wind, u, increases sharply with altitude (toward lower pressure) (Fig. 7.5). Hence the basic thermal state of the rotating atmosphere results in westerly winds that increase with height, and these reach a maximum, on average, near 30° latitude in the form of the meandering jet stream.

7.8 The Tropical Regions

In the tropics there is a rich variety of atmospheric phenomena that move energy around. Convection plays a major role and often is organized into various waves or patterns. Cumulus clouds can become quite strong and tall, forming cumulus congestus or even thunderstorms (cumulonimbus). However, much more powerful systems are those where the convection appears to self-organize into a more efficient mechanism for moving heat out of and away from the surface into the atmosphere. A lot of this organization occurs via the major convergence zones, the ITCZ, SPCZ and monsoonal troughs that together form the upward branches of the Hadley and Walker circulations. Mesoscale Convective Systems (MCSs) or Complexes (MCCs) are quite common. Over land and in conjunction with complex topography, wind shears can come into play and create supercell thunderstorms which may spawn tornadoes.

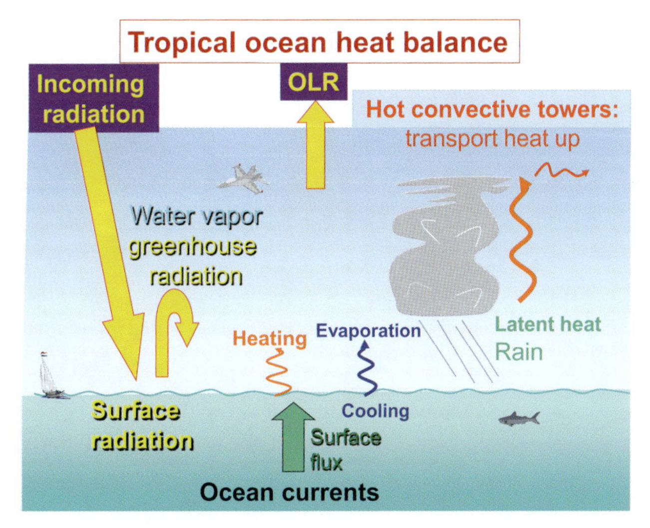

Fig. 7.9 A schematic showing some of the main energy processes operating in the tropics. Incoming radiation may be reflected by clouds or the surface, but that absorbed at the surface has a difficult time radiating back to space as infrared radiation because of the large water vapor presence in the atmosphere over the oceans. The main surface heat flux is evaporative cooling of the surface, that moistens the atmosphere, accompanied by a sensible heat turbulent flux. The moisture may fall out as rain in convection or organized tropical storms, which in turn move heat to the upper atmosphere above the water vapor layer, where it may be dispersed by winds or radiated to space.

Another major organized complex is the 30- to 60-day MJO, which takes place primarily from the tropical Indian Ocean to the western tropical Pacific Ocean and is the dominant form of intra-seasonal variation in the tropics. The MJO is characterized by an eastward progression of large regions of both enhanced and suppressed tropical rainfall. The anomalous rainfall is usually first evident over the western Indian Ocean and continues as the MJO propagates over the warm ocean waters of the western and central tropical Pacific. This pattern of tropical rainfall generally weakens as it moves over the cooler ocean waters of the eastern Pacific.

The following section introduces tropical storms, and a characteristic of all of these phenomena is that they move heat upwards, which stabilizes the atmosphere (see Fig. 7.9).

7.9 Tropical Storms

Various kinds of disturbances are generated in the tropics, often over land in association with complex topography (e.g., Africa) or as instability waves. These

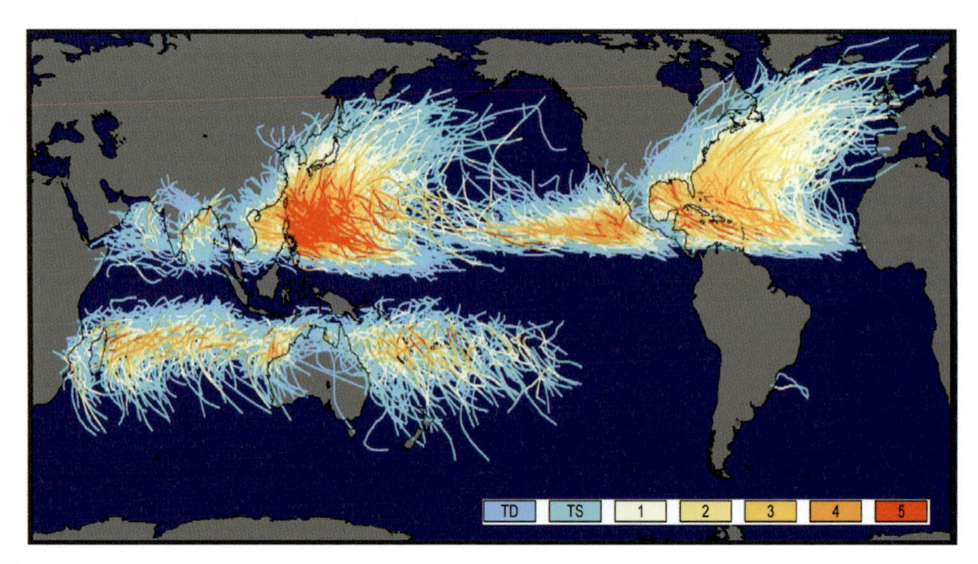

Fig. 7.10 Distribution of tracks of tropical storms and hurricanes through September 2006, with their strength category in color (lower key). Tropical depression (TD) 38 mph (61 km h^{-1}) or less. Tropical storm (TS) 39–73 mph (62–117 km h^{-1}). Hurricane category 1: 74–95 mph (119–153 km h^{-1}); 2: 96–110 mph (154–177 km h^{-1}); 3: 111–129 mph (178–208 km h^{-1}); 4: 130–156 mph (209–251 km h^{-1}); and 5: 157+ mph (252+ km h^{-1}) (Saffir–Simpson scale). After Trenberth (2007). © Scientific American, published with permission

can serve as a way for tropical convection to become organized into tropical storms. The disturbances begin the organization of thunderstorms that then act as a cooperative, and can spin up to become quite intense: they are labeled as *hurricanes* in the Americas, *typhoons* in the northwest Pacific, or *cyclones* in the Indian Ocean. They typically occur in deep summer in most of the deep tropics, or in the Indian Ocean before or after the monsoon season. The distribution of tropical cyclones is given in Fig. 7.10. Some of the following is adapted from Trenberth et al. (2018) and references therein.

Most theories of hurricanes are built on the general idea that they form only in regions where the sea-surface temperatures are above about 27°C (see Fig. 5.4). Hence, they form only in the tropics. The development of the vortex from the original loose disturbance requires positive reinforcement as the thunderstorms are organized and take the available warm moist air, lift it in deep convection and produce copious rainfall in the process. The rainfall in turn releases latent heat as the water vapor condenses into storm clouds, and this provides extra buoyancy for the air which rises even more. This in turn creates low pressure at the surface and encourages more warm moist air from the surrounding areas to flow toward the building storm, bringing in even more latent energy. Provided that the larger-scale winds above the surface are benign, which means an absence of wind shear in particular, the vortex can spin up in strength. Wind shear otherwise tears the incipient vortex apart.

As the vortex itself builds in strength, the surface pressure in the center of the storm drops and the winds circulating around the storm pick up, bringing even more moisture into the storm, and producing increased evaporation from the ocean in the vicinity. The whole storm complex itself becomes an entity as it develops an eye of clear sky surrounded by an eyewall of very intense thunderstorms, and a whole series of spiral arm bands that spiral moisture into the center of the storm.

From satellite, hurricanes are easy to see because they usually occur in relative isolation, surrounded by an area of relatively clear sky. The overall active rain and cloud region is where air is rising overall and it spreads out in the upper troposphere and lower stratosphere and subsides in the area around the visible storm. Hence, the surrounding clear areas are actually a key part of the storm and the tangible storm circulation is typically about 4 times that of the visible vortex. The average radius of a hurricane might be 400 km, but the circulation involved in such a storm extends to 1600 km or so, as the air from outside the vortex spirals in to help feed the storm with its fuel of warm moist buoyant air. Accordingly, the most vigorous hurricanes form away from land areas, as air from land that gets into a storm is apt to be dryer, undermining its ability to build. Strong winds over land also experience greater friction and surface wind stress, so that together these effects typically lead to weakening storms as they approach land. This does depend a lot, though, on the size and speed of the storm.

An example shows one day, September 6, 2017 (Fig. 7.11), that was especially active in the tropical Atlantic. The image contains three hurricanes: Katia, Irma,

Fig. 7.11 On September 6, 2017, hurricanes Katia, Irma, and Jose lined up across the Atlantic basin. The trio is visible in this NASA composite image (using several orbits), captured that day by the Visible Infrared Imaging Radiometer Suite (VIIRS) on the Suomi NPP satellite. Consequently, there are stripes of Sun glint. Courtesy NASA Earth Observatory. https://eoimages.gsfc.nasa.gov/images/imagerecords/90000/90918/hurricanes_vir_2017249_lrg.jpg.

and Jose. Katia is at far left near Central America. Irma became a category 5 storm and did considerable damage in Florida, in particular. Note the spiral arm bands and relatively clear skies surrounding each storm.

El Niño also affects which regions are favored and which are not (Fig. 7.12); see also Chapter 12 as to why. Generally, areas with higher SSTs are favored to be active. In El Niño conditions, the warmer central and eastern tropical Pacific favor more activity there at the expense of Atlantic activity. When activity is suppressed, it can lead to pent-up conditions that enable exceptional activity to break out once freed from the wind shear or subsidence. Activity in the Atlantic can become strong then in La Niña conditions, especially after a period of suppression, as happened in 1995.

Tropical storms and hurricanes spin up very strong winds which increase the surface evaporation (latent heat) by an order of magnitude. This cools the upper ocean through mixing heat downwards, and especially through evaporative cooling, and leaves a cold wake behind the storms. It creates a less favorable

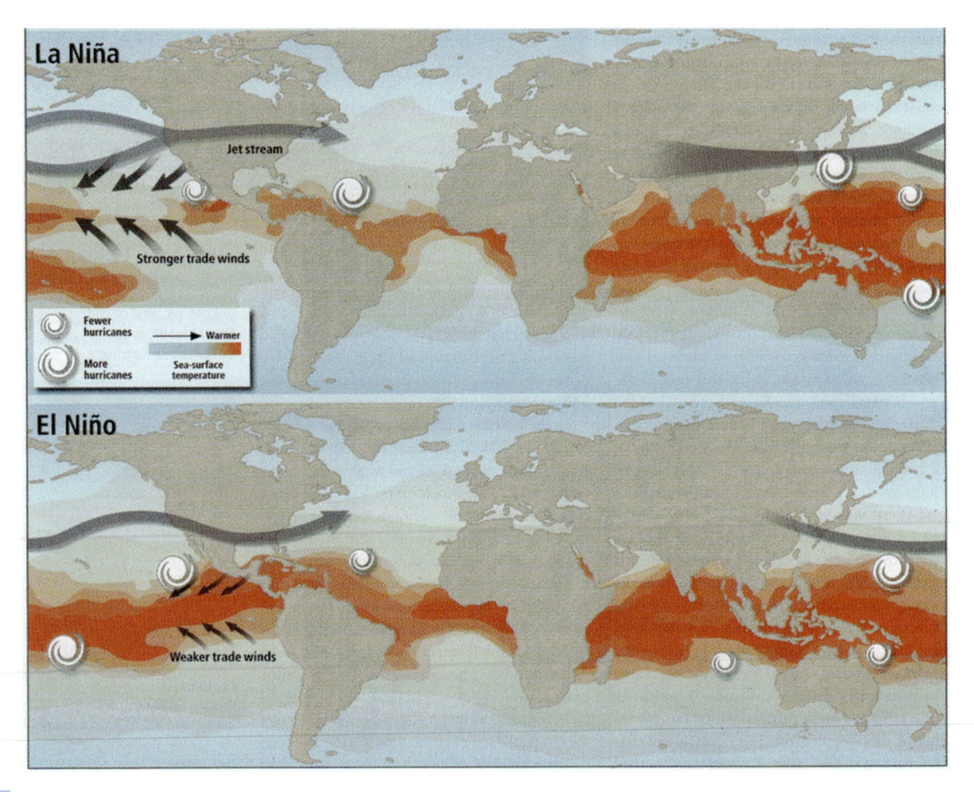

Fig. 7.12 Distribution of SSTs in La Niña and El Niño conditions, and associated tropical storm locations and tracks. Larger symbols mean more activity. From Trenberth (2007). © Scientific American, published with permission

environment in the immediate area for subsequent storms. The increased atmospheric moisture flows into the hurricane, often in spiral arm bands, and fuels the storm itself. Accordingly, it aids the strengthening of the storm if the storm keeps moving over virgin ocean, so that the storm can reach out and gather otherwise untapped moisture lurking in the atmosphere. As the intensity increases, the spiral arm bands become more circular and can form a new eyewall, forging an eyewall replacement, that frequently causes the storm to weaken momentarily but grow in size as the new eyewall has a much larger radius. A succession of eyewall replacements, as occurred in Irma for instance, can make for a very large hurricane.

Changes in tropical storms with climate change are included in Section 10.7.

7.10 Extratropical Cyclones and Anticyclones

A lot of heat energy is deposited into the subtropics via the subsidence in the downward branch of the Hadley Circulation. This heat is then carried away to cooler higher latitudes by mid-latitude storms that form polewards of the jet stream. In such storms the heat transport is side-by-side rather than overturning: warm moist air at one location is countered by cold dryer air at another longitude (Fig. 7.13).

The mid-latitude or extratropical *cyclones* (low-pressure areas or systems) and *anticyclones* (high-pressure systems) and their associated cold and warm fronts are examples of yet another instability in the atmosphere (called *baroclinic instability*). They arise from the equator-to-pole temperature differences and distribution of

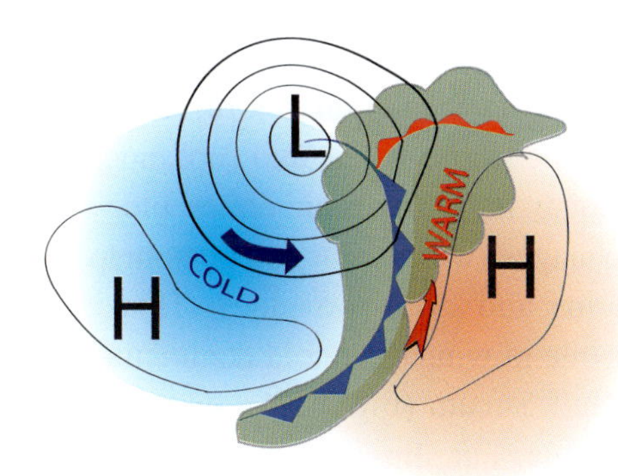

Fig. 7.13 Schematic of a mid-latitude northern hemisphere weather system of low-pressure extratropical cyclone **L**, high-pressure anticyclones **H**, and with the warm front (red), cold front (blue) leading warm (reddish) and cold (blueish) air. The strong upward motion and precipitation in the cyclone and fronts is schematically indicated in green.

heating (Figs. 1.2 and 5.2). The atmosphere attempts to reduce those temperature gradients by producing weather systems which have, in the northern hemisphere, southerly winds to carry warm air polewards and cold northerly winds to reduce temperatures in lower latitudes. In the southern hemisphere it is the southerlies that are cold, and northerly winds are warm. These weather systems are always present and migrate, develop, evolve, mature, and decay over days to weeks and constitute a form of atmospheric turbulence.

In terms of poleward transport of energy, the main transports at middle and higher latitudes occur in the form of sensible heat and potential energy which are combined as *dry static energy*, while at low to middle latitudes, latent energy also plays a major role, so that the combination of all three is *moist static energy* (discussed further in Section 9.2 and see Fig. 9.2).

Within a weather system, the boundary of a region where warm tropical or subtropical air advances poleward is necessarily a region of strong temperature contrast (Fig. 7.13). This boundary is called a warm front. As the warm air pushes cooler air aside, it also tends to rise, because warm air is less dense. Because the rising air also moves to regions of lower pressure it expands and cools, so that moisture condenses and produces extensive clouds and rain. The advancement equatorward of cold air occurs similarly along a cold front, but in this case the colder and therefore denser air pushes under the somewhat warmer air in its path, forcing it to rise, often causing convective clouds, such as thunderstorms, to form. The movement of warm air poleward and of cold air equatorward usually go together as part of the same system because otherwise air would pile up in some places, leaving holes elsewhere. Where the cold front has caught up with the warm front, the frontal system is referred to as occluded.

Weather systems over the oceans have a somewhat different character from those over land, because of the abundant moisture over the oceans which more readily allows clouds and rain to form. Over land, storms are often more violent, in part because the land can heat and cool much more rapidly than the ocean and also because mountain ranges can either block or create strong winds and wind-direction changes (wind shear) that can help facilitate the development of intense thunderstorms and even tornadoes. This is especially the case in spring and summer, while in winter the cold continental air moving over warmer waters such as in the Gulf Stream (Fig. 7.14), where the waters are quite warm, can trigger fields of cellular cumulus clouds as the warmer moist air rises. Often these are remarkably aligned into what are called "cloud streets." These occur when columns of heated air – thermals – rise through the atmosphere and carry heat away from the sea surface. The moist air rises until it encounters a warmer air layer (a temperature inversion) that acts like a lid. The inversion causes the rising thermals to roll over on themselves, forming parallel cylinders of rotating air. On

Fig. 7.14 This imagery from NOAA's Environmental Research Laboratory on October 11, 2019, has been modified to depict the locations of anticyclones (H) and extratropical cyclones (L) along with the location of a cold front (blue), warm front (red) and occluded front (purple) associated with the major cyclone off the east coast of North America. The cellular cumulus behind the cold front is symptomatic of the cold continental air over the relatively warm ocean. A second weather system exists farther west and some of the white area behind the L is snow on the ground, rather than cloud.

the upward side of the cylinders (rising air), water vapor condenses and forms clouds, and may produce showers. Along the downward side (descending air), skies remain clear.

The overturning cells and processes also develop in part to rearrange the winds and associated momentum balance of the atmosphere and maintain a geostrophic balance of broad westerlies in mid-latitudes. The Hadley Cell also transports momentum polewards in the upper branch and helps result in the strong subtropical jet stream near 30° latitude (Section 7.7). This is also maintained by poleward momentum transports by the mid-latitude eddies (cyclones and anticyclones); see Fig. 7.15. This occurs because the upper troposphere troughs and ridges tend to become oriented from southwest toward the northeast in the northern hemisphere and northwest to southeast in the southern hemisphere, in part because of the shape of the rotating Earth; if everything

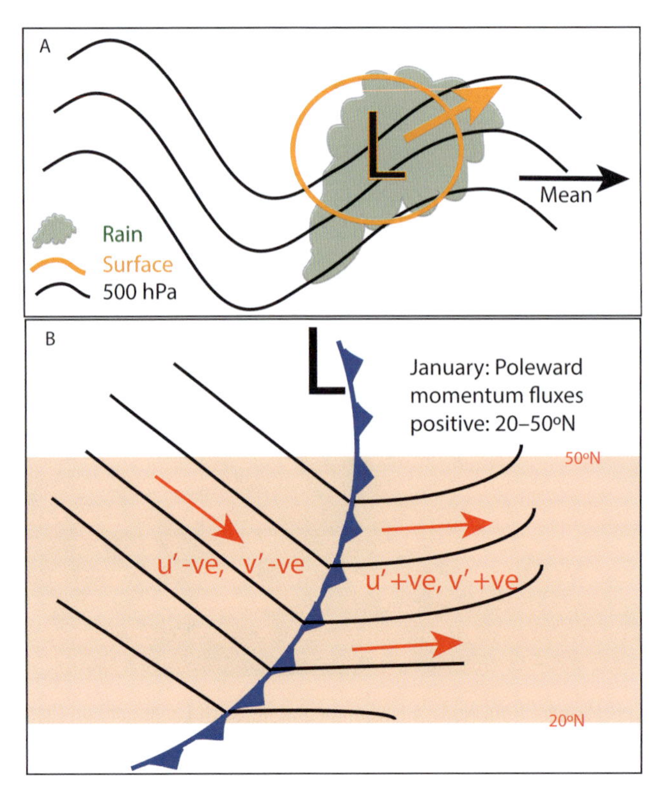

Fig. 7.15 The top panel schematically shows the location of a northern hemisphere surface low-pressure center in orange and its typical location relative to the flow in the mid-troposphere at 500 hPa level (black contours), so that they steer the low-pressure system northeastwards along with the accompanying rains. The lower panel shows the sea-level pressure field and cold front. If the departures from the average are denoted as ()′, then east of the cold front the westerlies (u′) are stronger than to the west, while the northerly component to the winds (v′) is strong to the west but southerly ahead of the cold front. Hence u′ and v′ are positively correlated on both sides of the cold front, and the westerly momentum is carried northwards. Note that the average could be the average zonally, around a latitude circle, or, because the system moves eastward, in time. (From Trenberth and Zhang, 2018.) © American Meteorological Society. Used with permission

moved at the same speed, the greater distances east–west at lower latitudes mean features lag behind.

The process of warm air rising and cold air sinking is pervasive in the atmosphere and is a vital part of the weather machine. As seen in Fig. 7.9, warmth is generally transferred from the surface to higher levels in the atmosphere, where the heat is eventually radiated to space. Further aspects of poleward energy transports by the atmosphere are given in Chapter 9.

References and Further Reading

Covey, C., P. J. Gleckler, C. Doutriaux, et al., 2016: Metrics for the diurnal cycle of precipitation: toward routine benchmarks for climate models, *Journal of Climate*, **29**, 4461–4471. doi: 10.1175/JCLI-D-15-0664.1.

Trenberth, K. E., 2007: Warmer oceans, stronger hurricanes. *Scientific American*, July, 45–51.

Trenberth, K. E., and D. P. Stepaniak, 2003a: Co-variability of components of poleward atmospheric energy transports on seasonal and interannual timescales. *Journal of Climate*, **16**, 3691–3705. doi: 10.1175/1520-0442(2003)016,3691:COCOPA.2.0.CO;2.

Trenberth, K. E., and D. P. Stepaniak, 2003: Seamless poleward atmospheric energy transports and implications for the Hadley circulation. *Journal of Climate*, **16**, 3706–3722. doi: 10.1175/1520-0442(2003)016,3706:SPAETA.2.0.CO;2.

Trenberth, K. E. and Y. Zhang, 2018: Near global covariability of hourly precipitation in space and time. *Journal of Hydrometeorology*, **19**, 695–713. doi: 10.1175/JHM-D-17-0238.1.

Trenberth, K. E., L. Cheng, P. Jacobs, Y. Zhang, and J. Fasullo, 2018: Hurricane Harvey links to ocean heat content. *Earth's Future*, **6**, 730–744, doi: 10.1029/2018EF000825.

8 The Dynamic Ocean

The Earth is "The Blue Planet" and all peoples depend on the ocean. The ocean is interconnected in the climate system through exchanges of water, energy and carbon, but moderates climate with its thermal inertia. Climate change threatens marine productivity, oxygen for ocean organisms, the carbon cycle and ocean acidification.

8.1 Ocean Heat Content

The oceans are wet, and play a major role in the hydrological cycle. They also have a high heat capacity. Section 6.2 provided the basic character of sea water and ocean structure. Figures 5.2 and 5.3 presented maps and the annual cycle of the mean surface air temperatures, and Fig. 5.4 gave the sea surface temperatures (SSTs). The latter are closely related to water vapor and precipitation in the atmosphere over the oceans. The SSTs are supported by the upper ocean temperatures and especially the mixed layer, which is the part of the ocean in direct contact with the surface through wind effects and heating from both radiation and temperatures in the atmosphere (see Fig. 8.8).

Although it is evident from the SSTs (Fig. 5.4) that the eastern sides of oceans are cool as currents flow equatorwards, while western sides of the main ocean basins are warm, with boundary currents (such as the Gulf Stream), the main SST gradients are north–south. Accordingly, zonal averages of the ocean temperatures are used along with averages across the different ocean basins to explore aspects of the mean (Fig. 8.1) and annual cycle of the ocean temperatures (Fig. 8.2).

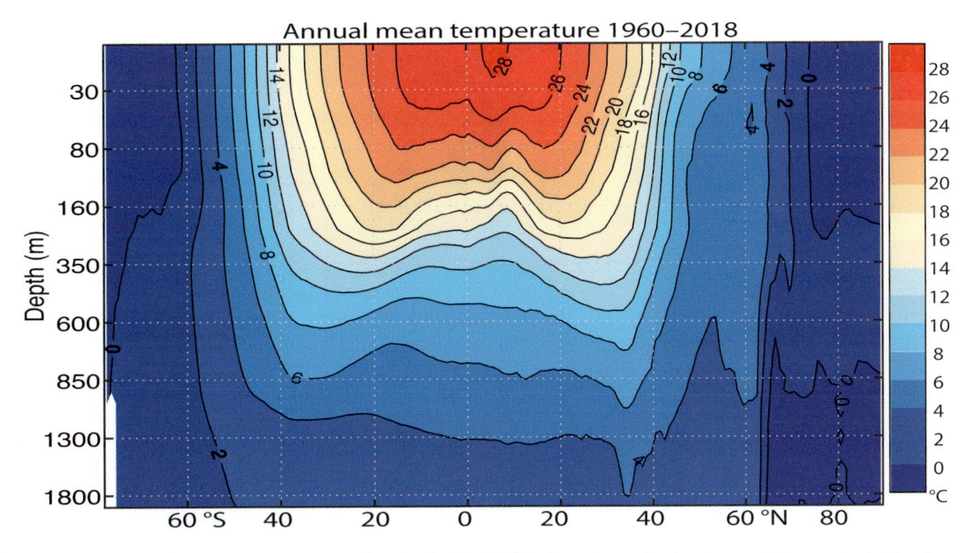

Fig. 8.1 Annual and zonal mean temperatures for 1960–2018 in°C, based on the Institute of Atmospheric Physics (IAP) ocean analysis (Cheng et al., 2017). Courtesy Lijing Cheng

The ocean heat content is the main reservoir for heat seasonally and for the changing climate, and the latter aspects are taken up in Chapter 14. The mixed layer is labeled as such because it has relatively uniform temperatures owing to mixing from mechanical wind effects and convection, and it sits on top of the abyssal ocean, separated by the *thermocline*. The annual mean zonal average of temperatures in the ocean (Fig. 8.1) blurs the thermocline somewhat because it varies with location. Figure 12.1 illustrates the east–west vertical temperature structure in the tropical Pacific, including the thermocline. Variations in the thermocline are huge in conjunction with ENSO (Figs. 12.5 and 12.8).

The mixed layer is deeper in winter than in summer (Fig. 8.2) and averages 50–100 m depth (Fig. 8.1). Temperatures tend to be somewhat higher in the northern hemisphere, or extend to slightly higher latitudes for a given value than in the southern hemisphere. However, the symmetry of Fig. 8.1 between the two hemispheres is remarkable given the differences in land versus ocean amount and the clear effects of western boundary currents such as the Gulf Stream and Kuroshio.

Surprisingly little attention has been paid to the annual cycle of ocean heat content (OHC). Near-surface ocean temperatures follow the SSTs. Given the large heat capacity of the ocean, it is not surprising that the seasonal heating from the Sun only slowly penetrates to greater depths, especially as warm waters at the surface create a stable layer that inhibits mixing. On the other hand, in the fall, as the SSTs cool, especially as vigorous weather systems come by and extract heat from the upper ocean, cool waters sit on top of warmer waters, which is an unstable configuration that triggers convection within the ocean. Accordingly, it is relatively easy to cool

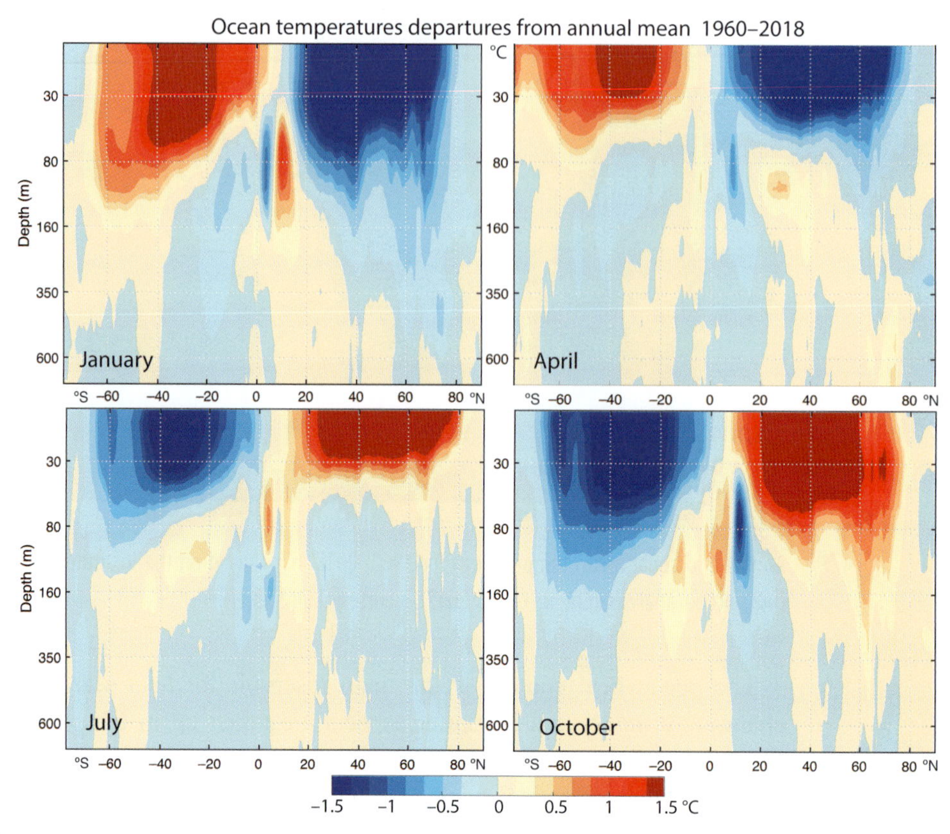

Fig. 8.2 Departures from the annual mean ocean temperatures for 1960–2018 based upon the ocean analyses from the IAP (Cheng et al., 2017) for January, April, July and October for the top 700 m. Courtesy Lijing Cheng

the deeper ocean but difficult to warm it. Horizontal ocean currents then play a role, as detailed in subsequent sections of this chapter and Chapter 9.

Ocean tides are a primary means of mixing the ocean as water sloshes around from the gravitational pull of the moon and Sun. Tides are long-period waves that move throughout the ocean but often form more complex progressions along coastlines. There they appear as a regular rise and fall. Because the Earth rotates through two tidal "bulges" every lunar day, coastal areas experience two high and two low tides every 24 hours 50 minutes. High tides occur 12 hours 25 minutes apart and it takes 6 hours 12.5 minutes for the water at the shore to go from high to low, or from low to high on average. In this process, the ocean moves waters over ocean bottom topography, which may cause mixing to occur at depth as well as near the surface. Tides help remove pollutants and circulate nutrients that ocean plants and animals need to survive, and hence they affect the reproductive activities of fish and ocean plants.

SSTs generally peak at the end of the heating season, in March in the southern hemisphere and September in the northern hemisphere. In Fig. 8.2 there are large

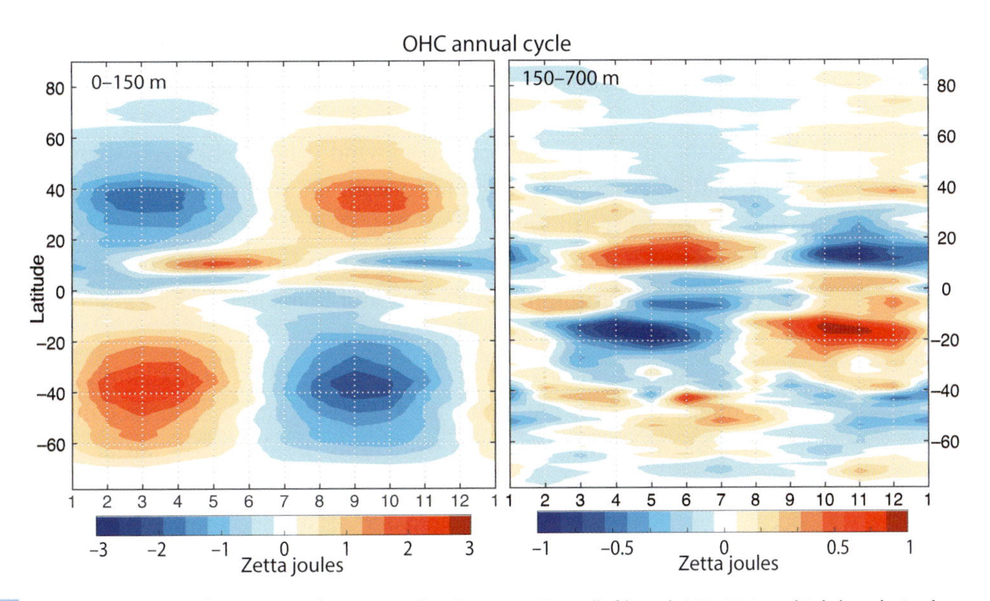

Fig. 8.3 The annual cycle of OHC vertical averages for the top 150 m (left) and 150–700 m (right) as latitude–time sections. The months of the year are labeled 1–12 on the bottom axis. Units are zettajoules (10^{21} joules). Courtesy Lijing Cheng

depths with ocean temperatures more than 1.5°C above the annual mean in January and April (note the color scheme in the figure saturates and does not show the detail of departures larger than ±1.5°C), and less than 1.5°C below the mean in July and October. The heat penetration exceeds 80 m depth in January but has already shrunk to 60 m or so by April in the northern hemisphere, while in the southern hemisphere the heat has penetrated to only about 40 m in July and is at 100 m by October. The picture in the deep tropics, between 5° S and 15° N is quite different, and is explored in detail in Chapter 12.

There is an annual cycle signal down to about 700 m depth in places, and the contrast between the upper layers 0–150 m versus below 150 m depth is highlighted in Fig. 8.3. The biggest signals below 150 m are from about 40° S to 30° N. From about 7° N (the average location of the ITCZ) to 40° S, the deeper ocean temperatures peak in October and are coolest in April, some 7 months later than for the layer above 150 m. The reverse also occurs in the northern hemisphere where the delay is 8 months between 10 and 30° N relative to the upper layers north of 20° N. These changes can only come about from substantial changes in ocean currents, which occur much more rapidly in low latitudes than at higher latitudes owing to the Coriolis parameter and how the ocean adjusts to perturbations (see Chapter 12 on ENSO for some aspects, and subsequent sections of this chapter). Hence, even in the upper 150 m (Fig. 8.3) between 5° S and 20° N there is perhaps an unexpected seasonal variation in ocean temperatures.

8.2 Ocean Currents

The ocean circulation is primarily wind-driven although there is a significant thermohaline component associated with internal changes in density arising from either temperature (thermo) or salinity (haline) effects. Tides can produce strong local currents, but they are reversible. Knowledge of the ocean surface currents is important for determining how ships and boats are carried along, and how heat is moved around, along with other materials.

At the surface, winds set up by the atmospheric circulation are dominated by atmospheric pressure gradients, the Coriolis force, and surface drag, and hence the geostrophic wind relation is considerably modified by surface friction (Sidebar 7.2). As the winds blow over the ocean, they create a surface wind stress on the ocean (Fig. 8.4) that causes a transfer of surface momentum, energy and water between the ocean and atmosphere. Both sensible heat exchanges (fluxes) and evaporative moisture fluxes occur.

The momentum exchanges are inherent in the surface wind stress, measured in newtons per square meter, as the ocean tries to speed up to match the atmospheric wind but is encumbered by the huge mass below. The wind stress depends strongly on the surface wind, although it is really related to the difference in velocity between the atmosphere and the ocean, and hence the ocean current also plays a small role. It also depends upon a drag coefficient and density, and the drag

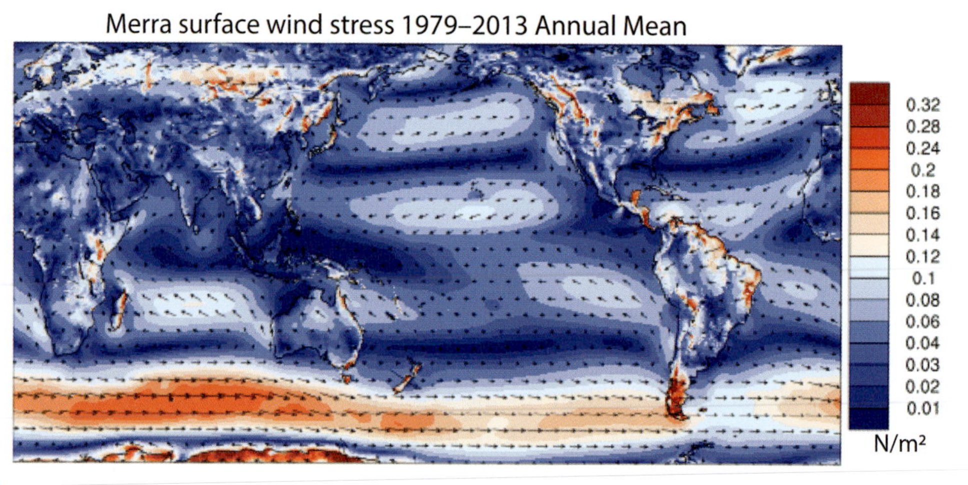

Fig. 8.4 Annual mean surface wind stress for 1979–2013 based upon NASA MERRA atmospheric reanalysis data. The units are N m^{-2}. Wind stress is a vector quantity and has both direction and magnitude, shown by the arrows with the net magnitude shaded in colors. Because of topography, values are messy over land, and largest values occur for the westerlies over the southern oceans, North Pacific and North Atlantic, and for the subtropical easterly trade winds.

coefficient quantifies how effective the drag or resistance of the slower body is to not speed up to the velocity of the faster-moving fluid. Empirically, it varies with wind speed. In light winds it depends a lot on small-scale turbulence. The ocean applies the reverse stress on the atmosphere, and hence in general the ocean acts to slow down the surface winds, while the atmosphere tries to drive ocean currents. Often the small ocean current speed is neglected in computing the wind stress.

Because of Earth's rotation (Sidebar 7.1), the frictional component of the vertically averaged transport of water is to the right in the northern hemisphere and to the left in the southern hemisphere. If one seeks a mathematical solution that results in frictional forces balancing Coriolis forces, then the result is an Ekman spiral with depth, in which the surface current is strongest and 45° to the right (northern hemisphere) or left (southern hemisphere) of the surface wind vector, and the turn increases with depth as the current slows (Fig. 8.5). Over the entire depth of the Ekman spiral, which typically runs about 100 m, the transport is 90° to the right (northern hemisphere) or left (southern hemisphere). Hence the frictional component of surface flows is called the Ekman transport.

To understand the main ocean currents over the open ocean (Fig. 8.7 given later), however, it is necessary to deal with the *vorticity*, or curl, as a measure of the local rotational character of the fluid, combined with the Coriolis component, and the critical quantity is the curl of the wind stress. The oceanographer H. Sverdrup showed that over the open ocean, the meridional transport of mass in the ocean is governed by the curl of the wind stress. In general, the atmosphere, through the prevailing westerly winds, acts to drive the ocean eastwards in middle and high

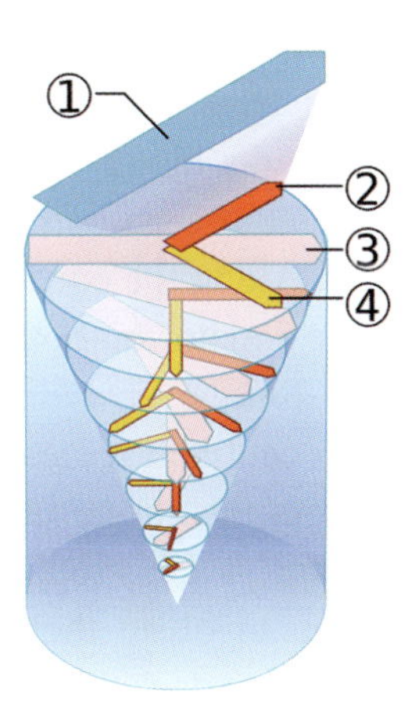

Fig. 8.5 The Ekman spiral in the northern hemisphere. Given the surface wind ① (blue arrow), then the force on the ocean ② (red arrow) is parallel, while the Coriolis force ④ (yellow arrow) is to the right, and together these result in the direction of the surface current ③ in pink. However, the current and therefore the Coriolis force drop with depth and the result is the Ekman spiral clockwise but dropping in magnitude with depth. https://en.wikipedia.org/wiki/Ekman_spiral#/media/File:Ekman_spirale.svg.

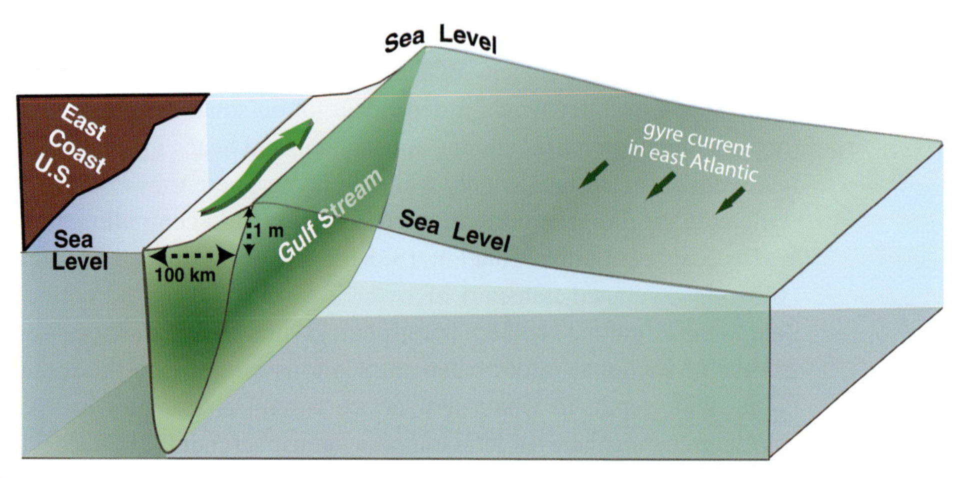

Fig. 8.6 Simplified schematic of the Gulf Stream off the east coast of the United States. A short north–south section is indicated showing the strong deep current. The sea level rises sharply by 1 m or more across the Gulf Stream, then gradually falls off east of there to northern Africa and the Iberian Peninsula across the Sargasso Sea, where a weaker return flow occurs that includes the Canary Current on the eastern side of the Atlantic Ocean. The stronger the current, the steeper the slope.

latitudes, and westward in the easterly trade winds, and hence the water transport has an equatorwards component in the westerlies, that converts into an equatorward current in the eastern ocean basins. The way the circulation is completed is through the development of relatively strong western boundary currents that return the waters to higher latitudes. These western boundary currents spin up sufficiently to experience strong lateral boundary friction that allows the total vorticity to be contained, and this works along western ocean boundaries. The results are the observed narrow western ocean currents, such as the Gulf Stream and Kuroshio (Fig. 8.7 given later). Along the equator, the prevailing easterly trade winds drive waters away from the equator on both sides, causing upwelling of cooler subsurface waters to occur. Similar upwelling occurs from equatorward-directed winds off the west coasts of continents, creating cold eastern boundary currents.

However, away from the strong influences of boundaries and surface friction, the ocean currents are in geostrophic balance, and hence a pressure gradient develops in the ocean to offset the Coriolis force. Because the ocean is fairly incompressible, one way it does this is by changing the sea level, and hence higher sea levels develop in association with the subtropical ocean gyres, and sea level is much higher at Bermuda than along the eastern United States coast, for instance, in association with the Gulf Stream (Fig. 8.6). In general, sea level is slightly more than 1 m lower along the coast than on the east side of the Gulf Stream. However, this also means that fluctuations in the Gulf Stream result in sea-level fluctuations along the coast, and a weakened Gulf Stream therefore means higher coastal sea

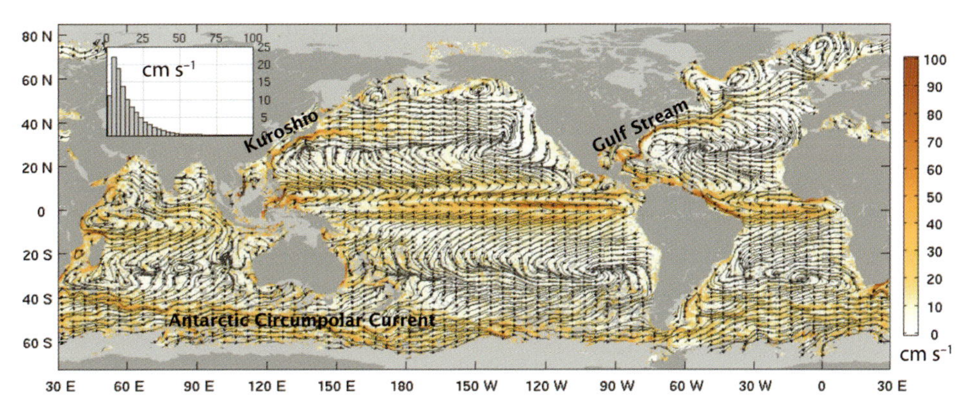

Fig. 8.7 The annual mean surface ocean currents in cm s^{-1}. Values are based upon satellite-tracked drogued drifter buoys, compiled by Lumpkin and Johnson (2013). Streamlines are drawn to indicate directions. The speed is in color. Inset (top left) shows histogram of mean current speed (cm s^{-1}, horizontal axis, from 0 to 100 at 3.125 cm s^{-1} intervals). The major currents (Gulf Stream, Kuroshio, and Antarctic Circumpolar Current) are indicated.

levels and risk of coastal flooding. Changes in temperature of the ocean also change sea level along with changes in prevailing winds.

8.3 Overturning Currents

The term "thermohaline" circulation used to be commonly used to denote the overturning circulation that was supposedly driven by temperature and salinity effects through changes in the density of the ocean waters. More recently the term "meridional overturning circulation" has been adopted to recognize that the circulation is primarily wind-driven even though it may have an important thermohaline component. The best known and strongest is the Atlantic Meridional Overturning Circulation: AMOC. Often the three-dimensional ocean circulation is referred to as the "conveyer belt" because it transports heat and chemicals around the world. The winds not only drive the ocean currents but also play a major role in surface exchanges of heat and moisture, and thus in both temperature and salinity.

The overturning circulation is difficult to measure because it is quite weak, but its evidence becomes clear from trace constituents in sea water and also isotopic signatures such as those associated with past detonations of nuclear bombs. Most gases are soluble in water, and hence their concentrations may be able to be used to track certain water masses. For example, chlorofluorocarbons are man-made and hence tracking their path through the ocean provides excellent information on circulations and the age since the water was last at the surface.

Ocean water takes up carbon dioxide directly in an exchange between the air and sea water. Tiny marine plants use carbon dioxide for photosynthesis, and many microscopic marine organisms use carbon compounds to make shells. When these marine microorganisms die and sink to the sea floor, they take the carbon with them. Carbon dioxide and oxygen are roughly doubled in terms of saturation values at 0°C versus 24°C. In one sense, warming the waters "boils off" the gases. Oxygen is typically close to or even above saturation value near the surface owing to bubbles of air and mixing. But as oxygen-rich waters sink, not only is the source cut off, but oxygen is consumed by organisms, depleting amounts.

As large values of carbon dioxide are observed deep in the North Atlantic, but not the Pacific, it is inferred that waters not only sink relatively quickly in the North Atlantic but also spread southward, as part of the AMOC. A particular isotope of carbon, C_{14}, is radioactive and produced naturally in the atmosphere by cosmic rays and especially from nuclear explosions above ground. The main isotope, C_{12}, is stable and makes up about 99% of carbon, while C_{14} decays with a half-life of about 5730 years. Using the known decay rates of C_{14} to C_{12}, it is possible to estimate the time since the water was at the surface. Along with salinity, such tracers can be used to identify water masses and flows, and their sources. Results show that relatively few parts of the ocean form deep water, in the North Atlantic and around parts of Antarctica, such as the Weddell Sea. The latter is referred to as Antarctic Bottom Water, which forms with the help of freezing ice that ejects brine, making for cold, salty, and dense waters.

The overturning circulation is mostly a direct circulation and hence transports heat down the temperature gradient. This is the case in the North Atlantic, with the warm northward flow in the Gulf Stream and a much deeper and colder flow southwards on the western side of the ocean, also as a boundary current. The heat transport is efficient because of the large temperature contrasts between the surface and deep ocean, whereas the gyre (horizontal) transports also contribute but the return flow in the east is not as cold. However, the AMOC extends well into the southern hemisphere, and hence the northward heat transport there is opposite in direction to the surface temperature gradients.

The meridional heat transport therefore is a vital part of the overall energy balance of the planet and is explored in Chapter 9.

8.4 Surface Energy Exchanges

Although the forms of energy vary considerably in the atmosphere, in the ocean the main form of energy is the ocean heat content (measured in zettajoules, 10^{21} J; Figs. 8.3 and 14.6). Kinetic energy plays a role at times, but oceans currents are typically measured in cm s^{-1} (Fig. 8.7), and hence the kinetic energy is relatively very small (fluctuations are order 10^{17} J). Latent energy plays a major role where sea ice forms and melts. Because the ocean is essentially incompressible, changes in altitude or depth do not alter the potential energy much because a mass equal in size has to replace the mass moving up or down. Accordingly, the exchanges with the atmosphere can be broken up mostly into radiative fluxes, sensible heat fluxes, and the evaporative/latent heat fluxes. Kinetic energy fluxes involve exchanges of momentum via the surface wind stress. The sensible heat fluxes mostly involve exchange of heat by air and water at different temperatures, but there is also a small component from precipitation: the temperature of raindrops may differ from that of the ocean, and there may be snow or other solid forms of precipitation. These latter components have mostly been neglected as tiny in past studies.

Many instruments have been deployed on platforms and ships to measure the turbulent surface fluxes, and most commonly these make use of an empirical bulk flux formulation that for sensible heat depends upon the difference in air versus water temperature, the density of the surface air, and an exchange coefficient. For evaporation the form is similar except the saturation specific humidity (that depends on the SST) versus the atmospheric humidity is used. The exchange coefficients may depend on wind speed and/or stability of the atmosphere, but appear to vary considerably in space and time. Attempts to measure the fluxes directly involve complex sensitive instruments that record every second to capture the puffs of wind and their associated temperature and moisture content. Accordingly, they are done only in a research setting for limited periods of time. These form the basis for the exchange coefficients, but they may not capture the full range of variability on all timescales.

The proof of these concepts comes when a fully coupled atmosphere–ocean is considered. Then the heat and moisture lost by the ocean must be matched by that gained by the atmosphere, and hence by carrying out complete heat or moisture budgets, it is possible to determine how closely the flux formulation is closed. Unfortunately, results show that the closure is not good enough to resolve the important changes associated with climate variability and change, and errors over the global ocean are typically several tens of watts per square meter. However, such fluxes are very useful because they vary enormously in space and time, often by hundreds of watts per square meter, and hence they reveal where and when the main exchanges take place.

This discussion also highlights a more fruitful way of proceeding by utilizing the fully coupled framework, inventorying the energy in the atmosphere and oceans continually, and ensuring that the changes match and are equal but opposite. This approach was impossible prior to about 2005 because the analyses of energy in both atmosphere and ocean domains were not accurate enough, except for climatological values that were

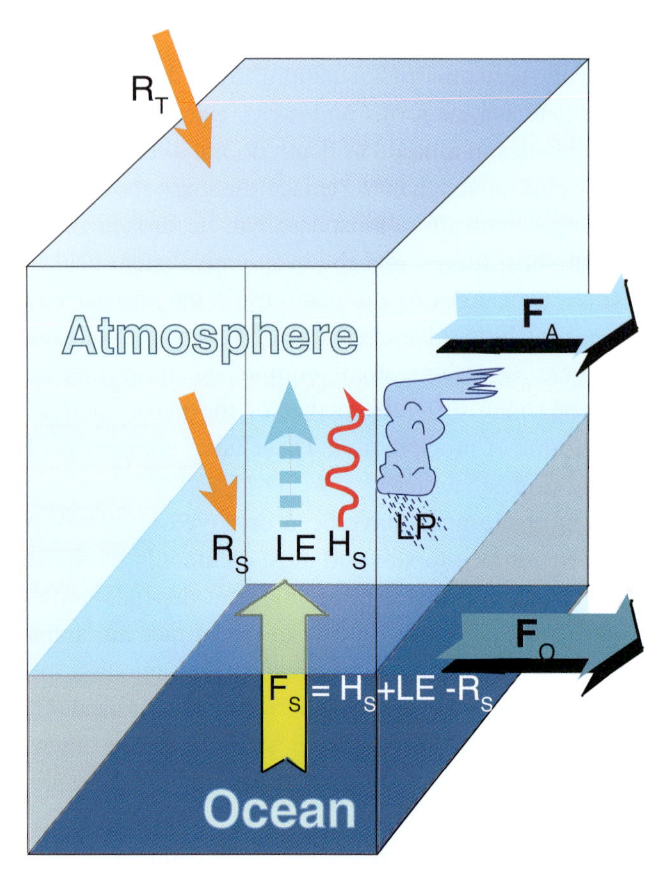

Fig. 8.8 Schematic diagram of the main energy terms in the vertically integrated atmosphere (top) and ocean (bottom) and how they balance. Main terms include radiation at the top and surface R_T and R_S, surface sensible heat flux H_S, and surface latent heat flux LE. The last three terms combine to give the surface flux F_S (also a small precipitation enthalpy term is included). Latent heat is realized in the atmosphere as precipitation LP. The vector transports of total vertically integrated energy in the atmosphere F_A and ocean F_O are indicated. Trenberth et al. (2019). © American Meteorological Society. Used with permission

averaged over many years. With high-quality atmospheric and ocean reanalyses of all available data to produce gridded fields, this diagnostic task has become possible. The framework is given in Fig. 8.8 and details are in Sidebar 8.1.

Individual monthly means of the net surface flux of energy can be readily computed, and annual mean values (Fig. 8.9) and seasonal means for December–February and June–August (Fig. 8.10) display warm colors to signify a heat flux out of the ocean into the atmosphere, while blue colors mean heat is going into the ocean. The pronounced seasonal cycle is dominant, and in the summer season (DJF in the southern hemisphere, JJA in the northern hemisphere) there is a very strong flux of heat into the ocean. In contrast, in winter (JJA in the southern hemisphere,

Sidebar 8.1: Atmospheric and Ocean Energy and Moisture Budgets

The atmospheric energy and moisture budgets can be performed with knowledge of the top-of-atmosphere (TOA) radiative fluxes (Fig. 5.8), measured by satellite, plus the full computation of all the energy transports and changes in heat in the atmosphere, to give the implied surface fluxes as a residual (Fig. 8.8). With R_T as the net radiation at TOA (T for top), $\boldsymbol{F_A}$ as the vertically integrated vector transport of atmospheric energy (**A** for atmosphere), and A_E as the vertically integrated energy in the atmosphere, then

$$\partial A_E/\partial t = R_T + F_s - \nabla \cdot \boldsymbol{F_A} \tag{8.1}$$

F_s is the net surface flux (s for surface) $F_s = H_s + LE - R_s$, where H_s is the surface heat flux, E is the surface evaporation, L is the latent heat of vaporization, and R_s is the surface radiation. Here the exchanges related to precipitation have been ignored. Note that it is the divergence, given mathematically by the ∇ operator, of the atmospheric energy transport that comes in here. The divergence relates to the change in horizontal energy transport along its path and implies that a vertical component at top or bottom is adding to or subtracting from the transport.

Then for the ocean (O),

$$\partial O_E/\partial t = -F_s - \nabla \cdot \boldsymbol{F_O} \tag{8.2}$$

and so the ocean heat content O_E is governed by the convergence or divergence of heat in the ocean and the surface flux. Note that summed together for the atmosphere and ocean, Eqs. (8.1) and (8.2) give

$$\partial(A_E + O_E)/\partial t = R_T - \nabla \cdot (\boldsymbol{F_A} + \boldsymbol{F_O}) \tag{8.3}$$

and the surface flux term cancels.

Similarly, a budget for moisture in the atmosphere can be done.

$$\partial A_m/\partial t = L(E - P) - \nabla \cdot \boldsymbol{F_m} \tag{8.4}$$

where A_m is the column-integrated atmospheric moisture (total column water vapor plus very small liquid and ice components), $\boldsymbol{F_m}$ is the vertically integrated transport of moisture by winds, and P is the precipitation.

Now Eq. (8.1) can be rewritten as

$$\begin{aligned}\partial A_E/\partial t + \nabla \cdot \boldsymbol{F_A} &= [(R_T - R_S) + H_s + LP] + L(E - P)\\ &= Q_1 - Q_2\end{aligned}$$

where

$Q_1 = (R_T - R_s) + H_s + LP$ is the net heating of the atmosphere from radiation, sensible heating and latent heating, and

$Q_2 = L(P - E)$ is the net latent heating due to the moisture flux out of the atmosphere.

Hence the atmospheric changes in storage of energy or transport of energy can be considered as either the difference in fluxes between the top and bottom of the atmosphere or as the total atmospheric heating

Sidebar 8.1: *(cont.)*

plus latent heating, where the latter is the difference between P and E. Hence this second term is related to the hydrological cycle in Eq. (8.4).

This framing of the problem of energy flows allows the surface flux of energy F_S to be computed from the other three terms in Eq. (8.1), and this is viable because of the global atmospheric analyses that exist, given the terms on the left-hand side of Eq. (8.1), while the R_T is obtained from radiation measurements at the TOA. Then given F_S, Eq. (8.2) can be used with estimates of changes in ocean heat content to compute the ocean heat divergence by ocean currents, a term not otherwise available.

Most of these terms can now be estimated quite accurately, although adjustments are required to satisfy global integral constraints, and thus F_S and E can be computed as residuals from the atmospheric analyses. Then given the change in ocean heat content and the surface flux term, the vertically integrated ocean heat divergence can also be computed as a residual. This enables the movement of heat around in the ocean to be computed, whereas it is difficult to measure otherwise owing to the small velocities in the ocean that are not measured, except at a few particular locations.

Actually, the horizontal transports of energy and its components, and also moisture, can be computed reasonably well, but taking the divergence often means computing small differences from large quantities, and the errors become significant on small scales. However, for monthly averages, on 1000 km scales or more, the estimated random errors are less than 10 W m^{-2}, and because they are random, the errors cancel out as larger and larger regions are considered. Globally, of course, the divergences of the transports are zero.

DJF in the northern hemisphere) there is a huge flux of energy out of the ocean into the atmosphere.

In particular, in winter in the northern hemisphere, when strong westerly winds bring cold dry air from the major continents out over the oceans, there are huge heat fluxes from the vicinity of the Gulf Stream, off North America, and the Kuroshio, off east Asia, into the atmosphere. These are often episodic as storms occur in the right place and surface fluxes can easily exceed 1000 W m^{-2} for a few hours. The surface heat flux is primarily in the form of evaporation, moistening the atmosphere and evaporatively cooling the ocean, but the sensible heat flux is also substantial. The response is often to form strong cumulus complexes in spectacular patterns in the atmosphere, called cumulus streets because they tend to align with one another parallel to the wind direction. An example is shown in Fig. 7.14 for October 11, 2019, although the convection is much more extensive in mid-winter. Such large cooling of the ocean lowers the SSTs, and hence lowers subsequent fluxes.

The predominant pattern is one of the excess solar radiation in summer going into the ocean as heat, and then ocean heat returning to the atmosphere in winter

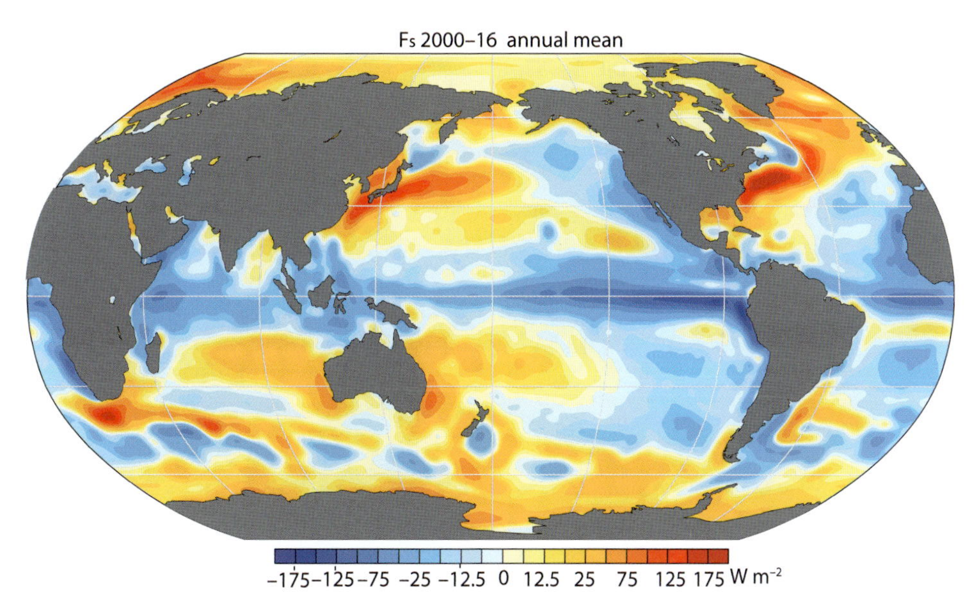

Fig. 8.9 Mean annual surface energy flux for 2000–16 computed as a residual of the atmospheric energy budget in W m^{-2}. Positive values are upward out of the ocean. Trenberth et al. (2019). © American Meteorological Society. Used with permission.

(Figs. 8.2, 8.3, and 8.10). It is this which gives rise to a maritime climate whereby the temperature extremes on nearby land are modulated by ocean effects, and the seasons are somewhat delayed on land.

Nevertheless, when the annual mean is computed (Fig. 8.9), there are strong patterns of surface fluxes that remain, although the magnitudes are reduced considerably relative to the seasonal extremes (Fig. 8.10). The wintertime heat losses off east coasts of northern continents remain evident. There is a general predominance of blue colors, signifying heat going into the ocean throughout the tropics and eastern oceans. The latter tend to be the location of subtropical anticyclones. However, the main source region for heat into the ocean is truncated polewards of about 15° latitude in part because of strong evaporative cooling associated with trade winds and extratropical winds. The patterns revealed in Fig. 8.9 are real. They also occur in comprehensive climate system models (such as that at NCAR) and come about because of ocean currents and the ocean bottom topography. For instance, the blue patch southeast of New Zealand lies over the Campbell Plateau (see Fig. 9.5), a somewhat shallower part of the Southern Ocean that diverts the Antarctic Circumpolar Current farther south.

Hence the blue areas in Fig. 8.10 are where heat enters the ocean, and warm colors are where heat comes out. For a stable climate these must balance through transports of heat by ocean currents from the blue- to orange-colored regions.

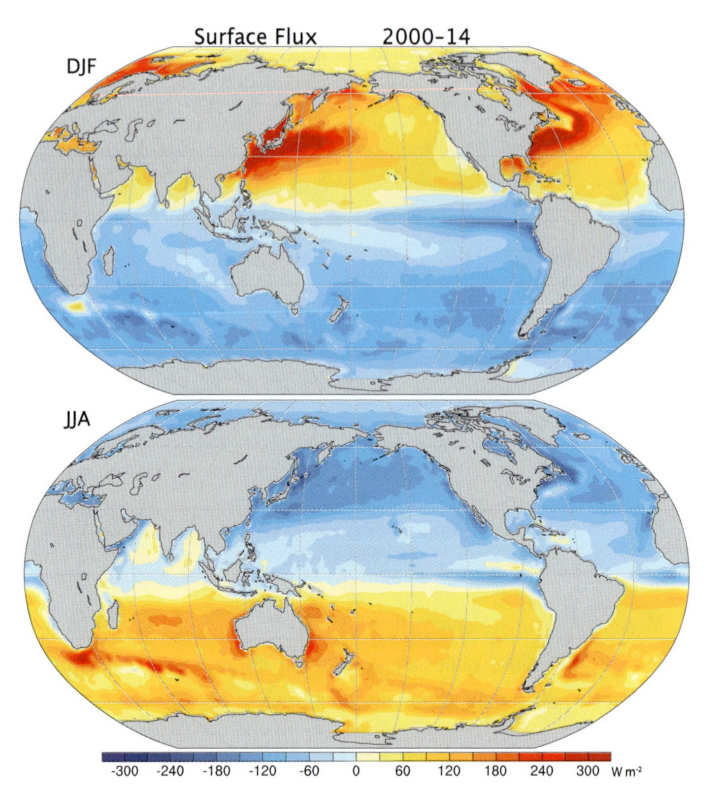

Fig. 8.10 Mean surface energy flux for 2000–14 for December–January–February (top) and June–July–August (bottom) computed as a residual of the atmospheric energy budget in W m^{-2}. Positive values are upward out of the ocean.

While some balances can be achieved relatively locally, it is readily seen that the Arctic is a sink for heat, and so too is the North Atlantic, and these cannot be completely compensated for by the ocean heating in the tropical Atlantic. Instead, there has to be a heat transport by ocean currents from the Pacific and Indian oceans, into the South Atlantic, and a northward transport of heat throughout the Atlantic Ocean (Fig. 9.1). This is a key part of the role of the AMOC. Although these aspects have generally been established for some time, it is only recently that they have been quantified for the mean annual cycle and interannual time series.

The tropical Pacific is a large heat source region for the ocean while the largest sinks lie off the coasts of Asia in the Pacific and North America in the Atlantic in the vicinity of the Kuroshio and Gulf Stream currents, where the fluxes are largest in winter.

Another facet of the patterns of heating of the ocean is that they appear to be related to that of carbon uptake. Carbon uptake is affected by the chemical buffering of anthropogenic CO_2 and is quantitatively the most important oceanic process acting as a carbon sink. Carbon dioxide entering the ocean is buffered due to scavenging by the carbonate ions and conversion to bicarbonate, so that the resulting increase in gaseous sea-water CO_2 concentration is smaller than the amount of CO_2 added per unit of sea-water volume. This factor influences the ratio of carbon to heat uptake. However, because both carbon and heat are moved around by the ocean currents and mixing in common ways, the circulation component can be better determined. Nevertheless, model results suggest that in future, the heat-content increases may be dominated more by where the heat is added to the ocean.

8.5 Oceans and Climate Change

Earth is "The Blue Planet" and all peoples depend directly or indirectly on the ocean. The ocean supports unique habitats. As has been shown, the ocean is interconnected with other components of the climate system through exchanges of water, energy and carbon. Meridional transports of heat are substantial, and clearly anything else in the ocean is similarly transported, with prospects for changes as the climate changes. The ocean has tremendous thermal inertia and moderates atmospheric temperature changes, but also provides the primary memory of past changes.

New data and analyses show continuing record high sea levels and global ocean temperatures (Chapter 14); stronger coastal storm surges with hurricanes (Chapters 7 and 10) and risk of flooding; amplification of the existing salinity patterns and the hydrological cycle (Chapter 10); and increased stratification (Chapter 14).

Stratification refers to the layering of less dense, lighter fluid atop denser, heavier fluid and determines the stability of a fluid, be it a gas like the atmosphere, or a liquid like the oceans. As the planet is continually warmed, the surface and lower atmosphere warms faster than the air aloft and the atmosphere becomes less stratified and less stable, which promotes vertical motion, mixing and convection. In the oceans, the opposite effect occurs. The warm, light surface waters warm faster than the cold deeper water, since heat penetrates slowly into the depths of the ocean. Global warming makes the oceans more stable. Consequently, this inhibits the ability of the ocean to sequester heat and carbon dioxide, with implications for acidification (Chapter 6). The increased stratification means less oxygen for ocean organisms that affects the whole food web and marine productivity and poses risks to fish populations.

As the ocean warms and OHC reaches record levels year after year (Section 14.5) marine heatwaves are becoming a lot more common, often with devastating consequences for marine life. The best documented case has been for "The Blob" from 2015 to 2016 where over 4 million square kilometers of the North Pacific surface ocean became exceptionally warm. An estimated 100 million cod disappeared and the whole food web was decimated, from phytoplankton, zooplankton, krill, swarms of small fish, half a million birds, and hundreds of humpback whales. Similar kinds of disruption in ecosystems have occurred elsewhere in marine heatwaves, such as the south Tasman Sea in 2015–16. That marine heatwave has been linked to the super El Niño event in 2015–16 (see Chapter 12). Each marine heatwave influences the local energy exchanges with the atmosphere with consequences for local weather; in the tropics strong hurricanes like Harvey may result, as occurred from the marine heatwave warming of the Gulf of Mexico in 2017. However, it also means that the lifetime of each marine heatwave is finite, and the hot spots move around from one year to the next as the ocean relentlessly warms.

References and Further Reading

Bronseleiaer, B., and L. Zanna, 2020: Heat and carbon coupling reveals ocean warming due to circulation changes. *Nature*, **584**, 227–233. doi: 10.1038/s41586-020-2573-5.

Cheng, L., K. Trenberth, J. Fasullo, T. Boyer, J. Abraham, and J. Zhu, 2017: Improved estimates of ocean heat content from 1960–2015. *Science Advances*, **3** (3), e1601545. doi: 10.1126/sciadv.1601545. http://advances.sciencemag.org/content/3/3/e1601545.

Cheng, L., K. E. Trenberth, N. Gruber, et al., 2020: Improved estimates of changes in upper ocean salinity and the hydrological cycle. *Journal of Climate*, **33**. https://doi.org/10.1175/JCLI-D-20-0366.1.

Cornwall, W., 2019: In hot water. *Science*, **363**, 442–445. doi: 10.1126/science.363.6426.442.

Ezer, T., L. P. Atkinson, W. B. Corlett, and J. L. Blanco, 2013: Gulf Stream's induced sea level rise and variability along the U.S. mid-Atlantic coast. *Journal of Geophysical Research: Oceans*, **118**, 685–697. doi: 10.1002/jgrc.20091.

Lumpkin, R., and G. C. Johnson, 2013: Global ocean surface velocities from drifters: mean, variance, El Nino–Southern Oscillation response, and seasonal cycle. *Journal of Geophysical Research: Oceans*, **118**, 2992–3006, doi: 10.1002/jgrc.20210.

Trenberth, K. E., Y. Zhang, J. T. Fasullo, and L. Cheng, 2019: Observation-based estimates of global and basin ocean meridional heat transport time series. *Journal of Climate*, **32**, 4567–4583. doi: 10.1175/JCLI-D-18-0872.1.

9 Poleward Heat Transports by the Atmosphere and Ocean

The atmosphere and the ocean together transport heat and energy polewards to offset the Sun–Earth geometry of incoming and outgoing radiation. Meridional heat transports by the atmosphere and ocean are explored along with energy transports across the equator.

9.1 Poleward Energy Transports

As Earth rotates on its axis and revolves around the Sun, the basic geometry determines an excess of radiation from the Sun is received in the tropics, and there is a deficit in polar regions (Fig. 1.4). Contrasts in temperatures between the two regions would be far greater than observed (Chapter 5) were it not for the dynamic transport of energy polewards by the atmosphere and ocean. The contrast is greatest in winter, when the polar night sets in, with zero incoming radiation in the Arctic or Antarctic, and the contrast in summer is less in the northern versus southern hemisphere owing to the location of large continental land masses in mid- to high latitudes that more readily warm up than the oceans do. How and where the poleward energy transports occur greatly affect local climates.

As documented in Chapter 7, the atmosphere is primarily responsible for the energy transports that compensate for the net radiation imbalances. The total poleward energy transport is readily determined from the TOA radiation balance (Section 5.6). The atmospheric transports can be computed using global atmospheric reanalyses, and then the ocean transport may be computed as a residual (Fig. 9.1). For the ocean, the energy transport is a heat transport. The total meridional energy transports for the climate system including the atmospheric

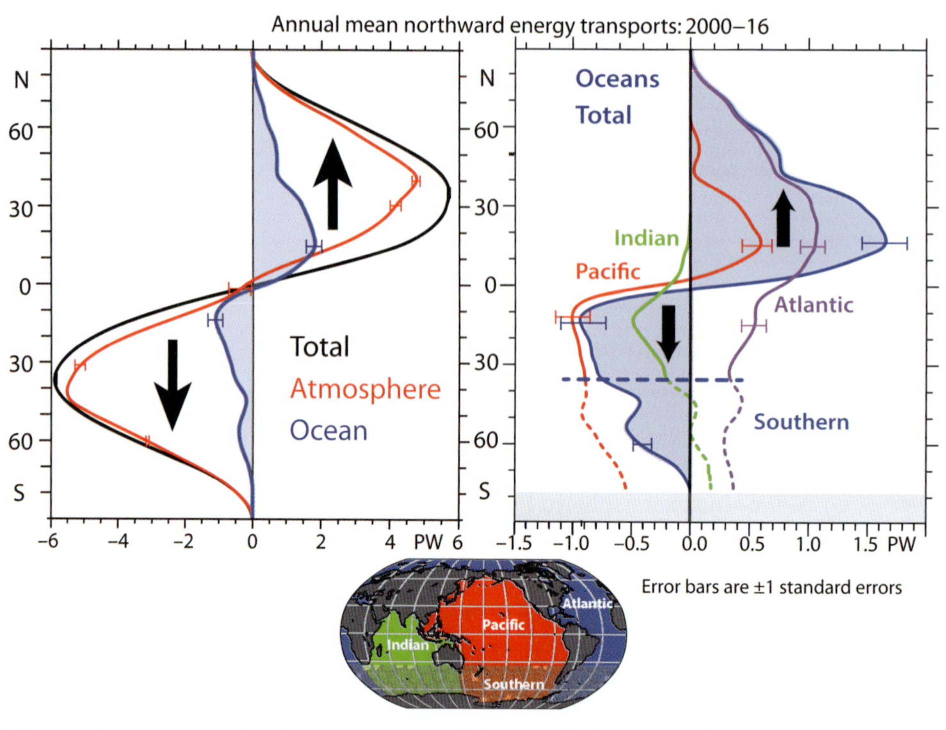

Fig. 9.1 Poleward heat transports for the climate system: (left) total (black line) based on TOA radiation measurements, the atmospheric transports (red) from atmospheric reanalyses, and the total ocean transport (blue) computed as a residual; (right) the ocean transport is broken down by oceans, as outlined in the panel below (not including Indonesian Throughflow effects). Here the Southern Ocean includes all ocean south of 35° S, the southern tip of Africa. From Trenberth et al. (2019). © American Meteorological Society. Used with permission

and ocean components have been well quantified since 2001, and the atmospheric transports have been reliably estimated since then, although earlier estimates were made using radiosonde[1]-based analyses dating from the 1970s.

Annual mean poleward transports of energy of 5.8–6 PW (petawatts) peak near 35° latitude, while the dominant atmospheric poleward transports peak near 41° latitude in each hemisphere (Fig. 9.1). Poleward ocean heat transports are dominant only between 0 and 12° N. At the peak of the total poleward transport, the atmospheric transport accounts for 78% of the total in the northern hemisphere and 96% in the southern hemisphere (Fig. 9.1).

[1] Radiosondes are expendable instruments attached to a weather balloon and launched once or twice daily at about 800 weather stations on land, mostly in the northern hemisphere.

The estimated meridional heat transports by the ocean can be examined in full and broken down into the annual long-term mean, the annual cycle and the anomalies, as the departures from the annual cycle (Section 9.3), and the individual ocean components of the total ocean meridional heat transport are given in Fig. 9.1.

9.2 Atmospheric Meridional Energy Transports

The atmospheric total energy transports can readily be broken down into components (Fig. 9.2) and how they vary with season (Fig. 9.3). The kinetic energy component is quite small and, although included in Fig. 9.2, is mostly neglected here. There are quite large variations from summer to winter in each hemisphere, seen in Fig. 9.3, especially in the northern hemisphere. The total poleward energy transport is dominated by the dry static energy component which depends on temperature transport, while the latent energy transport, associated with moisture movement, is dominant in the deep tropics. For the period 1979–2017, the total transport (Figs. 9.1 and 9.2) goes to zero slightly north of the equator where values change sign between hemispheres. Slightly farther north, at about 7° N, the location of the ITCZ, the three curves intersect in Fig. 9.2. North of there, the transport is northward for dry static energy. South of 7° N the transport reverses to become southward, consistent with Hadley Circulation transports. However, the

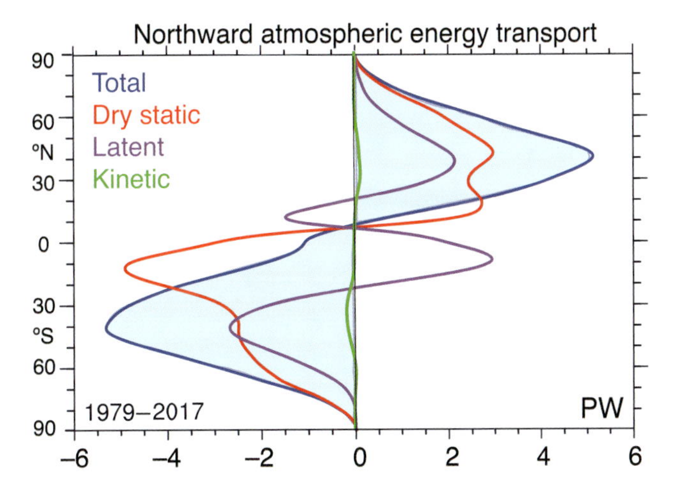

Fig. 9.2 Annual and zonal mean for 1979–2017 northward energy transport for the total atmospheric energy (blue), dry static (red), latent (purple) and kinetic (green) energy components; in PW. The transports are integrated from the poles. Negative values mean the transport is southwards. Based upon ERA-Interim reanalyses.

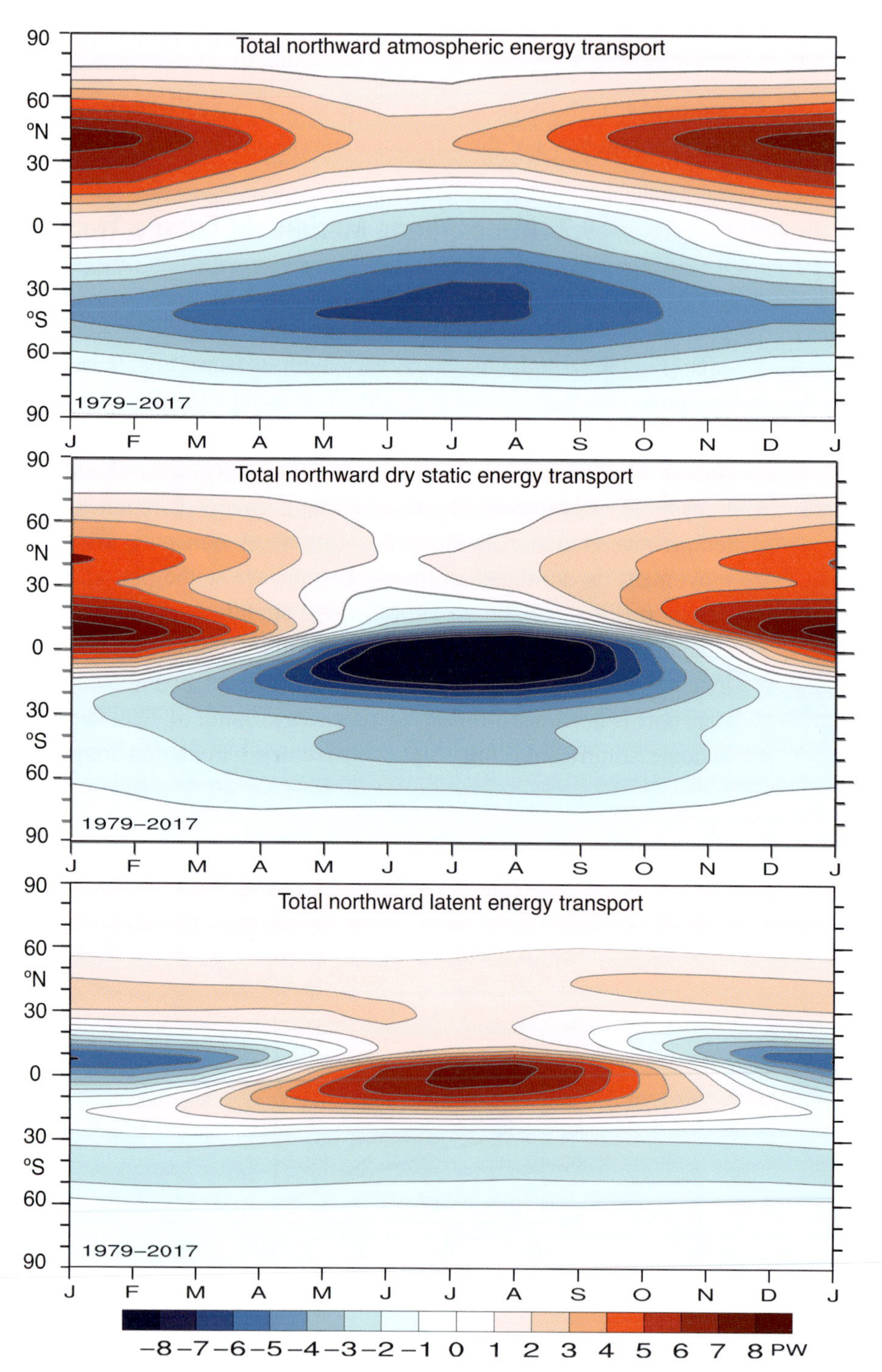

Fig. 9.3 Zonal mean northward atmospheric energy transport for 1979–2017 for the mean annual cycle, in PW. The atmospheric energy transports for the (top) total energy, (middle) dry static energy, and (bottom) latent energy, with shadings given in the key. The contour interval is 1 PW. Based upon ERA-Interim reanalyses.

moisture and dry static energy transports are in opposite directions between about 20° N and 20° S. They are dominated by the moisture in the lower part of the Hadley Circulation because moisture amounts drop rapidly with height in colder temperatures aloft. Associated northeasterly (northern hemisphere) and southeasterly (southern hemisphere) trade winds converge moisture towards the ITCZ, and thereby feed the Hadley Circulation with latent energy. The signature of the Hadley Circulation is that the moisture and associated latent energy transports are opposite to the meridional transport of the total energy, because the moisture feeds the latent energy into the rising branch where it is realized as heat (Fig. 7.4).

It is often useful to separate the total energy transports into the component associated with the mean flow versus that arising from transient storms, such as those in Figs. 7.13 and 7.14. The mean flow component is referred to as the "stationary" or "quasi-stationary" component and sometimes as the "standing" component, versus the "transient" component. Figure 9.4 presents the zonal mean transient poleward sensible heat fluxes as a function of altitude for the annual mean. As it deals only with temperature and does not include the specific heat factor, it does not have the units of an energy transport. What is noteworthy in Fig. 9.4 is where the transient transports occur, namely mid-latitudes and lower troposphere. There is a secondary maximum in the lower stratosphere (not fully shown). The difference between the transients and the total is the quasi-stationary component that dominates the tropics, while the transient component is dominant polewards of 30° latitude.

The Hadley Circulation dominates the quasi-stationary component but, in the northern hemisphere in particular, there is also a significant component from standing waves that remain more or less in place, anchored by the land–sea distribution and mountain ranges. While a characteristic of the overturning energy transports is that the latent energy component is opposite in sign to that of the dry

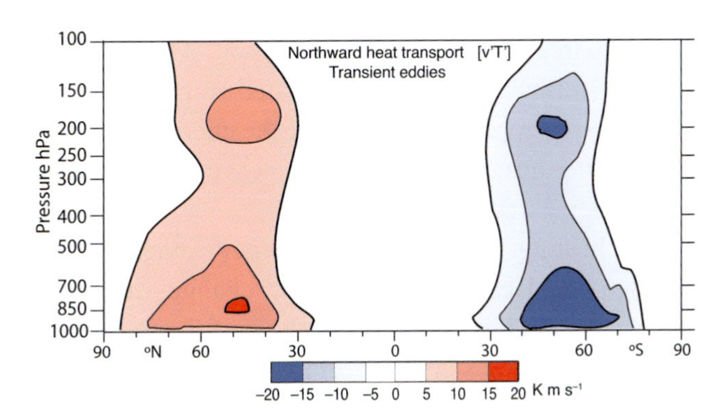

Fig. 9.4 Zonal mean northward transient heat transport as a function of latitude and pressure in K m s^{-1}. Based upon ERA-Interim reanalyses.

static and total energy transport, in contrast, for transient and standing waves, both latent and dry-static components contribute towards poleward transports, signifying that the circulation transports both heat and moisture in the same way, as illustrated in Fig. 7.13. The stationary component diverges energy out of the tropics but converges energy between about 22° and 50° latitude, especially for the dry static energy. The transient eddies pick up that energy and transport it polewards of 45° latitude to warm the higher latitudes (Fig. 9.4).

The seasonal variations in these transports (Fig. 9.3) show the smaller annual cycle in the southern hemisphere, where transient eddies are active year round, while they are much less active in summer in the northern hemisphere because land changes the overall patterns of heating. The dry static energy and latent energy are both active at the same times and almost the same locations (Fig. 9.3), except that the moisture transports peak at lower latitudes where it is warmer, and the atmosphere can hold more moisture.

The total atmospheric transport in the tropics is a fairly small residual from the cancellation between dry static and latent energies. The dominant processes are those associated with large-scale overturning: the global monsoon in which the Hadley and Walker circulations are embedded. Latent heat is released in the upward branch (Fig. 7.4) while subtropical cooling balances the warming from subsidence in the downward branch and is a combination of radiative cooling to space and cooling by transient waves that transport the heat to higher latitudes. These waves are called "baroclinic" because of the role of temperature gradients and stratification, and their link to heat transport. They are not only present in the climatological mean, but also in the variability on timescales longer than a month. The biggest perturbations occur with El Niño events; see Chapter 12. As the Hadley Circulation changes and the east–west Walker Circulation weakens, the locations of storm tracks change (Fig. 11.2).

Although the signature of the Hadley Circulation can be discerned from both the dry static and latent energy transports, it is remarkable that there is no sign of the Hadley Circulation in the total transport. This highlights the seamless link to mid-latitude storms and that the associated storm tracks depend on the warm subsiding air in the subtropics as a source of the energy that they in turn transport polewards. By drawing heat out of the subtropics, the extratropical storms actually help drive the Hadley Circulation.

The heating of the planet is largely externally controlled by the Sun–Earth geometry and can be modified by changes in albedo. Other considerations are the changes in heat storage and the partitioning of energy among the components of the climate system, especially how much heat is taken up by the ocean. Nevertheless, it is worth highlighting the importance of these results for understanding climate dynamics. The observed inverse and symbiotic relationship among the atmospheric quasi-stationary and transient waves should be obvious given the need to satisfy the heat budget, which can be thought of as a strong constraint. If one is strong then the other is weak. Hence this interaction contributes to the seamless nature of the atmospheric heat transports. But the constraint applies to the whole system and emphasizes the strong tendency for compensations to also occur among climate system components. If atmospheric poleward transports are strong, then ocean transports weaken, and vice versa (see Section 9.3).

9.3 Ocean Meridional Heat Transports

Ocean meridional heat transports (MHTs) have been deduced as a residual using energy budgets to produce latitude versus time series (see Sidebar 8.1). The top-of-atmosphere (TOA) radiation is combined with the vertically integrated atmospheric energy divergence from atmospheric reanalyses to produce the net surface energy fluxes everywhere. The latter is then combined with estimates of the vertically integrated ocean heat content tendency to produce estimates of the ocean heat divergence. In all results presented here, the OHC tendency used is from the ORAS5 (Ocean Reanalysis System 5) ocean reanalysis from ECMWF. It uses four-dimensional data assimilation in a comprehensive ocean model and, in addition to assimilating ocean soundings, it includes surface sea-level height from altimetry, SSTs and surface flux estimates. Five ensemble members are generated by perturbing both observations and forcing fields. However, prior to 2005, values are not as reliable, and values after 2015 were not reanalyses but operational analyses. As a result, biases are present, but they have been corrected by using the constraint that the global mean ocean change in heat must match the surface heat flux, that is in turn constrained by TOA radiation. Figure 9.1 presented the total meridional heat transports for the climate system, plus the atmospheric and ocean components.

Locally, the ocean heat transport consists of two major parts. The bigger part is referred to as "rotational" or nondivergent. This is the part where heat is carried around by ocean currents in an irregular circle but which ends up back where it started. This component is actually very large, especially for the Antarctic Circumpolar Current, for instance. The second is the divergent component that not only transports heat but moves it from one place to another and deposits it there, and this is what can be estimated using the residual technique. On the other hand, direct oceanographic estimates of temperatures and currents measure the whole transport, most of which is rotational and therefore does not contribute to the net movement of heat across regions, and hence even small uncertainties or errors can make huge differences in the inferred transports by each component. This is why the residual technique has a large advantage.

By taking the zonal integral across longitudes, the rotational component of heat transport can be mostly removed because of mass conservation. That is to say that the flow northwards in one area is compensated by a southward flow elsewhere. This is, of course, not quite true because the oceans are mostly connected (the nonconnected parts, like the Caspian Sea are isolated). It is desirable to isolate the Atlantic from the Pacific, for example. The narrowest link in the north is through the Bering Strait between Russia and Alaska joining the Pacific and Arctic Oceans, so that as long as that can be accounted for, such as by direct measurements, then the combined Arctic and Atlantic Ocean heat transports can be separated from those in the Pacific, at least as far as 35° S (the southern latitude of Africa). The net transport of mass and heat through the Bering Strait is quite small and two orders of magnitude smaller than the zonal mean.

It can then be useful to consider the Indo-Pacific oceans together. The Pacific is isolated in the northern hemisphere but is linked to the Indian Ocean via the Indonesian ThroughFlow (ITF) (Fig. 9.5). The ITF mass and heat transports are significant and there is a net mass transport of order 15 sverdrups ($10^6\ m^3\ s^{-1}$) from the Pacific into the tropical Indian Ocean, accompanied by about 1 PW of heat transport, and this flow must be compensated for by a return flow around Australia. In that sense, Australia acts as an island in the midst of the huge Indo-Pacific Ocean. Of course, there are many other small islands but these are of little consequence for mass and heat transports because they are sufficiently small. This is not the case for Australia (Fig. 9.5).

The actual movement of waters around Australia is quite involved and complicated. After transitioning through the Indonesian region, the flow southward and westward can occur almost anywhere in the Indian Ocean, but ultimately has to flow eastwards south of Australia, where it becomes part of the Antarctic Circumpolar Current and West Wind Drift (Fig. 9.5). Some of this flow runs into the Campbell Plateau and New Zealand, and moves northwards into the Tasman Sea, but some may transit all the way to South America. In both cases, the return northward flow is at a colder temperature than for the Indian Ocean southward flow, and hence there is also an energy transport involved. Therefore, the Indian and Pacific oceans are not separate in terms of the mass and heat transports. The actual energy transport has to account for both the warm ITF component polewards and the colder return flow equatorwards. Sometimes these components are referred to as "temperature transports" because they are not actual energy transports without consideration of the mass transport. The ambiguity of the temperature of the return flow and the complexity of the path make this difficult to tie down, and instead a simple reference temperature, such as 0°C, is sometimes used to define the temperature transport.

Using that approach, estimates have been made of the meridional ocean heat transport with and without an adjustment for the ITF for the Pacific and Indian

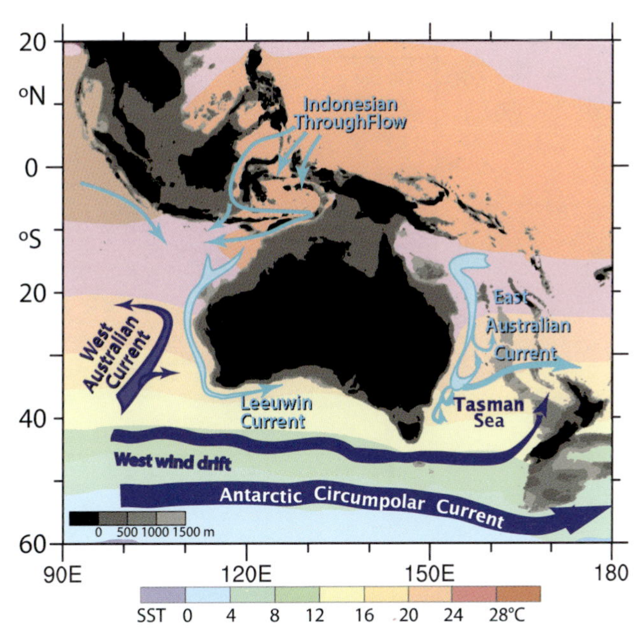

Fig. 9.5 The primary region of mass flows in the Indonesian ThroughFlow (ITF). The shades of gray show the bathymetry at the surface (black), 500, 1000, and 1500 m depth. The background field is the annual mean SST, scale at bottom. The arrows depict various currents: the cyan arrows show warm waters in the ITF, East Australian current, or Leeuwin current; the dark blue arrows show the Antarctic Circumpolar Current and the associated West Wind Drift, plus the West Australian current. The Tasman Sea lies between Australia and New Zealand. From Trenberth and Zhang (2019). © American Meteorological Society. Used with permission

Oceans, but there is always ambiguity arising from how the mass balance is treated. Hence there is considerable merit in treating the Indo-Pacific Ocean as a unit in terms of the meridional transports of heat. These arguments apply to the total field and affect the mean values and the mean annual cycle, but a good assumption is that the effects are systematic, and therefore the anomalies of heat transport from year to year can be readily isolated in each basin. For the mean MHT, the melting and freezing of sea ice have also been considered, although this is difficult owing to uncertainties in the volumes involved.

In Fig. 9.6 the global ocean MHT has quite a strong annual cycle in both hemispheres, although this is seen to come mostly from the Indo-Pacific. The largest annual cycle is from 5 to 10° N in the Pacific, which is also the location of the ITCZ over the oceans. Values range from more than 5 PW northwards in the southern summer, to 5 PW southwards in the northern summer. Peak values of the transports occur at the times of highest SSTs in each hemisphere, March in the

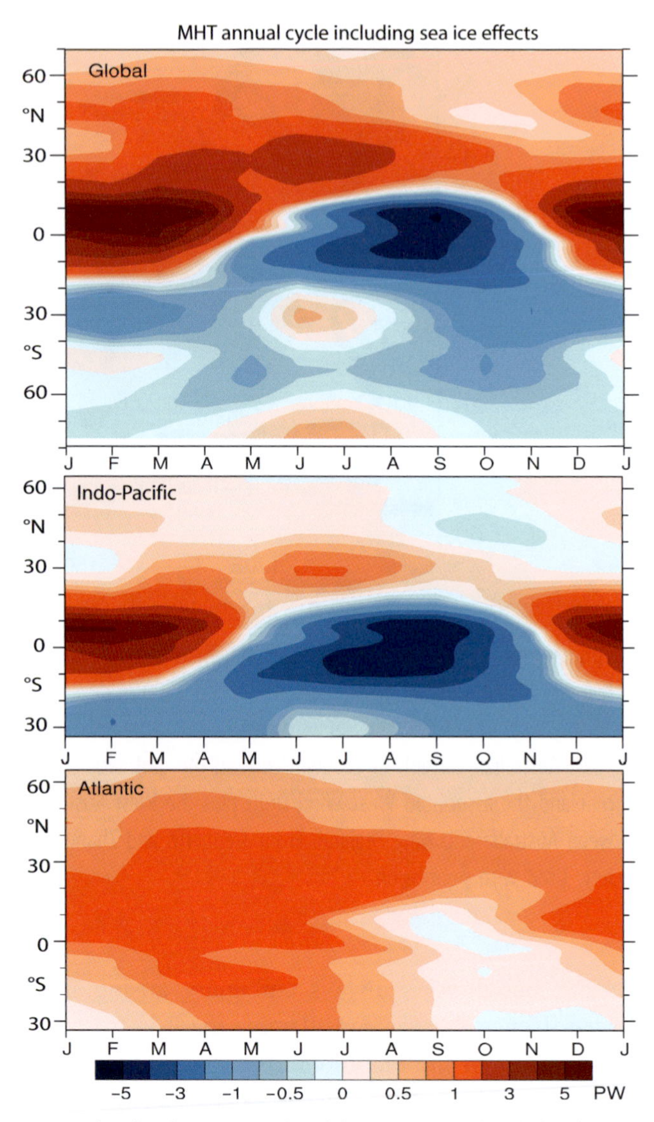

Fig. 9.6 The inferred mean annual cycle of mean meridional heat transport (MHT) by the oceans in PW for the global (top), Indo-Pacific (middle), and Atlantic (lower). Trenberth et al. (2019). © American Meteorological Society. Used with permission

south and September in the north. This is symptomatic of the buildup in heat and OHC through the long summers (Fig. 8.2), while the largest northern hemisphere heat losses by the ocean are in winter. Over the southern oceans, south of 35° S where the Antarctic Circumpolar Current dominates, the ocean heat transport is mainly polewards except for December–January–February near 45° S, when the Sun resides south of the equator. The reverse applies in the northern hemisphere, where only briefly, in August–September–October, is the ocean heat transport southwards at high latitudes, primarily in the Indo-Pacific.

For the Atlantic (Fig. 9.6) the computations include an allowance for the formation and melting of sea ice in the North Atlantic and Arctic. Heat moves northwards at all latitudes from January to August as the heat from the Pacific and Indian oceans is redistributed and offsets the heat losses in the Arctic and North Atlantic. Average heat transport is northwards in most months in the southern hemisphere in the Atlantic, and mainly from 0 to 15° N and south of 25° S in July to September there is a weak southward transport, evidently purging some of the summertime heat from the north. The MHT values are positive year-round from 10 to 50° N. MHT values are slightly higher for November through June at 26° N, and with values of 0.9 (July, September, October) to about 1.3 PW in March–April. These values are reasonably compatible with direct ocean heat transport estimates from the RAPID-MOCHA[2] moored array along about 26° N.

9.4 Ocean Poleward Heat Transport Variability

The 12-month running mean time series for the global ocean MHT and the anomalies, as departures from the mean annual cycle (Fig. 9.6), are given in Fig. 9.7. The total field emphasizes the poleward transports in both hemispheres. The anomalies, however, reveal considerable variability that is larger in the tropics, and strongest just north of the equator. The global ocean MHT time series (Fig. 9.7) may seem unduly spotty in the deep tropics, but this is real and mostly related to the El Niño phenomenon, and best deciphered by separating out the contributions from each ocean.

The time series for the Atlantic (Fig. 9.8) shows the northward transport at all latitudes (note the different color scheme here), but with largest values between the equator and 45° N. MHT tends to be somewhat higher in the first part of the record until 2009 when values drop precipitously, and with a smaller repeat in 2012. These fluctuations can be seen more clearly in the anomalies in MHT (Fig. 9.8), which are quite a lot weaker than for the Pacific. The large variability in the MHT zonal mean time series is primarily associated with the atmospheric circulation and, in particular, with modes of natural variability of the North Atlantic Oscillation (NAO) and the

[2] RAPID-MOCHA (Rapid Climate Change–Meridional Overturning Circulation and Heatflux Array) is an international oceanographic research program that deployed a series of moorings and instrumentation approximately along 26° N in March 2004 whose observations are combined with many others to estimate the MOC and its transports. https://www.rapid.ac.uk/rapidmoc/overview.php.

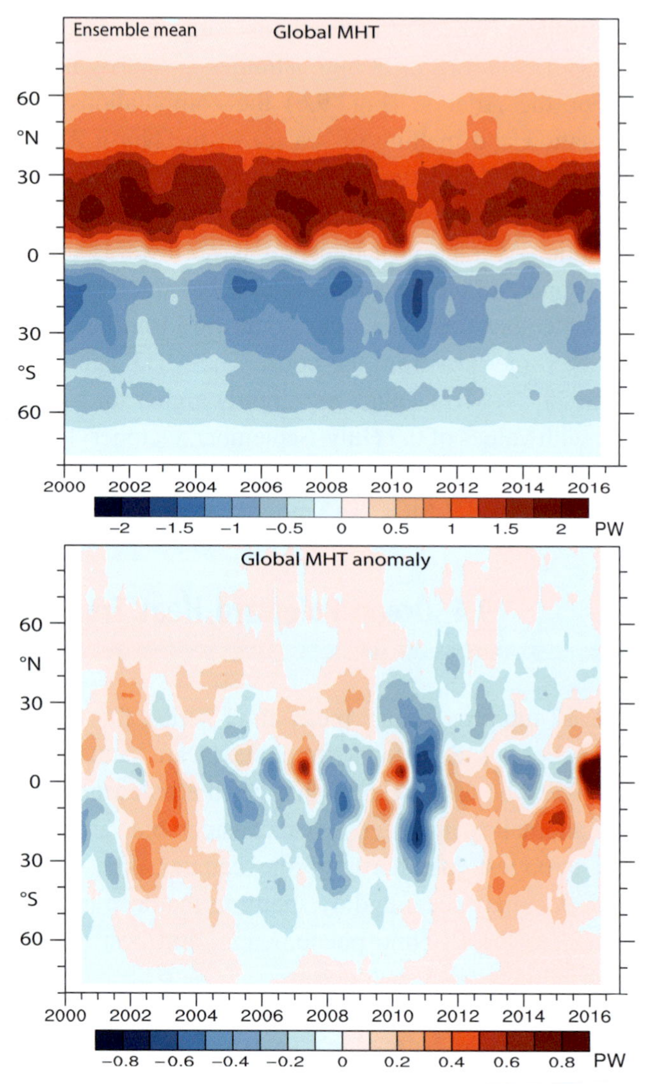

Fig. 9.7 Latitude–time series of zonal and mean MHT for the global ocean as the total (top) and anomalies (departure from annual cycle of Fig. 9.6) (below) in PW. Values are 12-month running means. Trenberth et al. (2019). © American Meteorological Society. Used with permission

closely related Northern Annular Mode (both described in Chapter 11). Positive correlations of the Atlantic MHT between the equator and about 45° N are verified, with the NAO leading by about 2 months. At higher latitudes between 50 and 60° N the correlation reverses and is marginally significant.

The large variability in the deep tropics for the global MHT time series was noted above, but most of this comes from the Indo-Pacific (Fig. 9.9) and is real. Indeed, the variability is associated with ENSO events, as seen in the bottom

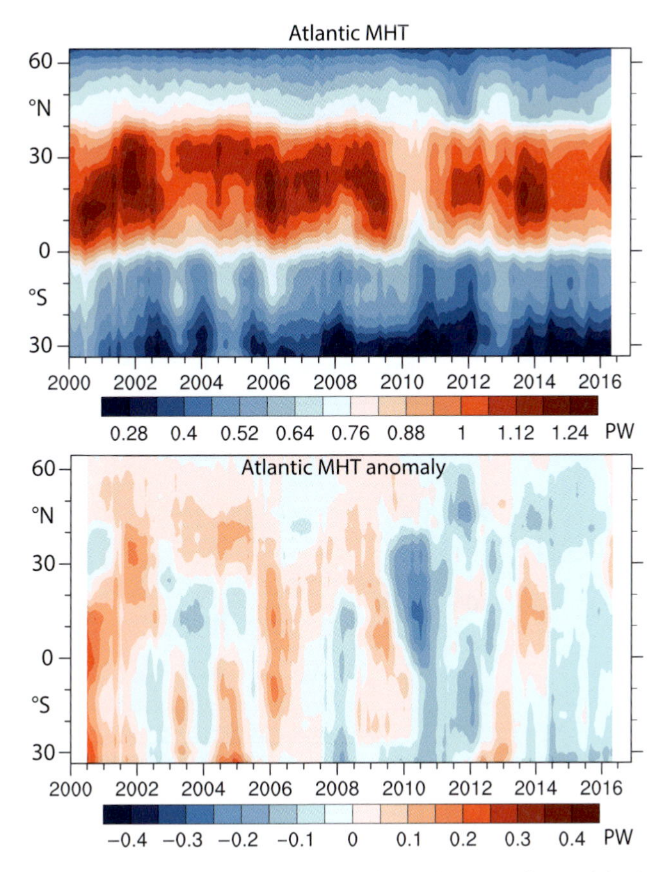

Fig. 9.8 Latitude–time series of zonal and mean MHT for the Atlantic Ocean as the total (top) and anomalies (below) in PW. Values are 12-month running means. Trenberth et al. (2019). © American Meteorological Society. Used with permission

panel of Fig. 9.9, which shows the key index of ENSO (see also Chapter 12). In the Indo-Pacific there is consistent northward MHT by the ocean north of about 5° N which is the location of the ITCZ. The largest value and farthest south extension of the northward MHT is in 2015–16 in association with the major El Niño event.

The Indo-Pacific has been split into the anomaly components from the Pacific and Indian Oceans (Fig. 9.9 lower panels). In El Niño events, the deep warm water stored in the Warm Pool in the tropical western Pacific thermocline becomes much shallower as warm waters surge eastwards. These warm waters deepen the thermocline in the central and then eastern Pacific and result in warmer waters being upwelled, greatly changing the SSTs. Subsequently, warm waters spread polewards along the coast of the Americas. In the northern hemisphere, warm waters can spread all the way to the Gulf of Alaska. Chapter 12 details El Niño aspects.

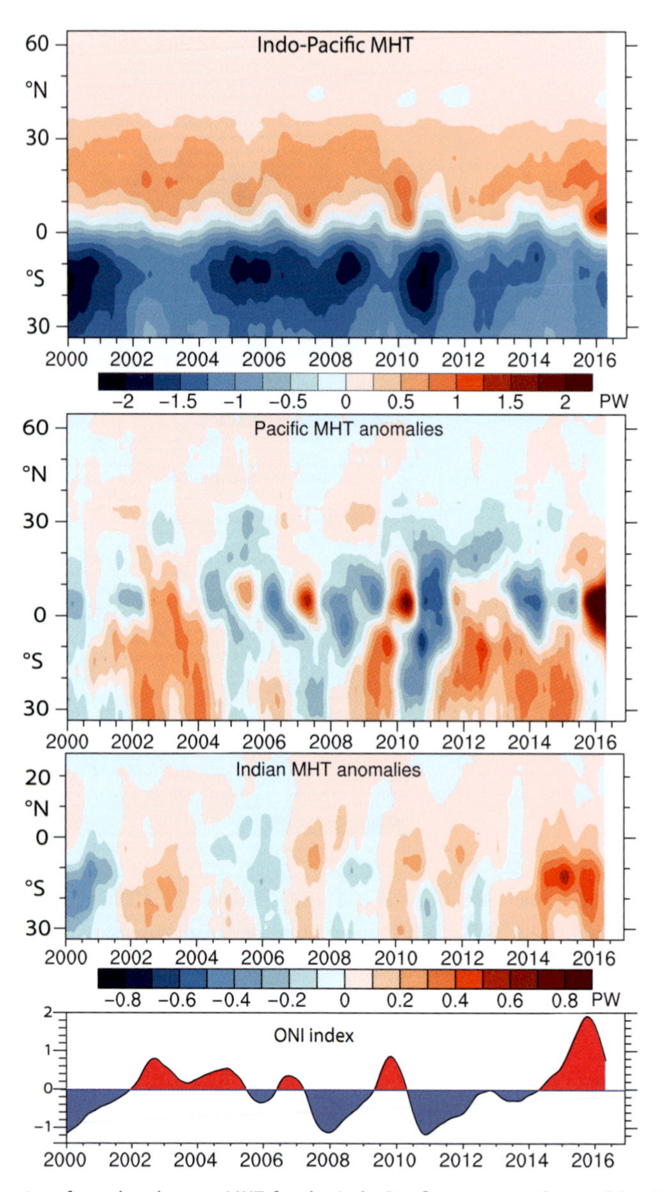

Fig. 9.9 Latitude–time series of zonal and mean MHT for the Indo-Pacific oceans as the total (top) and anomalies for each ocean: Pacific (upper middle panel) and Indian (lower middle panel) in PW. (Bottom) ONI is the Ocean Niño Index (Chapter 12). The ITF is included. Values are 12-month running means. Adapted from Trenberth and Zhang (2019) and Trenberth et al. (2019). © American Meteorological Society. Used with permission

The MHT anomalies (Fig. 9.7) as a whole are much stronger in the Pacific than the Indian Ocean (Fig. 9.10), and the main fluctuations are associated with ENSO (Fig. 9.9). In the southern hemisphere, the MHT variations are somewhat comparable in magnitude in the Indian and Pacific Oceans and it is mainly near the equator and

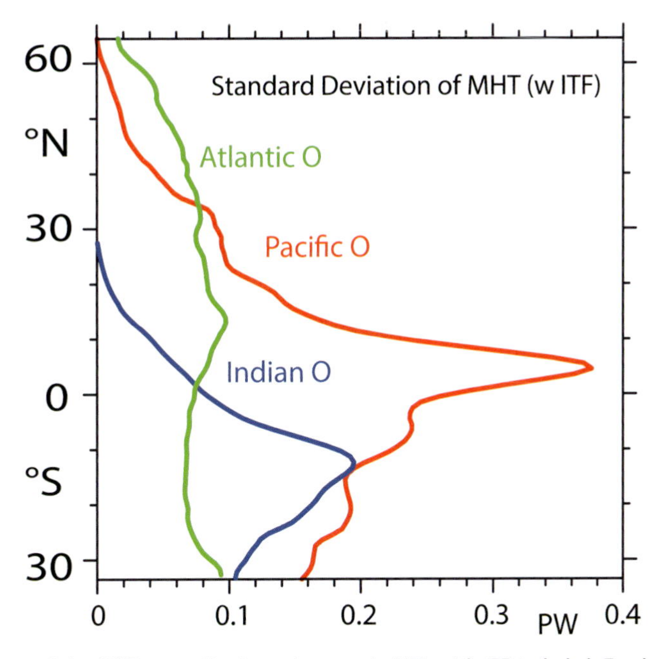

Fig. 9.10 Standard deviation of the MHT anomalies in each ocean in PW, with ITF included. Trenberth and Zhang (2019). © American Meteorological Society. Used with permission

peaking about 5° N that the Pacific MHT variability dominates. Moreover, the southern hemisphere anomalies at times reveal reverse changes in the two oceans, consistent with the role of the ITF. The ONI El Niño index leads MHT near the equator in the Pacific by 3–4 months, consistent with the ITF response. There are also intriguing relationships for Pacific MHT in the southern hemisphere from 20 to 30° S where fluctuations tend to lead El Niño by 8 months, and it also leads the MHT in the Indian Ocean in the same zone by 8–9 months. It appears that strong northward MHT near 25° S propagates northward to influence the equatorial region in the Pacific, especially after 2005. The weaker relationship prior to then may be a consequence of the absence of adequate Argo data to describe what has occurred (see also Fig. 14.3 and discussion). In the 2015–16 El Niño event, a lot of the northward MHT propagation comes from the Indian Ocean.

Much has been made of the Atlantic Meridional Overturning Circulation (AMOC) variations with regard to decadal and centennial climate variability, and indeed, the variations are larger in Atlantic versus the Pacific MHT north of 35° N, but the largest interannual variability in MHT by far arises in the Pacific Ocean tropics (Fig. 9.10), a new finding that is yet to be exploited for its predictability prospects. Because the time series (e.g., Fig. 9.7) are only from 2000 to 2016, they do not adequately sample the multi-decadal climate variability.

9.5 Interhemispheric Flows of Energy

There is considerable interest in the relative energy/heat budgets of the two hemispheres, and how much MHT there is across the equator. Discussions have dealt with the role of land distribution, the atmospheric versus the ocean contributions, and their implications for understanding the whole system. The trans-equatorial energy flow is complicated by the fact that the climate is changing, and it is essential to properly deal with the changes in heat storage. The new MHT estimates provide further insights into this issue.

A summary of the results (Figs. 9.11 and 9.12) provides the time series of MHT on the equator for the individual oceans, the global ocean, the atmosphere and the TOA radiation values. The ocean transports in the Pacific at the equator vary considerably and dominate the global ocean values. Of special note are the extremely large values in 2016, associated with the super El Niño event. Values range from −0.3 PW to 1.2 PW for the 12-month running mean. A range from −0.3 to 0.6 PW exists without 2016 included. Accordingly, the actual transports depend critically on the period sampled.

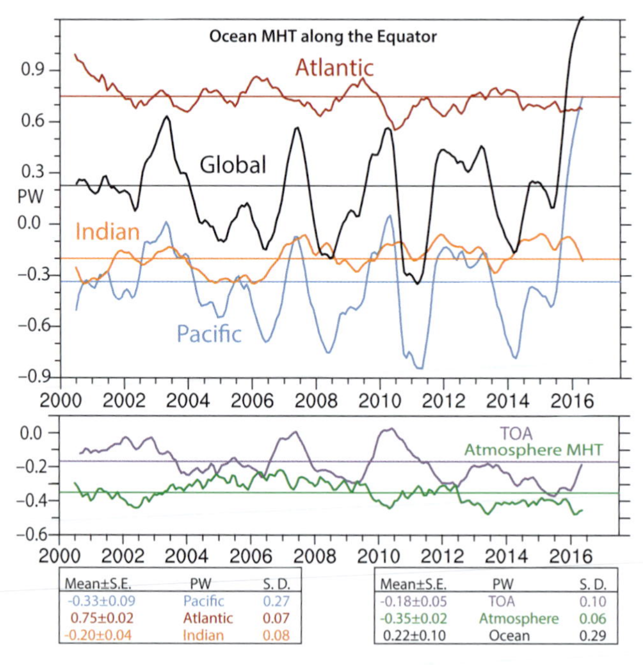

Fig. 9.11 Time series of MHT as 12-month running means across the equator for the oceans: (top) Global, Atlantic, Pacific, and Indian; and meridional energy transports at TOA and for the atmosphere (middle) in PW. At the bottom, the mean values are given along with the standard error of the mean, and the standard deviation in PW. Trenberth and Zhang (2019). © American Meteorological Society. Used with permission

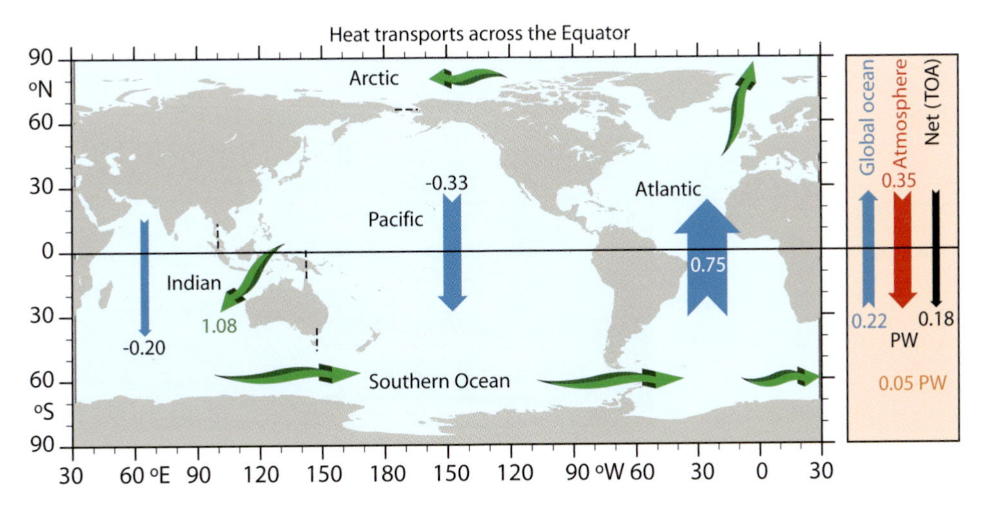

Fig. 9.12 Schematic of the ocean heat flows across the equator for 2000–16 as blue arrows, complemented by green arrows showing flows among the oceans. (Right) The global ocean, atmosphere, and totals are given in PW, along with the differential change in storage (orange) in the Southern Hemisphere. From Trenberth and Zhang (2019). © American Meteorological Society. Used with permission

Also of interest in Fig. 9.11, and summarized in Fig. 9.12, is that on the equator for 2000–16 the mean ocean MHT is −0.33 PW (i.e., southward) for the Pacific, −0.20 PW for the Indian, and +0.75 PW for the Atlantic associated with the AMOC, giving a net ocean MHT of 0.22 PW northward. In turn this heat is transported into the North Atlantic and into the Arctic Ocean, with a very strong annual cycle. The schematic green arrows in Fig. 9.12 illustrate the main movement of heat within the ocean in regions away from the equator.

However, the atmosphere has a distinct southward transport of energy of −0.35 PW, the ocean transport is northward at 0.22 PW, while overall the TOA values give −0.18 PW. Accordingly, there is an extra −0.05 PW equatorial transport that represents a heat flow into the southern hemisphere ocean and arises because of the much greater ocean area and volume, and relative uptake of heat in the southern hemisphere. Accordingly, while the global ocean energy budget is balanced by the TOA Earth's Energy Imbalance (EEI), this is not true locally or for each hemisphere alone. The value of 0.05 PW is equivalent to 1.1 W m^{-2} over the ocean and it is largely because of the greater area in the south that the imbalance arises. The value may be underestimated and could be as large as 0.1 PW. Clearly, the Southern Ocean acts as a giant sponge for heat, mixing it deep within the ocean (see Fig. 14.4).

The transports necessarily have to balance, and the TOA value (Fig. 9.13) is the sum of the atmosphere and ocean transports but with an extra 0.05 PW imbalance into the southern oceans included. These equatorial energy transport values differ considerably from previous estimates because of the different period used – the

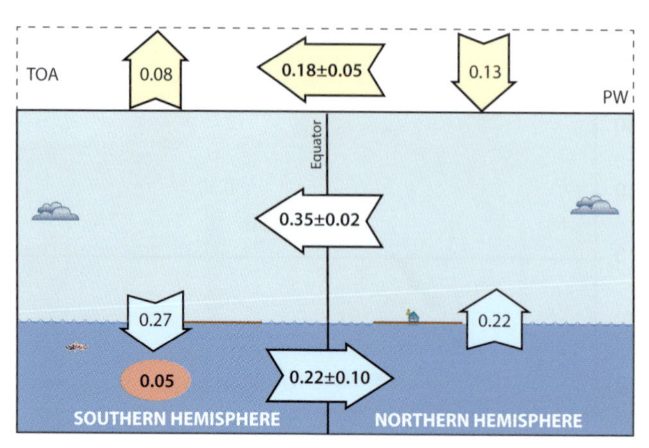

Fig. 9.13 Interhemispheric transports in PW. The atmosphere (light blue), land (brown), ocean (blue), and TOA are indicated. At the TOA (yellow arrows) the imbalance reflects the larger component into the southern hemisphere ocean (red bubble). The net anomalous surface energy fluxes for the oceans are also given. The global mean EEI does not affect transports but its regional differences do. Trenberth and Zhang (2019). © American Meteorological Society. Used with permission

large perturbations in 2016 – and the previous failure to account for the differential uptake in the OHC in the two hemispheres. Significant differences also arise from the revised atmospheric energy budget computations and the different ocean datasets used, because several ocean analyses have been shown to be quite deficient as they do not conserve energy. The results highlight the need to fully take account of the variability.

The MHTs for the global ocean, Pacific and Indian Oceans vary somewhat in sync, as does the TOA MHT. The Atlantic MHT variations along the equator are not only small, they are also mostly not significantly correlated with others. The MHTs in the Indian and Pacific Oceans seem to vary together (Fig. 9.13), with peaks more or less coinciding from 2002 to 2015, but that relationship is countered by the huge ITF effects in 2016 that go in opposite directions, although those effects are much greater farther south. Most of the anomalous ITF heat advection is compensated by the meridional temperature transport anomalies across the southern Indian Ocean boundary during both El Niño and La Niña.

References and Further Reading

Oort, A. H., 1971: The observed annual cycle in the meridional transport of atmospheric energy. *Journal of the Atmospheric Sciences*, **28**, 325–339,

Oort, A. H., and T. Vonder Haar, 1976: On the observed annual cycle in the ocean–atmosphere heat balance over the Northern Hemisphere. *Journal of Physical Oceanography*, **6**, 781–800.

Peixoto, J. P., and A. H. Oort, 1992: *Physics of Climate*. New York: American Institute of Physics, 520pp.

Trenberth, K. E., and J. M. Caron, 2001: Estimates of meridional atmosphere and ocean heat transports. *Journal of Climate*, **14**, 3433–3443.

Trenberth, K. E., and Y. Zhang, 2019: Observed inter-hemispheric meridional heat transports and the role of the Indonesian Throughflow in the Pacific Ocean. *Journal of Climate*, **32**, 8523–8536, https://journals.ametsoc.org/doi/pdf/10.1175/JCLI-D-19-0465.1.

Trenberth, K. E., Y. Zhang, J. T. Fasullo, and L. Cheng, 2019: Observation-based estimates of global and basin ocean meridional heat transport time series. *Journal of Climate*, **32**, 4567–4583. doi: 10.1175/JCLI-D-18-0872.1.

10 The Changing Hydrological Cycle

Water is essential for life. The water cycle is described along with issues for models and understanding impacts. Changes occurring in the water cycle with climate change are addressed along with changing storms, hurricanes, and extremes of drought and floods. Reasons why the extremes of floods and droughts increase, and the societal needs are outlined.

10.1 The Global Water Cycle

The hydrological cycle fundamentally involves evaporation (E) of moisture from the land and vegetation or ocean surface into the atmosphere, and back again as precipitation (P) (Fig. 10.1). Evaporation produces evaporative cooling at the surface and moistens the air. It includes transpiration from plants in which water enters the atmosphere through the tiny stomata in leaves as photosynthesis occurs. Together these are called evapotranspiration. Precipitation and the relationships to humidity and sea surface temperatures (SSTs) were introduced in Section 5.4. Water vapor is moved around by the atmosphere, and the precipitation occurs elsewhere, often in preferred locations such as mid-latitude storm tracks or tropical monsoons and convergence zones.

The global energy balance at Earth's surface (Fig. 3.1) has a net of 160 W m^{-2} energy absorbed at the surface from the Sun. A net 57 W m^{-2} is radiated back to

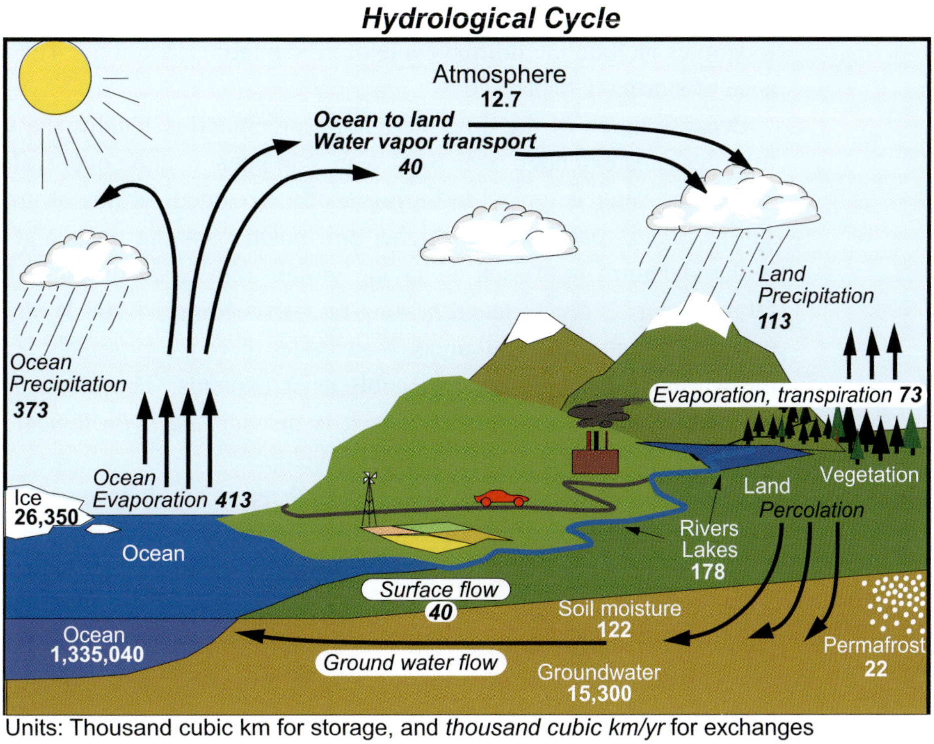

Fig. 10.1 The hydrological cycle. Estimates of the main water reservoirs, given in plain font in 10^3 km^3, and the flow of moisture through the system, given in slant font (10^3 km^3 yr^{-1}), equivalent to exagrams (10^{18} g) yr^{-1}. From Trenberth et al. (2007). © American Meteorological Society. Used with permission

space as infrared energy, and 20 W m^{-2} heats the atmosphere directly as thermals. The dominant compensation and cooling at the surface are from evaporative cooling, estimated globally as 82 W m^{-2}. This is slightly higher than implied by the global mean precipitation estimate of 2.69 mm day^{-1} from the Global Precipitation Climatology Project (GPCP) (for 1979–2018) in Table 5.2, since the average latent heating under most conditions is about 29 W m^{-2} per mm day^{-1} of precipitation. Hence GPCP is likely missing a small amount (in heavy events or in mountainous areas, perhaps).

Accordingly, the global hydrological cycle plays a major role in the surface energy budget overall. Much more than the global number is the fact that the structure of the release of latent heat into the atmosphere from precipitation varies enormously in space and time, and much more than radiation. In several places (Chapter 5; and see especially Chapters 11 and 12) it is noted that the structure of the atmospheric heating is dominated by precipitation, which leads to the striking patterns of variability and propagation of energy into remote regions through teleconnections (Chapter 11). The variability in time is even more startling. Examples include the diurnal cycle of precipitation (Section 7.4), and the importance of storms (Chapter 7).

Because much of the time there is no precipitation locally, the variations start from zero and extend to over 1000 W m^{-2} in latent heating over scales of hundreds of km; the latter is typical in hurricanes for extended periods of time. In thunderstorms, values can be much higher still but for smaller areas and shorter times. A thunderstorm is classed as having violent rain when the rainfall rate exceeds 50 mm $hour^{-1}$, giving latent heating rates exceeding 34 700 W m^{-2}, but usually not for long, and for small areas. No wonder there are extremely large updrafts as the warm heated air rises and cools as it expands. These rough estimates are presented simply to emphasize that it is through the hydrological cycle that the main extremes in the climate system develop.

The water cycle is also involved in energy in other ways because of the formation of clouds, and the radiative influences of both clouds and water vapor. As noted in Chapter 3, water vapor is the single most important natural greenhouse gas in the atmosphere (Fig. 3.2) and constitutes a strong positive feedback to anthropogenic climate forcing from carbon dioxide (CO_2). The water vapor feedback (Section 13.2.3) is critically important in understanding past and future climate change and its global and regional impacts. Clouds, introduced in Section 3.3, absorb and emit thermal radiation and have a blanketing effect similar to that of greenhouse gases (Fig. 3.2). But clouds are also bright reflectors of solar radiation and thus also act to cool the surface, leading to strong cancellation between the two opposing effects of shortwave and longwave cloud heating.

The hydrological cycle involves the continuous circulation of water and is driven by exchanges of energy. An estimate of the global hydrological cycle in units of 10^3 km^3 per year for 1988–2004 (Fig. 10.1) is that evaporation over the oceans (413) exceeds precipitation (373), leaving a net of 40 units of moisture transported from oceans onto land as water vapor. On average this flow must be balanced by a return flow over and beneath the ground through river- and streamflows, and subsurface groundwater flow. Consequently, precipitation over land exceeds evapotranspiration by this same amount (40). The oceans make up about 70.8% of Earth's surface versus 29.2% for land. The area of the oceans is roughly 240% that of land, and hence the average precipitation rate over the oceans exceeds that over land by 72% (allowing for the differences in areas). The average lifetime of a molecule of water vapor in the atmosphere is about 9 days. It has been estimated that on average over 80% of the moisture precipitated out comes from locations over 1000 km distant, highlighting the important role of the winds in moving moisture around.

Water vapor may be transported perhaps thousands of kilometers before it is involved in clouds and precipitating systems. Condensation of atmospheric moisture produces clouds or fog, tiny droplets that have little or no fall velocity and which come in many shapes and sizes and involve not just water droplets, but also ice particles of various kinds (e.g., snow crystals: dendrites, prisms, needles, and plates). Precipitation occurs when the clouds build to a point where the droplets become large enough to fall either as rain, snow, or other forms of ice (hail, graupel, etc.). With condensation, as clouds form with liquid water as part of the precipitation process, the latent energy that went into evaporation in the first place is given back, and this adds to the heating of the atmosphere and may make the air more buoyant so that it precipitates even more.

On land, once precipitation has fallen, the surface water substance may sit there in the case of snow, or it may freeze into ice in ponds and lakes. More commonly, it may percolate and infiltrate into the soils, run off into streams, rivers and lakes, and eventually find its way to the ocean, both on the surface and below the surface as groundwater. The surface water weathers rocks, erodes the landscape, and replenishes subterranean aquifers. As precipitation over land generally exceeds evaporation, there is usually runoff, even if it does not all get to the ocean, except in very dry areas where it all may evaporate.

Precipitation varies enormously in space and time (see Section 5.5 on how often it rains). Efforts have been made to determine how representative values are at one location and time. For spatial scales greater than 25 km based on hourly data, correlations between one location and others fall off with distance quite quickly, with an e-folding distance (the distance the correlation falls to 37% of the original value; see Box 10.1) of about 130 km, and varying from 100 to 160 km, depending on latitude. The highest values are for the main ocean storm track regions in southern winter. For ocean regions in the eastern tropical and subtropical Pacific and Atlantic there are much shorter e-folding decorrelations of less than 80 km, and this is also true in nearly all land areas, with some regions having an e-folding decorrelation of only 50 km. That is to say, precipitation is extremely spotty over land in summertime when convection dominates.

Box 10.1: E-folding Time or Distance

When a variable (such as precipitation P) has a rate of change in either space or time that depends on the amount of the variable present, the values fall off exponentially. The exponential constant $e \approx 2.718\ldots$. For example, if the rate of change of P is $\partial P/\partial t = -aP$, where t is time, and a is a constant parameter, then the solution is $P = P_0 e^{-at}$, where P_0 is the initial value. Then values drop off at an e-folding rate of a. That is to say, the values drop by a factor of e and fall to about 37% of the initial value after time $1/a$. The exponential function e^x forms the basis for natural logarithms and is very common in mathematics.

In terms of the covariability in time, the e-folding time for decay of correlations ranges from 6 to 9 hours over the oceans but only 2 to 5 hours over land. These numbers relate to the lifetime and movement of the associated weather systems. The small scales in both space and time of precipitation mean also that the latent heating is often isolated and localized, and does not result in cooperative complexes that can drive larger-scale circulations in the atmosphere. Chapter 12 on El Niño provides the best example where cooperation does take place. On the other hand, once the precipitation has fallen on land, it becomes integrated in catchment areas and river basins, and flows together into rivers and lakes. Given a widespread outbreak of storms, this amalgamation is what can trigger a flood. Over oceans, precipitation becomes fresh water that is mixed with the salty waters and thereby integrated into the net salinity of the ocean.

10.2 The Surface Freshwater Flux

The net surface freshwater flux is the difference between evaporation E and precipitation P, as $E - P$ into the atmosphere, or $P - E$ into the ocean. There are very distinct patterns of E and P in the tropics and subtropics. In particular, the strongest evaporation occurs in the subtropics, where large semi-permanent subtropical anticyclones reside over the oceans year-round although with a seasonal cycle. The relatively clear skies allow solar radiation through to the surface, and a lot of that heat goes into evaporation. The resulting water vapor flows out of the anticyclones both equatorwards and to higher latitudes. As noted in Chapter 7, southeast trade winds in the southern hemisphere, and northeast trade winds in the northern hemisphere carry moisture equatorwards and westwards where they feed either the ITCZ that resides about 5–10° N year-round over the Pacific and Atlantic or the SPCZ in the southern hemisphere (Fig. 5.1).

Over the Indian Ocean there is very strong seasonality associated with the monsoons, and the moisture flows into the monsoon rains. The atmospheric moisture transported to higher latitudes feeds the mid-latitude storm tracks, where copious rains occur in frontal systems – often now called atmospheric rivers, because the moisture is transported large distances and precipitation is focused into a narrow-elongated region. Because the locations of precipitation and evaporation differ, and rivers on land pour fresh water into the ocean, strong differences can develop in the ocean surface salinity, and ocean currents have to compensate by transporting relatively fresh water to some areas and salty water to others. Hence the oceans have their own hydrological cycle in addition to their role in surface evaporation, supplying moisture to the atmosphere and thus to land areas.

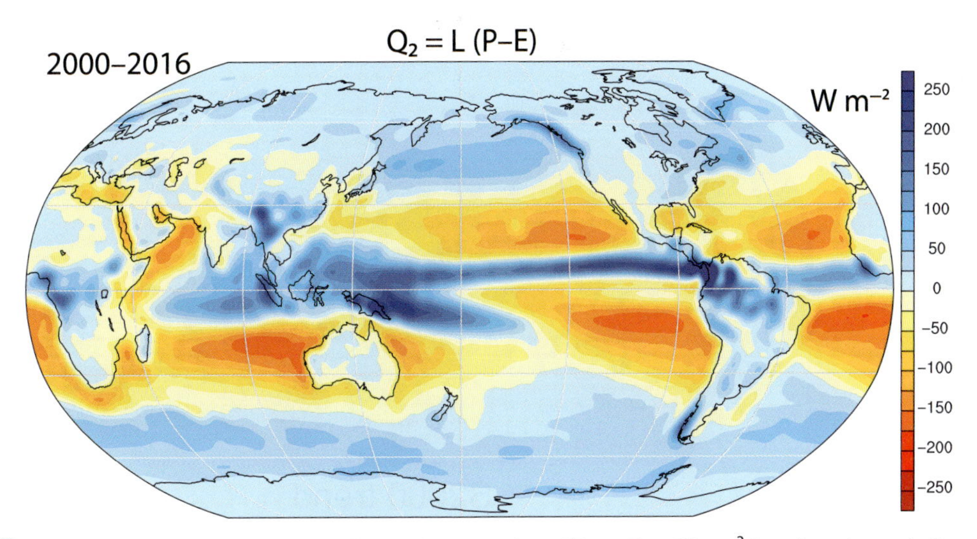

Fig. 10.2 The total annual mean latent heat flux in the atmosphere $L(P - E)$ in W m^{-2} based on the vertically integrated moisture divergence and tendencies for 2000–16 using ERA-Interim reanalyses. Divide by 29 to get mm day^{-1} (145 W m^{-2} is 5 mm day^{-1}). Trenberth et al. (2019). © American Meteorological Society. Used with permission

Figure 10.2 shows an estimate of $P - E$ expressed as a latent heating for the annual average for 2000–16. Over land, unless there are rivers and lakes present, P should exceed E so that colors should be a bluish shade. Small errors exist in some locations, such as Australia, where the model used to assimilate the data had a spurious source of moisture (essentially, even though evaporation took place, the soil moisture was restored artificially but not through rainfall). The orange colors highlight the dominance of evaporation especially over the subtropical oceans, while the ITCZ and SPCZ show up clearly as regions with an excess of P over E. In middle to high latitudes, P is greater than E, indicating that some sources of moisture come from the subtropical evaporation.

Very little precipitation falls in the subtropical high-pressure zones in the downward branch of the Hadley Circulation (Figs. 5.5 and 5.6) so that regions like the subtropical Pacific are actually a form of desert in spite of the surface waters. The same latitudes over the continents is where the true deserts form with little or no vegetation: the Sahara (North Africa), Kalahari and Namib (southern Africa), Gobi and Arabian/Syrian (Asia), Mojave (North America), Atacama (South America), Great Victoria and others (Australia). Polar regions are also deserts in terms of precipitation.

Rather than simply show the surface salinity, Fig. 10.3 presents the vertically integrated total salinity for the top 2000 m of the ocean (the change pattern is

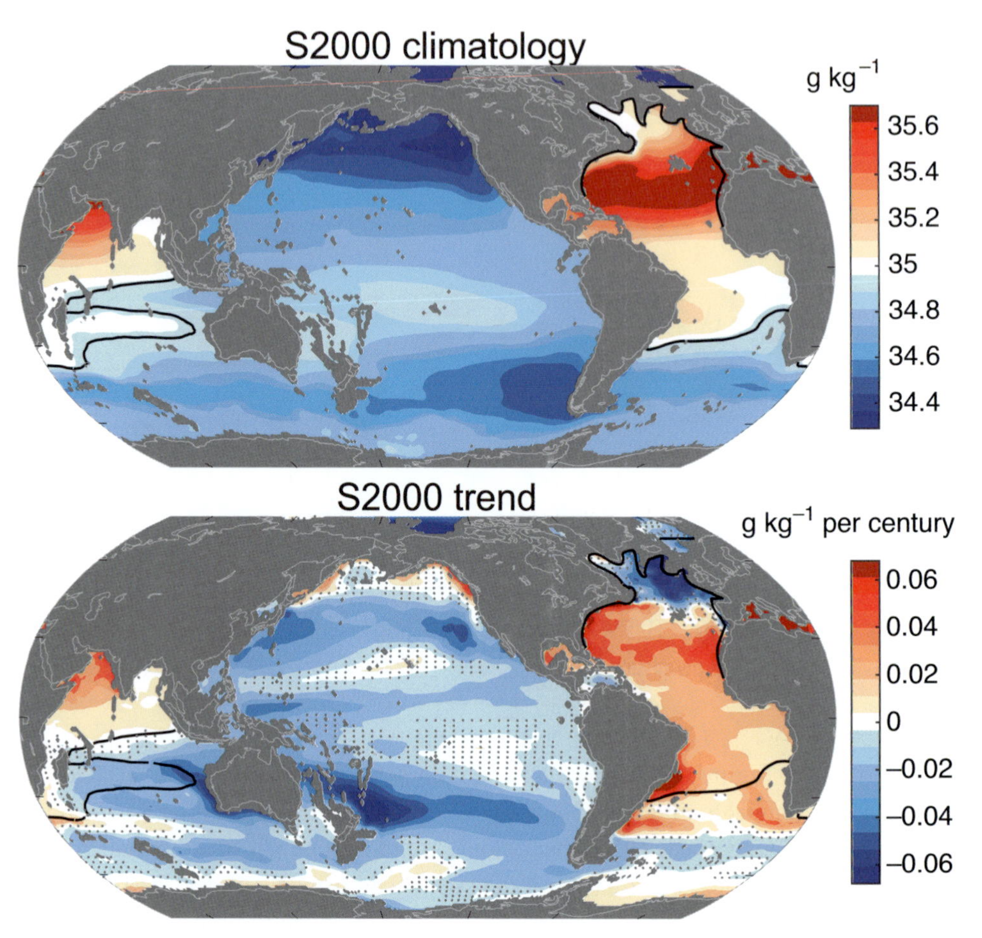

Fig. 10.3 Ocean salinity averaged over the top 2000 m for the climatology 1960–2005 (top) and the trend from 1960 to 2017 (bottom). Stippling indicates regions where trends are not significant. The thick black contour shows the median salinity value from the top panel. Adapted from Cheng et al. (2020). © American Meteorological Society. Used with permission

discussed below). $P - E$ is a primary driver of the salinity, but the latter can readily be mixed in the vertical, and the 0–2000 m salinity is more robust than surface salinity because it is less sensitive to shallow pools of fresh water brought about by rainfall. There is a dominance of evaporation ($E > P$) over the tropical and subtropical Atlantic (Fig. 10.2), and moist northeast trade winds readily transport water vapor across the Central American isthmus into the Pacific. However, there is no compensating return moisture flow in the westerlies in mid-latitudes because the Rockies in the northern hemisphere and the Andes in the southern hemisphere form major barriers for transport of moisture. The result is an excess of precipitation in the Pacific compared with evaporation, and a more saline Atlantic and fresher Pacific as a whole (Fig. 10.3). This is also in spite of the large freshwater transport into the tropical Atlantic from the Amazon and Mississippi rivers because the source of that water is the Atlantic and rainfall east of the two major mountain ranges noted above. Similarly, strong monsoonal winds carry moisture from the tropical Indian Ocean into Southeast Asia, where some runoff feeds rivers that flow into the west Pacific Ocean.

10.3 Climate Model Performance

Climate models do not perform very well in simulating the hydrological cycle. The overall cycling of water and patterns of precipitation are very broadly correct, but the details are not. The models prefer to have a double ITCZ – one in each hemisphere – and a much too extensive cold tongue in the tropical Pacific so that the SPCZ is quite poor in its configuration. Difficulties also exist in simulating monsoon rains. Most models have a hydrological cycle that is too intense, and precipitation amounts are too high, compared with observations. However, the mean amount can often be tuned. Even bigger issues occur with the character of precipitation. Intermittency is poorly modeled and has been given little attention. In nearly all models, the simulated frequency of precipitation is too high, and the intensity is too low, so that sometimes the rainfall in a model is described as "perpetual drizzle." These shortcomings relate to the inability to accurately simulate the weather disturbances in the model and associated clouds that result in rain. They are most readily determined by scrutinizing the diurnal cycle of precipitation (Section 7.4). Most models have not simulated realistic tropical storms and hurricanes, and convective cloud systems are generally not explicitly included.

All of these issues stem from the fact that convection occurs at scales too small to simulate explicitly and thus has to be parameterized. Convection is set to occur to prevent instabilities in the model atmospheric temperature structure in the vertical. Nevertheless, the evidence suggests that the model onset of convection is generally premature compared with nature, so that model atmosphere instability fails to build up as observed, and the resulting intensity is too weak (Fig. 10.4). These shortcomings are almost universal, and they have become acceptable to the modeling community. Some improvements have occurred in recent years, but the issues have not been resolved. Therefore, models do not provide a very reliable basis for examining hydrological cycle change, as the climate changes. Attribution is an ongoing issue. Hence it is a challenge to reliably inform an adaptation strategy and predict extreme events.

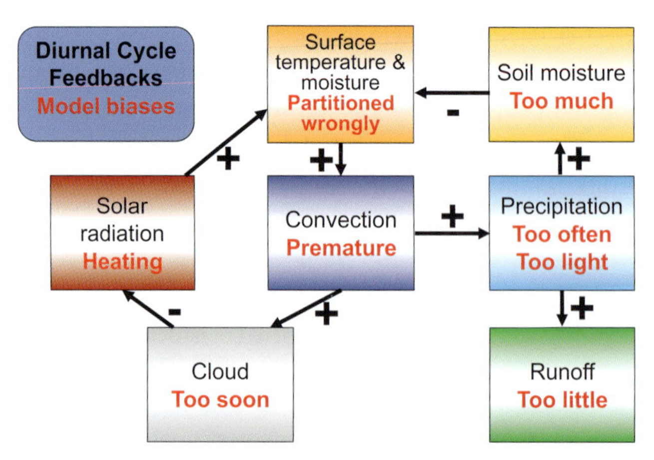

Fig. 10.4 Key feedback mechanisms involved in the diurnal cycle are given along with the model biases (in red). Convection relates to the stability of the column. Adapted from Trenberth et al. (2003). © American Meteorological Society. Used with permission

10.4 The Changing Water Cycle

Theory predicts an enhancement of the global hydrological cycle in response to global warming, including the increase of both evaporation E and precipitation P, as well as increases in precipitation intensity. Because the amount of moisture in the atmosphere is quite small, any increase in global evaporation has to be matched by an increase in global precipitation. But this is certainly not true locally.

Under a warming climate, the global precipitation amount changes are governed by energy changes and evaporation, which models suggest is about 1–3% per degree Celsius surface warming, although with considerable uncertainty. One consideration is that any extra latent heat release also has to be balanced by radiative cooling, and radiative forcing from increased greenhouse gases affects this process. However, some aerosols act to short-circuit the water cycle, which otherwise acts to transfer energy through evaporation and subsequent condensation to the lower atmosphere via latent heating. Carbonaceous aerosols, which are typically brown or black, instead block the surface heating and heat the atmosphere directly, taking precipitation out of the loop. Other aerosols, such as sulfate aerosols that form a whitish veil, also block the surface heating and thus evaporation, but instead of locally heating the atmosphere, the radiation is reflected and cooling results. Hence the actual outcome depends on observed emissions, and the global change in precipitation amount can be less than the projected 1–3% per degree Celsius under global warming.

In contrast, given upward motion in the atmosphere, the intensity of precipitation depends upon the amount of moisture available. As temperatures rise with global warming by 1°C, the average near surface moisture-holding capacity goes up by about 7% from the C–C relation (Section 5.3). This increases atmospheric demand for moisture from the surface and dries the land and vegetation. In most places where moisture availability is not an issue, such as the oceans and near-coastal regions, the actual atmospheric moisture is apt to go up by the C–C amount, thereby feeding 7% more moisture into any storms or weather systems, and hence it rains harder, as is observed. Or it snows harder, provided that temperatures are low enough. The availability of moisture from the surface is largest where SSTs are highest, while interiors of continents are prone to be dry (see Section 5.3).

A major consequence of the differences between the 7% for intensity versus the 1–3% for amount is that the frequency and duration of precipitation events must decrease in some sense. Indeed, there are many studies which show that precipitation intensity has increased at the C–C rate or higher. Values can easily be higher because of feedbacks on the atmospheric dynamics in some phenomena. The reason is that latent heat released by the extra precipitation enhances the upward motions and thus the low-level moisture convergence, potentially doubling the increment. The outcome depends a lot on where the main precipitation and associated latent heating occur relative to the storm center, and in hurricanes the latent heating directly intensifies storms. For thunderstorms, the effect can also add to intensity. In extratropical storms the influence is less, as heating contributes more to changes in motion and direction of development of the storm. Other factors, such as a larger size and longer lifetime can further enhance the storm precipitation, although the increased vertical transport of heat is apt to stabilize the atmosphere and ultimately limit the outcomes. Moreover, at some point the storm may run out of moisture supply because storms act to dry out the atmosphere. Several studies of particular events have demonstrated that with today's warming, there have likely been increases in heavy precipitation of order 20–30% compared with pre-industrial conditions (an example is given in Section 15.2).

As temperatures rise, the likelihood of precipitation falling as rain rather than snow increases in areas where temperatures are near freezing, especially in spring and autumn – at the beginning and end of the snow season. Such changes have already been observed over land in the mid- and high latitudes of the northern hemisphere, leading to increased rains but reduced areas of snowpack, consequently resulting in diminished water resources in summer, when they are most needed. However, snowfalls can be heavier in mid-winter as long as temperatures are still below 0°C but higher than they used to be.

The small spatial scales of precipitation and its intermittency in time make it difficult to accurately observe these trends. Trends in global precipitation are reasonably well established from GPCP and are likely the best-known hydrological atmospheric change, but the global mean *P* trends since 1979 have been negligible and instead global *P* is dominated by variability, in particular through ENSO. Hence there is no clear sign of the expected increase in global amount and perhaps a good reason is because of the effects of aerosols and their changes over time.

There is extra energy in dry regions, where it is sunny, but the locally evaporated moisture is carried away by the winds to places where it is precipitating. Higher temperatures mean increased atmospheric demand for moisture, and increased evapotranspiration in the dry areas exacerbates aridity and droughts. Meanwhile, in precipitation locations, it rains (or snows) harder and there is an increased risk of flooding. Because, with warming, more precipitation occurs as rain instead of snow and snow melts earlier, there is increased runoff and risk of flooding in early spring, but increased risk of drought in summer, especially over continental areas.

The strong dependence on the local environment for the moisture supply for precipitation has several implications. Firstly, the environment has changed because of the warmer oceans and less ice, so that it is order 1°C warmer and with 5–15% more moisture in the environment as a resource for all storms; see also Section 15.2. Offsets have occurred locally where air pollution has played a role in blocking the Sun. Secondly, because precipitation has very strong structures (see Fig. 10.2 for instance), while *E* is more uniform, precipitation changes in intensity are manifested most where it already precipitates and it is expected that there ought to be a strong amplification of the pattern of $E - P$, the net surface freshwater flux. These statements also depend on the somewhat uncertain effects of changes in aerosol. Hence, overall these environmental changes lead to the "*wet get wetter*" and the "*dry get dryer*" regions as a basic concept for how precipitation patterns change, especially over oceans. This is also referred to as the "*rich get richer and the poor get poorer.*" This pattern is simulated by climate models and is projected to continue into the future.

This has obvious consequences for salinity because the absolute changes are greatest where values are already large (Fig. 10.3). Taking advantage of the integrating effects of the ocean, by essentially treating the ocean as a big rain gauge, several studies of the limited number of longer-term observations of changes in surface ocean salinity have suggested a "*fresh gets fresher, salty gets saltier*" pattern in line with expectations based on changes in rainfall (wet get wetter, dry get dryer). New research has analyzed the top 2000 m of the ocean to show these changes from 1960 to 2017 versus the mean conditions for 1960–2005 (Fig. 10.3). Note the already salty Atlantic Ocean is getting saltier. Similarly, there is a marked contrast between the northern and the rest of the Indian Ocean that is also growing in magnitude. The Pacific everywhere is getting fresher. The pattern enhancement of both $P - E$ and salinity is an amplification of the climatology (Fig. 10.3).

However, with more precipitation per unit of upward motion in the atmosphere, i.e. "*more bang for the buck*," atmospheric circulation weakens, causing monsoon winds to falter. The result is heavier monsoon rains, because the greater amount of moisture more than compensates for the weaker circulation overall, and it is the combination of the two factors that matters physically. However, the monsoon is less reliable.

10.5 Extremes of Rain and Floods

A figure summarizing the changes in the atmospheric water cycle (Fig. 10.5) highlights the interactions between changes in temperature versus evaporation, but with the environmental moisture increasing associated with the C–C relationship. This is what leads to increased intensity of weather systems, including all of those listed, because they feed on the environmental moisture. Increased lifetime duration and size are other important factors commonly present.

Unfortunately, global datasets on extremes are often spatially incomplete and maybe of poor quality and too short to be useful in really determining the odds, even as the odds are changing. Extremes are also poorly simulated in most climate models. However, there is evidence from precipitation data over land that the extremes of precipitation are generally increasing. Storms, whether individual thunderstorms, extratropical rain or snowstorms, or tropical cyclones, supplied with increased moisture, produce more intense precipitation events. Such events are observed to be widely occurring, even where total precipitation is decreasing: "*it never rains but it pours!*" This increases the risk of flooding. Whether flooding results or not depends on the degree of mitigation and drainage systems.

In the tropics and subtropics, precipitation patterns are dominated by shifts as SSTs change, with El Niño a good example for interannual variability. The volcanic eruption of Mt. Pinatubo in 1991 led to an unprecedented drop in land precipitation and runoff, and to widespread drought, as precipitation first shifted from land to oceans and evaporation faltered, before precipitation decreased as a whole, providing lessons for potential impacts from possible geoengineering.

Frequently, different extremes are manifested spatially, with one area drenched while another region not that far away is under drought and wildfire. A case in point may be the heavy flooding rains in Jakarta and Indonesia in early January 2020, while Australia was under major heatwaves and bushfires. However, it can also happen temporally, as was the case in Japan in July 2019, when there was the heaviest rainfall in over 35 years and a death toll of well over 200 from July 2 to 9, followed later in the same month by an extreme heatwave and temperatures over

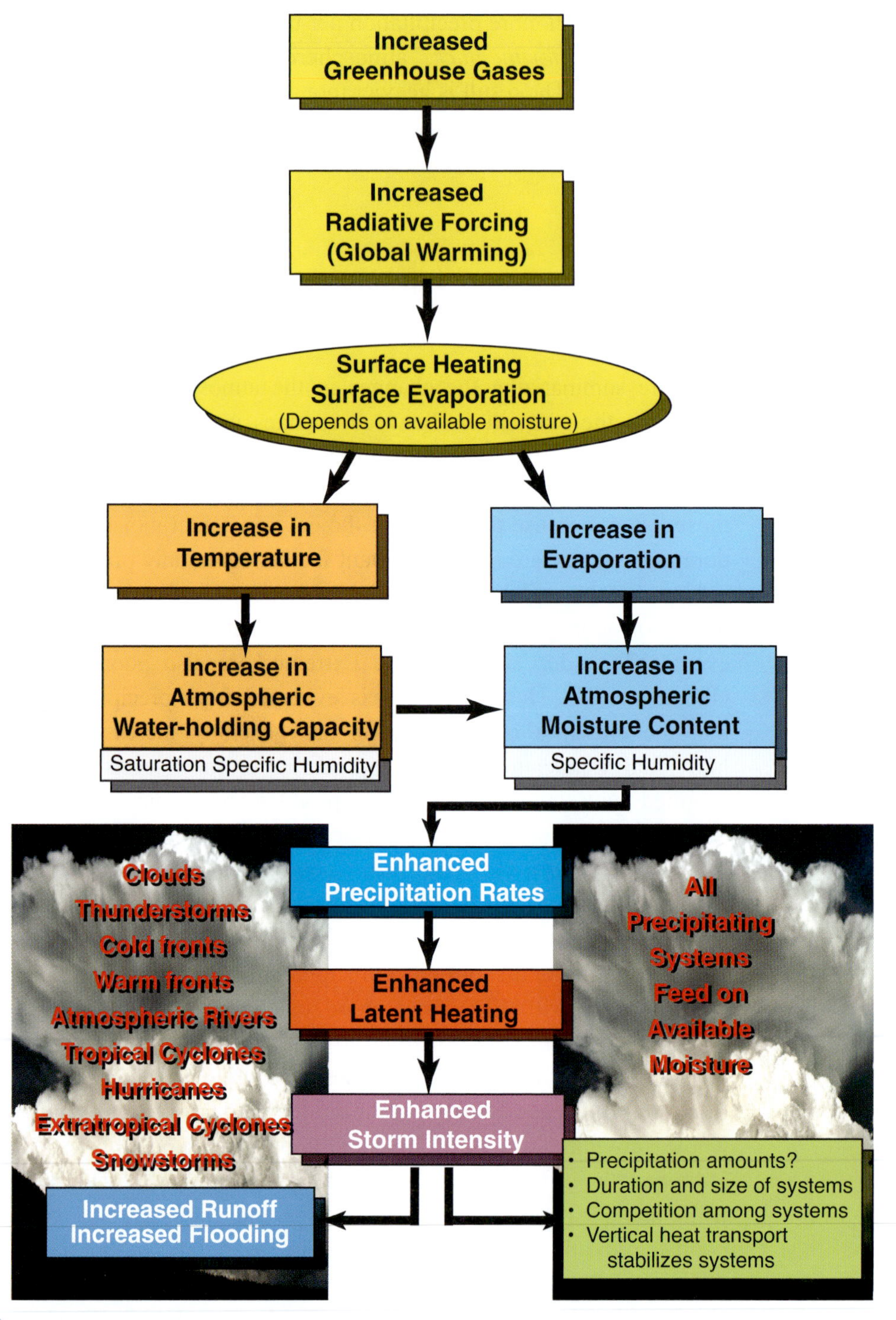

Fig. 10.5 Schematic flow of the influences arising from increasing greenhouse gases on precipitation leading to increased intensity, that in turn leads to enhanced storm intensity and perhaps size and lifetime. A short list is given of some phenomena involved and all feed on the environmental moisture content. The green box lists some extra factors. Adapted from Trenberth (1998)

40°C for the first time in many places (including Tokyo), with over 80 deaths and 22 600 hospitalized with heat stroke.

10.6 Drought and Wildfires

Drought occurs naturally and its incidence varies widely, both spatially and in time. Good reliable metrics of drought are difficult to come by, however, and partly stem from different definitions of drought. Drought is generally defined as a prolonged absence or marked deficiency of precipitation that results in water shortage for some activity. It is also defined as a period of abnormally dry weather sufficiently prolonged for the lack of precipitation to cause a serious hydrological imbalance. It is useful to recognize the following:

- *Meteorological drought* is mainly a prolonged deficit of precipitation.
- *Agricultural drought* relates to moisture deficits in the topmost 1 m or so of soil (the root zone) that impacts crops.
- *Hydrologic drought* is related to below-normal streamflow, lake and groundwater levels.

These differences emphasize the relative roles of precipitation, evapotranspiration (ET) and runoff in drought caused by climatic factors, and the timescales increase as surface and hydrological factors come in. More generally, water availability is a societal and environmental concern, which also brings in the demand side, and thus there are other possible definitions related to water scarcity.

A number of simplified indices of drought have been developed, and they have strengths and weaknesses. A drought index is, by design, a simplified measure of drought severity and development. Often the viability of a drought index depends on the availability of quality datasets that allow it to be computed. An index has to be calculable easily with limited data available from observations. Even now there are many comprehensive land surface models that cannot simulate the land surface water balance well for many regions and cannot easily be used to calculate a drought index, because reliable forcing data to drive them does not exist for the real world. Precipitation measurements are widely available but good coverage is essential. Several indices depend only on precipitation. The disadvantage is that they ignore the ET and runoff aspects. Some indices use a highly idealized potential evapotranspiration, which depends on temperature of the lower atmosphere, and these are essentially an index of atmospheric demand for moisture. Because they do not account for water availability, they do not reliably depict drought.

An index that has merit is the Palmer Drought Severity Index (PDSI) because it attempts to calculate the water balance using a two-layer bucket-type model. It is named after Wayne Palmer who developed it in 1965. The PDSI uses readily available temperature and precipitation data to estimate relative dryness. Normalization in the Palmer model minimizes the errors associated with some earlier assumptions made by Palmer, and the actual evaporation is often determined to a large degree by the availability of soil moisture (and thus is affected by precipitation), not only by potential evapotranspiration, over many land areas. The Palmer model uses a top and a lower soil layer to consider how soil moisture storage and runoff is generated and when the soil storage reaches its capacity. However, it has recently been criticized for using a simple water balance model, yet the indices used to replace it have far greater flaws. More recent PDSI developments have included a "self-calibrating" aspect to allow for spatial differences, and various improvements have been made in how the potential evapotranspiration is calculated. Even so, there is a risk of overestimating PDSI changes unless allowance is also made for the effects of the vegetation response to increasing carbon dioxide on ET.

Discrepancies in drought time series relate not only to the indices used but also the data used. Accurate attribution of the causes of drought also requires accounting for natural variability, especially El Niño–Southern Oscillation effects, owing to the predilection on average for wetter land during La Niña events. As it uses temperature data and a physical water balance model, the PDSI can capture the basic effect of global warming on drought through changes in potential evapotranspiration and it has been reasonably successful at quantifying long-term drought. The PDSI is highly correlated with streamflow. The PDSI has also been widely used in tree-ring based reconstructions of past droughts in North America and other regions.

Increased heating from global warming may not cause droughts, but when droughts occur they are likely to develop sooner and become stronger, as discussed in Section 2.4. Hence droughts occur naturally, but they are now more intense, set in a little quicker, and increase the risk of wildfire (or bushfire) and heatwaves.

According to the World Wildlife Fund early in 2020, from 1979 to 2013 the global fire season length increased on average by 19%. Factors included the persistent hot dry weather related to climate change, as well as human factors such as land conversion for agriculture and poor forest management. In April 2020, the number of fire alerts across the globe was up by 13% compared to 2019 – which was already a record year for fires. Extensive fires have been taking place in the Amazon rainforests and in the Pantanal Region (shared by Brazil, Bolivia, and Paraguay), the planet's largest tropical wetland. In response to changes in Brazilian government policies that encourage deforestation, settlers have cleared lands for agriculture and cattle grazing. According to Brazil's National Institute for Space Research (INPE), there were 6803 fires in July 2020 vs. 5318 in 2019, threatening biodiversity, ecology and indigenous peoples. During 1972–2018, California experienced an 8-fold increase in summertime forest-fire area and this was very likely driven by drying of fuels due to increased atmospheric aridity promoted by human-induced warming. The warm season increased temperatures by 1.4°C and significantly decreased humidity. In the northern summer of 2020, many major wildfires in the western parts of the United States, especially in California, Oregon, and Colorado, exceeded previous record values and, while forest management is no doubt a factor, climate change effects have played a major role (Section 2.4).

10.7 Changes in Tropical Storms

Hurricanes and typhoons are natural phenomena. Hurricanes keep tropical oceans cooler as a consequence of their strong winds that increase evaporation, and they also mix heat to greater depths within the ocean. Indeed, they are the only viable means for this to happen in the deep tropics over the ocean in summer (see Fig. 7.9), where the cold outbreaks in middle latitudes fail to reach.

It is generally expected that storm and hurricane activity are affected by climate change, because of higher ocean heat content (OHC), sea levels, and SSTs (Section 14.5), and therefore all storms occur in a warmer and moister environment, increasing precipitation and thus latent heat release. The resulting environment invigorates hurricanes (or typhoons/tropical cyclones) to make them more active.

The activity can be manifested in several ways: (i) more storms, (ii) more intense storms, (iii) bigger storms, (iv) longer-lived storms, and (v) much greater rainfall and risk of flooding. The heavier more extensive rains combine with all other options. There is speculation about changes in tracks with climate change, but the tracks are predominantly determined by the weather situation, and thus the location of high- and low-pressure systems, cold fronts, troughs and ridges, and the jet stream. However, the mix in how activity is manifested can vary a lot and confounds tracking just one parameter. Data on all of these parameters are of decidedly mixed quality, making estimates of some long-term changes uncertain.

Many theoretical and modeling studies have suggested that there may instead be fewer storms, in part because of changes in atmospheric stability and heating profiles in the atmosphere with increased carbon dioxide, although this is very much a model result, and in part because a few bigger storms can replace a lot of smaller storms in terms of their impact on the ocean. Increases in intensity, size and duration mean that more heat comes out of the ocean, leaving a cold wake behind and this works against increases in numbers.

Nevertheless, increased activity does not occur everywhere at once. Even though SSTs are generally increasing everywhere, some spots become warmer than others and activity tends to be favored in those regions. In turn the storms mix up and deplete heat from the ocean through evaporative cooling and wipe out what was a hot spot. Preferred areas of activity together alter the Hadley and Walker circulations through large-scale overturning in the atmosphere of the tropics. In subsidence regions, the atmosphere is anticyclonic and stable, and convection is suppressed. In regions between the general cyclonic activity regions and the anticyclones, the low and upper level winds move in opposite directions, creating vertical wind shear which tends to tear incipient vortices apart. Just where the strong wind shear occurs depends upon both the rotational (e.g., geostrophic) and divergent (non-geostrophic) part of the flow, and hence the real situation involves more than just overturning cells. What it means is that there is a competition throughout the tropics for where the main tropical cyclone activity will occur.

Hurricanes are normal events in summer, with an average of 6 hurricanes and 12 named storms in the Atlantic. However, in 2017 there were 17 named storms and 10 hurricanes, 6 of which were classed as "major." The Atlantic hurricane season in 2017 broke numerous records during the course of the well-above-normal activity, especially with the tremendous damage from Harvey, Irma, and Maria. Irma broke many records including the longest lifetime as a category 5 storm, and it generated the most Accumulated Cyclone Energy of any storm in the tropical Atlantic. Irma was unprecedented in the way it straddled Florida affecting the whole peninsula as it moved northwards. Damages from Harvey and Irma were in the hundreds of billions of dollars. Nevertheless, even greater damage occurred both from Irma and then Maria in the Caribbean Islands and Puerto Rico.

In 2018–19, hurricanes Florence, Michael and Dorian in the Atlantic, cyclones Idai, Kenneth and Fani in the Indian Ocean, and typhoons Mangkhut and Hagibis in the Pacific Northwest stand out as examples that have broken records. Like Harvey, the size and duration of Florence was remarkable given the proximity to land, with exceptional rainfalls and flooding as a result. Heavy rains can extend far inland, as far as Canada for east coast United States storms such as Irene in 2011 and Isaias in 2020. On the other hand, Michael moved rapidly and made landfall as a category 5 storm, so that the main damage was from wind and storm surge.

Globally, in 2019, there were 96 named tropical storms, well above the 1981–2000 average of 82. In the Atlantic in 2020, the higher OHC and SSTs triggered a much greater number of storms (Fig. 10.6). This figure features five tropical storms all at once, two of which created havoc as they made landfall in Bermuda (Paulette) and on the Gulf Coast (Sally). In 2020 the hurricane season in the Atlantic may illustrate the case for greater numbers of storms, because once again (the first time was 2005), the large number of tropical storms pushed the naming well into the Greek alphabet (which means more than 22 storms as some

Fig. 10.6 GeoColor visible satellite image of an Atlantic Ocean packed with five tropical cyclones at 10:20 a.m. eastern daylight time, Monday, September 14, 2020. Image credit: RAMMB/CIRA/Colorado State University http://rammb-slider.cira.colostate.edu/

letters are not used), and the name Zeta has been used for only the second time (number 27). In Southeast Asia, Vietnam suffered tremendous flooding from four tropical cyclones, such as typhoon Molave on October 29, 2020, and super typhoon Goni hit the Philippines in early November as one of the strongest to make landfall ever.

These events would not have happened the way they did without human-induced climate change and warmer oceans. It is not just the higher SSTs that play a role but the warming of the upper ocean through the top 150 m or so that matters, because these strong storms mix up the ocean and normally create a cold wake behind them, but the wake is not as cold as it used to be. Warnings about the risk of more intense hurricanes have been clear for many years, and especially since 2005 (the year of Katrina) but the preparedness for these events has been inadequate.

10.8 Water, Climate, and Society

Water for society has multiple dimensions (Fig. 10.7). Water is irreplaceable and non-substitutable. It is more than just another natural resource. Indeed "*Water is Life.*" Often overlooked is the fact that water is a tremendous solvent for many

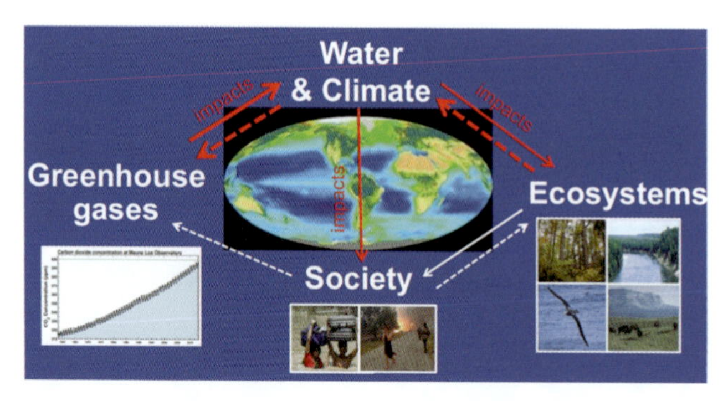

Fig. 10.7 Schematic to illustrate the interactive nature of water, climate and society. The dashed arrows suggest feedback effects.

things and thus widely used in cleaning. There are many competing demands for water. These range from drinking water, irrigation of crops, hydroelectric power, stream and river life (including fowl and fish), industrial uses, and cleaning. The global water challenges include unsustainable use of water, including declining groundwater levels and aquifers, and dry rivers; increasing competition for water resources, at local, regional, and international scales; degradation of water quality from over-extraction and pollution; as well as flood and drought risk.

A well drilled into an aquifer is called an artesian well. Drilling for water in one location and tapping water elsewhere through lateral pipes has led to stealing of waters that has, on occasion, extended across national borders. The increasing demand from growing populations, economic development, and agriculture threatens the future of water availability. These threats are compounded by environmental change, including land use and management, and climate change. A result is decreasing water security, increased international tensions, and local arguments over water rights. Accordingly, armies of water rights lawyers exist.

Examples of regional problems include the Aral Sea drying up; arsenic in water in Bangladesh; blue-green algae in Lake Winnipeg, Canada; and excessive nutrients and algal blooms in the Great Lakes. Other problems stem from the phenomena in Section 10.4, and, in 2019–20 alone, major floods occurred in Pakistan and India (September 2019), UK (October 2019), Venice (November 2019), Jakarta (January 2020), Michigan (May 2020), South China, Japan, and New Zealand (July 2020); while drought and wildfires in 2019 and 2020 especially afflicted California, Brazil, and Australia. As noted above for 2017–18, tropical storms/hurricanes caused damage in Africa, India, Bangladesh, the Philippines, China, Vietnam, United States, Bermuda, and several Pacific Islands.

In extratropical mountain areas, the winter snowpack forms a vital resource, not only for skiers, but also as a freshwater resource in the spring and summer as the snow melts. Warming leads to earlier melt of the snow that exists, and greater evaporation and ablation (the combined processes of sublimation, fusion or melting, and evaporation which remove snow or ice). There are several documented cases where reduced snow (versus rain) and earlier spring melt contributed to diminished snowpack and, later, low soil moisture, which likely contributed substantially to the subsequent widespread intense drought, heatwaves and wildfires because of the importance of recycled moisture. Recycling is where rainwater is evaporated only to fall again in the same catchment, and it is much more important in summer, because the transport of moisture by large extratropical storms diminishes.

Changes in extremes matter most for society and human health. Ironically, both ends of the hydrological cycle are becoming more extreme. The warming climate brings higher temperatures and heatwaves, along with wildfires and drought at one extreme, but more intense precipitation, increased heavy rains and risk of flooding, and more intense storms at the other. The impacts vary greatly depending on the infrastructure and management systems in place. Consequences vary enormously with country and location (Fig. 10.8). Increased flooding risk can be mitigated by good water management practices. Droughts may be mitigated by irrigation and use of stored water in reservoirs and as groundwater. The biggest challenge for water managers is how to save the excess water from times of heavy rain for the times when there is not enough.

Extremes may or may not be rare – some places expect extremes because of their location. However, extremes often break records and result in values recorded outside of previous experience. The result is that things break as thresholds are crossed – these may relate to engineering decisions in structures based on obsolete data that failed to account for climate change. An example is the devastation and loss of lives with hurricane Katrina in New Orleans in late August 2005 as the levees and floodwalls along Lake Pontchartrain failed. The consequences are that the damage costs are greatly amplified, perhaps by one or two orders of magnitude, and there is loss of life and limb. Then there is the proverbial "*straw that breaks the camel's back*": a modest increment by climate change has devastating consequences. The extreme nonlinearity of such events makes them very difficult to deal with and plan for, and economists generally cannot cope well with such "catastrophes," as they are called.

The prospects are for major shortages and water stress (Fig. 10.8) in many water-scarce areas by 2050. For many developing nations, water-demand increases due to population growth and economic activity have a much stronger effect on water stress than climate change. By 2050, economic growth and population change alone can lead to an additional 1.8 billion people living under at least moderate water stress, with 80% in developing countries. The strongest climate impacts on water stress are projected to be in Africa, but strong impacts also occur over Europe, Southeast Asia, and North America (Fig. 10.8). The

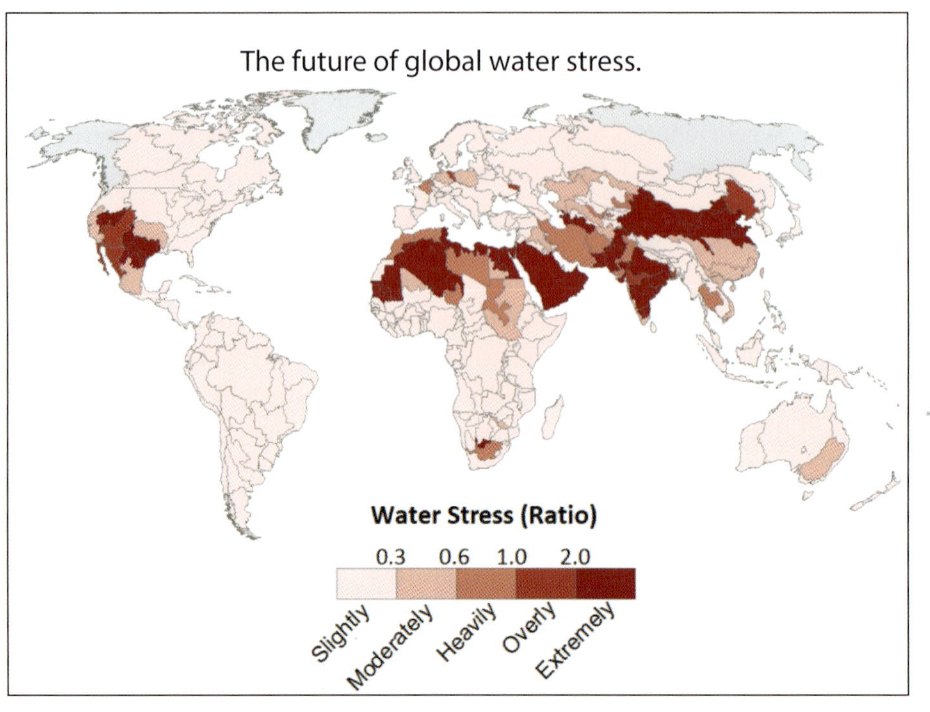

Fig. 10.8 Water stress. Shown is the global distribution of water stress index (WSI) by assessment subregions. These simulated values are an average for the years 1981–2000 of an historical baseline simulation. The shading levels also denote stress categories: WSI < 0.3 is slightly exploited, $0.3 \leq$ WSI < 0.6 is moderately exploited, $0.6 \leq$ WSI < 1 is heavily exploited, $1 \leq$ WSI < 2 is overly exploited, and WSI ≥ 2 is extremely exploited. From Schlosser et al. (2014). Reprinted with permission

combined effects of socioeconomic growth and uncertain climate change lead to a 1.0–1.3 billion increase of the world's 2050 projected population living where total potential water requirements will consistently exceed surface water supply. This raises the total water-stressed population to 5 billion out of the expected 9.7 billion population.

Water will be a primary pressure point for society in the future.

References and Further Reading

Cheng, L., K. E. Trenberth, N. Gruber, et al., 2020: Improved estimates of changes in upper ocean salinity and the hydrological cycle. *Journal of Climate*, **33**, doi: 10.1175/JCLI-D-20-0366.1.

Dai, A., 2011: Characteristics and trends in various forms of the Palmer Drought Severity Index (PDSI) during 1900–2008, *Journal of Geophysical. Research,* **116**, D12115, doi: 10.1029/2010JD015541.

Schlosser, C. A., K. Strzepek, X. Gao, et al., 2014: The future of global water stress: an integrated assessment, *Earth's Future*, **2**, 341–361, doi: 10.1002/2014EF000238.

Trenberth, K. E., 1998: Atmospheric moisture residence times and cycling: implications for rainfall rates with climate change. *Climatic Change*, **39**, 667–694.

Trenberth, K. E., 2011: Changes in precipitation with climate change. *Climate Research*, **47**, 123–138. doi: 10.3354/cr00953.

Trenberth, K. E., and A. Dai, 2007: Effects of Mount Pinatubo volcanic eruption on the hydrological cycle as an analog of geoengineering. *Geophysical Research Letters*. **34**, L15702, doi: 10.1029/2007GL030524.

Trenberth, K. E., and J. Fasullo, 2007: Water and energy budgets of hurricanes and implications for climate change. *Journal of Geophysical Research*, **112**, D23107, doi: 10.1029/2006JD008304.

Trenberth, K. E., A. Dai, R. M. Rasmussen, and D. B. Parsons, 2003: The changing character of precipitation. *Bulletin of the American Meteorological Society*, **84**, 1205−1217. doi: 10.1175/bams-84-9-1205.

Trenberth, K. E., L. Smith, T. Qian, A. Dai, and J. Fasullo, 2007: Estimates of the global water budget and its annual cycle using observational and model data. *Journal of Hydrometeorology*, **8**, 758–769. doi: 10.1175/JHM600.1.

Trenberth, K. E., A. Dai, G. van der Schrier, et al., 2014: Global warming and changes in drought. *Nature Climate Change*, **4**, 17–22, doi: 10.1038/NCLIMATE2067.

Trenberth, K. E., L. Cheng, P. Jacobs, Y. Zhang, and J. Fasullo, 2018: Hurricane Harvey links to ocean heat content. *Earth's Future*, **6**, 730–744, doi: 10.1029/2018EF000825.

Trenberth, K. E., Y. Zhang, J. T. Fasullo, and L. Cheng, 2019: Observation-based estimates of global and basin ocean meridional heat transport time series. *Journal of Climate*, **32**, 4567–4583. doi:10.1175/JCLI-D-18-0872.1.

11 Teleconnections and Patterns of Variability

The links between weather regimes and teleconnections affect how climate change is manifested. The main patterns of variability are described, along with the role of sea surface temperatures.

11.1 Weather Regimes

The weather systems undergo incredibly rich variety and never repeat. If left alone in the atmosphere, mid-latitude storms would continually march eastwards, modulated by very large-scale atmospheric waves – planetary waves – set up by the land–sea differences and continental mountain ranges. The latter are mainly in evidence in the northern hemisphere and vary from summer to winter because the mid- to high-latitude land is colder than the adjacent seas in winter, but warmer in summer. Nevertheless, this idealized state seldom exists and frequently the weather systems get stuck into a certain repeating pattern of sorts, called weather regimes.

Low-pressure cyclones may continually develop in preferred regions and track in certain directions, and even though each one is different, they constitute a distinct cluster that differs from the above picture. So-called blocking anticyclones may continually develop and intensify in preferred regions. Most commonly, the reason for such preferred patterns, or modes of variability, to exist is because of interactions of the atmosphere with the oceans or underlying surface that features anomalous conditions. Over land, this could be large wet areas or snow-covered regions. Over oceans it is because of anomalous SSTs and perhaps sea ice. How long such preferred patterns last depends then on how well the unusual surface conditions are sustained: for example, how much water lies around on land; how deep the snow is, or how well supported the SSTs are by underlying ocean heat content anomalies.

Of course, if there is persistent low pressure in one region, there must be compensating high pressure somewhere else. Alternatively, if the air is rising in one area, it must be subsiding somewhere else. This means that anomalous conditions of one kind in one area may well be accompanied by anomalous conditions of a different kind somewhere else. These are teleconnections. They come as distinctive preferred patterns, but ones that can switch in sign, so that the high- and low-pressure regions are reversed.

The question of how such teleconnection patterns change as the climate changes is important. Because the patterns relate to the underlying large-scale atmospheric circulation, any teleconnection changes will depend on the magnitude and shape of those changes. Current understanding is that circulation changes are relatively small from global warming and often lost in the noise of natural variability, so that the actual changes can only be sorted out using climate models averaged over large ensembles well into the future. Instead, it is best to consider the teleconnections as largely unchanging in pattern and location, with the main changes instead being more likely in the amount of time spent in each phase. This is why it is important to understand the current teleconnections and their triggers, if they have any.

11.2 Teleconnections

Teleconnections are linkages in climate variability across great distances. Implied in the term teleconnection is that there is a physical reason for the simultaneous variations, often of opposite sign, over distant parts of the globe. A primary reason for teleconnections in the extratropics is the presence of Rossby waves in the atmosphere, which naturally feature sequences of high- and low- pressure systems. In the tropics they may also originate from the monsoonal-type overturning circulations that link the upward and downward branches, and thus in association with variations in the Hadley Circulation, Walker Circulation, and monsoons. However, in most cases of quasi-stationary atmospheric teleconnections, there is a source region of heavy rainfall and atmospheric heating that initiates the perturbation, and the latent heat energy is dispersed and propagated away from the region, often far afield, to where it can be dissipated perhaps by radiating energy to space. The biggest and best example of such interactions with global consequences is the El Niño phenomenon (Chapter 12), which gives rise to the strongest covariability over large distances through the Southern Oscillation: a global-scale pattern in sea-level pressure, and hence surface winds, which is treated in detail in the next chapter. Teleconnections can also occur through the ocean owing to exchanges between the Indian and Pacific Oceans via the Indonesian ThroughFlow.

Accordingly, these teleconnection processes are a means of dispersing energy and they are especially relevant when large-scale marine heatwaves develop, as is increasingly the case as the oceans warm up. In the tropics, the temperature of the surface waters is readily conveyed to the overlying atmosphere, and because warm air is less dense it tends to rise whereas cooler air sinks. As air rises into regions where the air is thinner, the air expands, causing cooling and therefore condensing moisture in the air, which produces rain (Fig. 11.1). Low sea-level pressures set up over the warmer waters while higher pressures occur over the cooler regions in the tropics and subtropics, and the moisture-laden winds tend to blow toward low pressure so that the air converges, resulting in organized patterns of heavy rainfall. The rain comes from convective cloud systems, which preferentially occur in the convergence zones, the ITCZ and SPCZ, and which can be moved around somewhat by changes in SSTs. Often the most intense cluster of convection forms a convective cloud system (CCS) or complex (CCC) that is a phenomenon in its own right, and in turn these may develop into tropical storms.

In the tropical atmosphere, anomalous SSTs force anomalies in convection and large-scale overturning with subsidence in the descending branch of the local Hadley Circulation which is primarily in the winter hemisphere. The resulting strong upper tropospheric divergence in the tropics and convergence in the subtropics act as a Rossby wave source (Fig. 11.1). The so-called Rossby waves, after Carl-Gustav Rossby, enable the propagation of energy away from the region. The waves are quasi-stationary provided their source continues and their influence can be felt on a timescale of a week or longer in quite remote parts of the hemisphere. The climatological stationary planetary waves and associated jet streams can make the total Rossby wave sources somewhat insensitive to the position of the tropical heating that induces them and thus can create preferred teleconnection response patterns, such as the Pacific–North American (PNA) pattern.

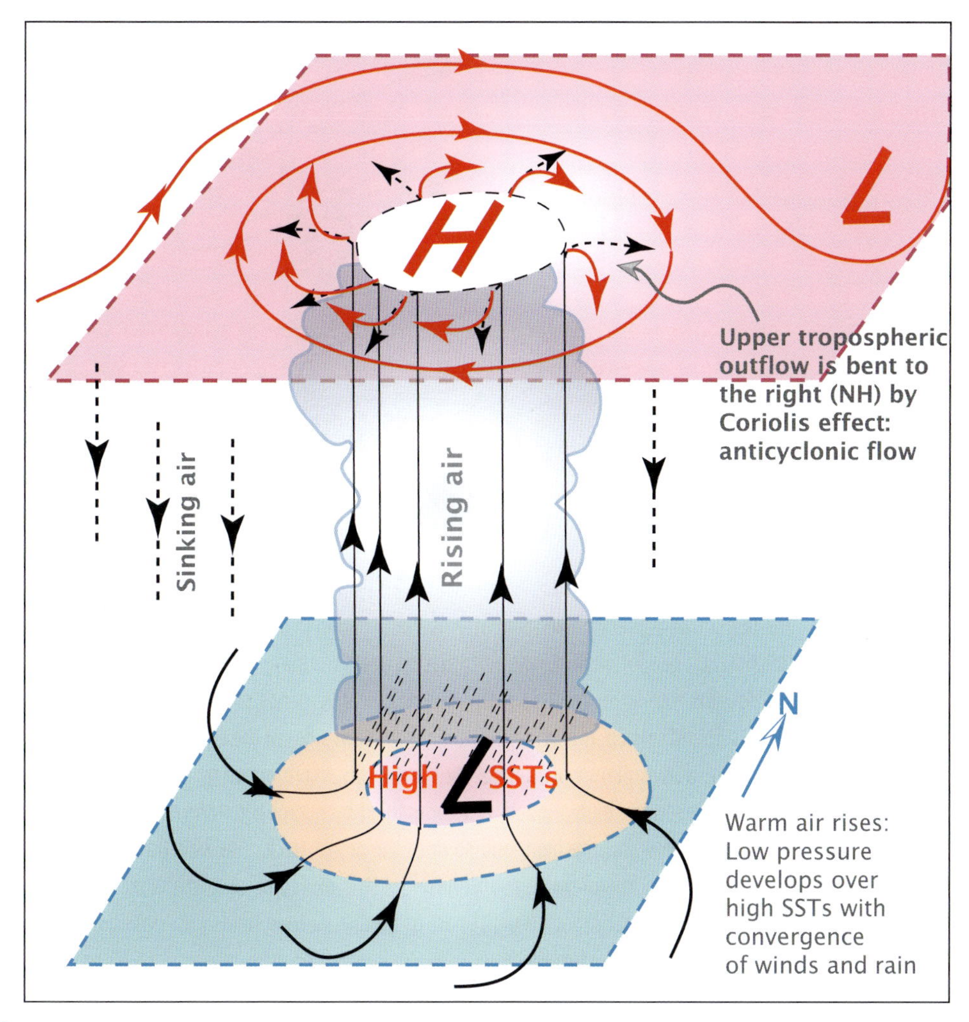

Fig. 11.1 Over the tropical Pacific Ocean where high SSTs exist, warm moist air rises, creating low surface pressure and convergence of moisture. In this depiction for the northern hemisphere, widespread convection leads to strong latent heating of the atmosphere and divergent outflow in the upper troposphere and lower stratosphere. The Coriolis force acts on the outflow, creating anticyclonic conditions and thereby spinning up a Rossby wave downstream (red). Energy propagates downstream and polewards at the group velocity of the wave. The mirror image happens in the southern hemisphere, but propagation is limited in summer conditions, and the main effects are seen in the winter half year.

The resulting wave pattern in the atmosphere, in this case the northern hemisphere (Fig. 11.2), alters the jet stream and preferred storm tracks. In the subtropics, the storm tracks change in such a way that the transient disturbances (including storms) feed back and influence the outcome through

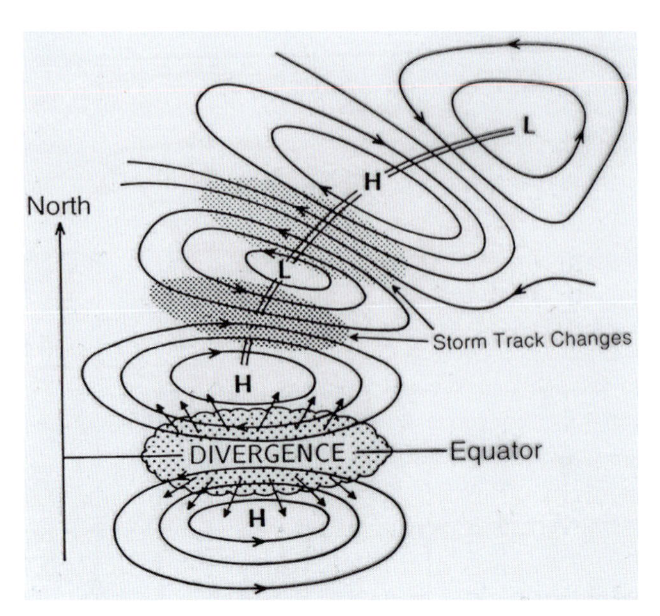

Fig. 11.2 The upper tropospheric wavetrain patterns that develop in the northern hemisphere in winter in the stream-function and geopotential height perturbation fields resulting from a strong large-scale convective event on the equator. The latter results in strong upper air divergence of air that is affected by the Coriolis force (Fig. 11.1) and sets up anticyclonic conditions on both sides of the equator. That high-pressure region in turn sets up the downstream wavetrain of lows and highs accompanied by changes in storm tracks. A stronger jet stream between the high and low is accompanied by a stronger storm track, while the jet stream weakens between the low and the high, and so does the storm track. The wavetrain tends to follow a great circle path, modified by the prevailing subtropical jet stream. From Trenberth et al. (1998). (The stream-function depicts the rotational [nondivergent] part of the actual flow fields.) © 1998 by the American Geophysical Union. Published with permission

changes in heat, vorticity and momentum transports, as well as changes in heating through changes in rainfall-induced latent heating. These factors mean that it is not just the wave source energy that comes into play, but the redistribution of energy throughout much of the hemisphere. These aspects can also be interpreted as interactions with annular modes (described below). The large natural variability means that results are not deterministic but continually evolve, and many climate model–based experiments have clearly shown the dominant role of the tropical and subtropical effects, while also indicating, however, that large ensembles are essential to sort out signal from noise, possible only with models.

Sidebar 11.1 in this chapter provides the definitions of the indices of the various modes and patterns of variability featured here.

Sidebar 11.1: Indices of Natural Variability Patterns of the Climate

A number of teleconnection patterns have historically been defined from either station data (SOI, NAO, SAM) or from gridded fields (NAM, AMO, PDO/NPI and PNA), or both: indices based upon regional patterns (multiple grid points) are more robust to small-scale perturbations, but the simpler station-based data can usually be extended further back in time. See www.cgd.ucar.edu/cas/catalog/climind/.

- **Atlantic Multi-decadal Oscillation (AMO) Index**. The time series of annual mean SST anomalies averaged over the North Atlantic (0–60° N, 0–80° W) as departures from the global mean SST.
- **Indian Ocean Dipole (IOD)** given by the **Dipole Mode Index (DMI)** represented by anomalous SST gradient between the western equatorial Indian Ocean (10° S–10° N, 50–70° E) and the southeastern equatorial Indian Ocean (10° S–0° N, 90–110° E).
- **North Atlantic Oscillation (NAO) Index.** The difference of normalized MSLP anomalies between Lisbon, Portugal and Stykkisholmur, Iceland, extends back in time to 1864.
- **Northern Annular Mode (NAM) Index.** The amplitude of the pattern defined by the leading empirical orthogonal function of winter monthly mean NH MSLP anomalies poleward of 20° N. The NAM has also been known as the Arctic Oscillation (AO), and is closely related to the NAO.
- **North Pacific Index (NPI).** The average MSLP anomaly in the Aleutian Low over the Gulf of Alaska (30–65° N, 160° E–140° W) and an index of the PDO.
- **Pacific Decadal Oscillation (PDO) Index.** The pattern and time series of the first empirical orthogonal function of departures from the global mean SST over the North Pacific north of 20° N. The PDO broadened to cover the whole Pacific Basin is known as the **Inter-decadal Pacific Oscillation (IPO).** They exhibit virtually identical temporal evolution.
- **Pacific–North American (PNA) Index.** The mean of normalized 500 hPa height anomalies at 20° N, 160° W and 55° N, 115° W minus those at 45° N, 165° W and 30° N, 85° W.
- **Southern Annular Mode (SAM) Index.** The difference in average MSLP between SH middle and high latitudes (usually 45° S and 65° S), from gridded or station data, or the amplitude of the leading empirical orthogonal function of monthly mean SH 850 hPa height poleward of 20° S. Also known as the Antarctic Oscillation (AAO).
- **Southern Oscillation Index (SOI).** The MSLP normalized anomaly difference of Tahiti minus Darwin, further normalized by the standard deviation of the difference. Available from the 1860s. Darwin can be used alone, as its data are more consistent than Tahiti prior to 1935. It is closely linked with El Niño and SST indices (see Chapter 12).

11.3 Annular Modes

There are some preferred patterns of weather that can develop without very much influence from outside of the atmosphere. The first class are called annular modes:

the Northern Annular Mode (NAM) in the northern hemisphere – also misleadingly called the Arctic Oscillation; and the Southern Annular Mode (SAM) in the southern hemisphere – also misleadingly called the Antarctic Oscillation. Fundamentally, these are hemispheric variations in the prevailing latitude of the main belt of zonal westerly winds and their associated jet streams and storm tracks. Here the term "storm tracks" refers to the prevailing track of cyclonic and anticyclonic perturbations and associated weather systems. In this case, the storms are associated with Rossby waves, and hence feature both a high and low perturbation, but they also play a key role in feeding on the temperature gradients and systematically moving heat polewards and upwards (see Chapter 7). The NAM has a presence year-round but is strongest in winter and weakens in summer and shifts polewards. The SAM is always present. In the positive phase, the westerlies and storm tracks are shifted polewards: northward in the NAM or southwards in the SAM. For instance, in the SAM positive phase, the westerlies are stronger from 45 to 60° S and weaker farther north.

The NAM is strongest in the Atlantic sector, where it has its own name: the North Atlantic Oscillation (NAO). Reasons for this are the planetary-scale waves which mean that the main storm tracks in the northern hemisphere occur over the oceans. The equivalent to the NAO in the southern hemisphere is not present because of the absence of the zonal asymmetries set up by continents: the dominant land masses of the southern continents are in the subtropics or tropics.

The NAM, SAM, and NAO all are natural modes of variability. They occur in the atmospheric circulation of climate models where the SSTs are specified at climatological values, and hence do not vary. Nevertheless, small subtle influences from land or ocean may nudge these patterns into one phase or another. Also, they may relate to the variations in north–south heating in the atmosphere, although there are strong feedbacks from the mid-latitude eddies in terms of both heat and especially momentum transports, that help sustain these patterns.

Some other teleconnection patterns also form naturally through an initial disturbance in one location that then sets up an atmospheric wavetrain into more remote areas. They can also be triggered by interactions with the surface.

11.4 SAM

The SAM is associated with synchronous pressure or height anomalies of opposite sign in middle and high latitudes, and therefore reflects changes in the main belt of subpolar westerly winds (Fig. 11.3, top). Enhanced Southern Ocean westerlies occur

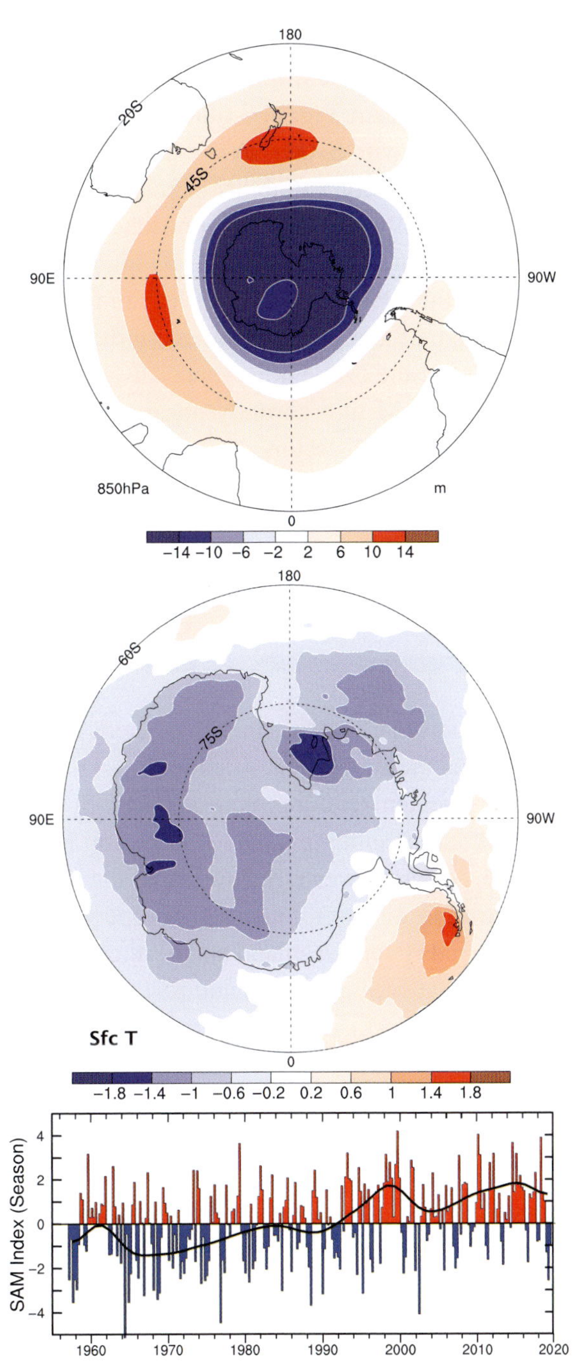

Fig. 11.3 (Top) The SAM geopotential height pattern as a regression based on the SAM time series for seasonal anomalies at 850 hPa. (Middle) The regression of changes in surface temperature (°C) over the 23-year period (1982–2004) corresponding to a unit change in the SAM index, plotted south of 60° S. Values exceeding about 0.4°C in magnitude are significant at the 1% significance level (adapted from Kwok and Comiso, 2002). (Bottom) Seasonal values of the SAM index calculated from station data (updated from Marshall, 2003; https://legacy.bas.ac.uk/met/gjma/sam.html). The smooth black curve shows decadal variations.

in the positive phase of the SAM. Accordingly, the SAM is also an index of the polar vortex in the southern hemisphere, and during the positive phase cool air is trapped south of the jet stream: Fig. 11.3 (middle) shows the temperature anomalies south of 60° S associated with SAM. SAM contributes a significant proportion of southern mid-latitude circulation variability on many timescales and is the leading mode in an analysis of monthly mean *global* atmospheric mass, accounting for around 10% of total global variance. The SAM index (Fig. 11.3, bottom) reveals a general increase beginning in the 1960s consistent with a strengthening of the circumpolar vortex and intensification of the circumpolar westerlies that has been associated with the development of the ozone hole, especially in the southern summer. Only in spring 2016 did the SAM turn abruptly negative in association with large decreases in Antarctic sea ice in all sectors. Nevertheless, although a very strong negative SAM occurred in November 2016 (−3.12), the overall high values of SAM have continued.

11.5 NAO

The NAO refers to changes in the atmospheric sea-level pressure difference between the Arctic and the subtropical Atlantic (Fig. 11.4). It exerts a dominant influence on winter surface temperatures across much of the northern hemisphere, and on storminess and precipitation over Europe and North Africa. When the NAO index is positive, enhanced westerly flow across the North Atlantic in winter moves warm moist maritime air over much of Europe and far downstream, while stronger northerly winds over Greenland and northeastern Canada carry cold air southward and decrease land temperatures and SST over the northwest Atlantic. Temperature variations over North Africa and the Middle East (cooling) and the southeastern United States (warming), associated with the stronger clockwise flow around the subtropical Atlantic high-pressure center, are also notable. Positive NAO index winters are also associated with a northeastward shift in the Atlantic storm activity, with enhanced activity from Newfoundland into northern Europe and a modest decrease to the south. More precipitation than normal falls from Iceland through Scandinavia during high NAO index winters, while the reverse occurs over much of central and southern Europe, the Mediterranean, parts of the Middle East, the Canadian Arctic, and much of Greenland.

From the time series of the NAO (Fig. 11.4) it can be seen that the NAO was preferentially positive especially from 1981 to 2009 versus 1951 to 1980 (Fig. 11.5), and so was the NAM. This highlights some significant decadal variability that may also be influenced by global climate change of some sort, likely involving lower stratospheric and ozone changes.

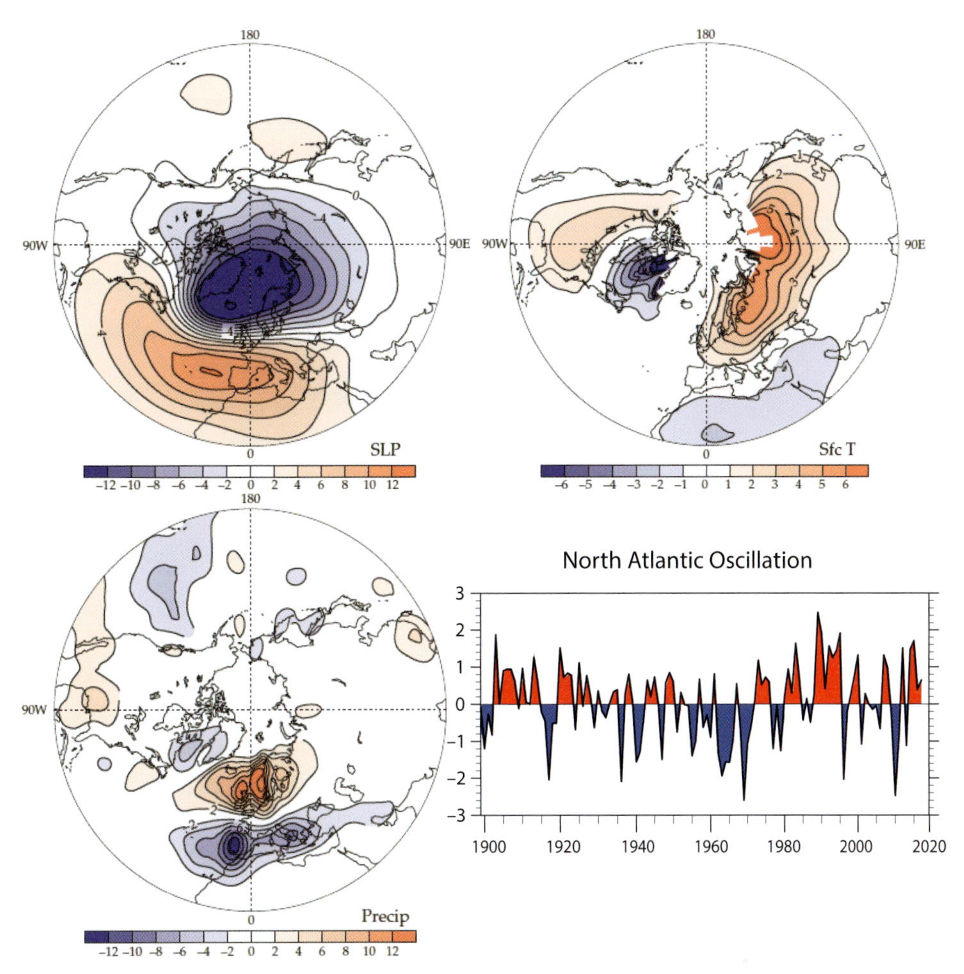

Fig. 11.4 Changes in northern winter (December–March) surface pressure, temperature, and precipitation corresponding to a unit deviation of the NAO index. (Top left) Mean sea-level pressure (0.1 hPa). (Top right) Land-surface air temperature and SST (0.1°C; contour increment 0.2°C): regions of insufficient data (e.g., over much of the Arctic) are not contoured. (Bottom left) Precipitation for 1979–2009 based on global estimates (0.1 mm day^{-1}; contour interval 0.6 mm day^{-1}). (Bottom right) Station-based index of winter NAO. The indicated year corresponds to the January of the winter season (e.g., 1990 is the winter of 1989/90). Adapted and updated from Hurrell et al. (2003)

11.6 PNA

Perhaps the best known example of a teleconnection, aside from the NAO and ENSO (Chapter 12), is the Pacific–North American (PNA) pattern, that occurs mainly in winter in the northern hemisphere. It features four centers of action,

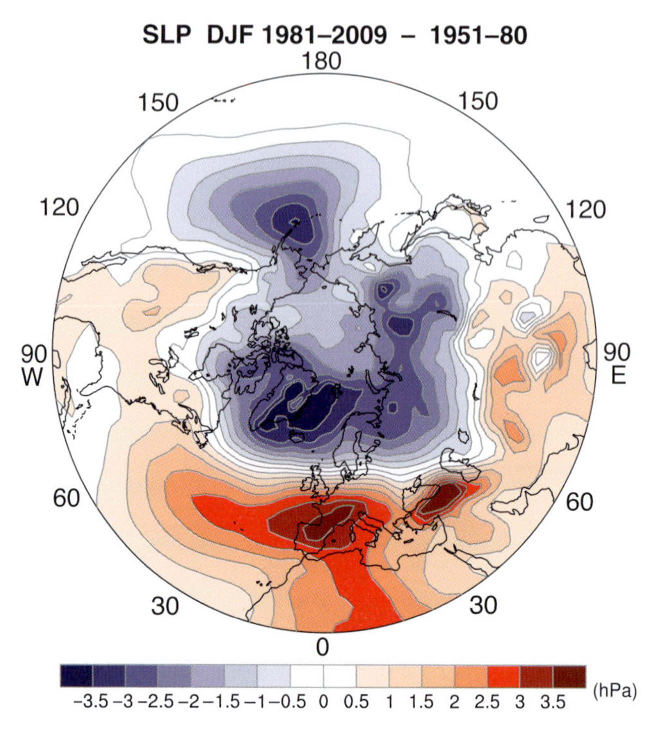

Fig. 11.5 For northern winter (DJF), the average sea-level pressure difference between 1981–2009 versus 1951–80. It illustrates large multi-decadal variability dominated by a positive phase of the NAM and the NAO. (See also Fig. 11.4 bottom right.). Adapted from Hurrell and Trenberth (2010)

which is rare. A positive PNA teleconnection pattern in the middle troposphere coincides with higher-than-normal pressure near Hawaii and over the northwestern United States and western Canada, while pressures are typically lower-than-normal over the central North Pacific and the southeast United States (Fig. 11.6). Variations in the PNA pattern represent changes in the north–south migration of the large-scale Pacific and North American air masses, storm tracks and their associated weather, affecting precipitation in western North America and the frequency of Alaskan blocking events and associated cold air outbreaks over the western United States in winter. Changes in the PNA can be triggered at any point, but a favored region is the tropical node where variations in convection with ENSO play a role, and hence the PNA is often part of the response to ENSO.

11.7 PDO and NPI

Decadal to inter-decadal variability in the atmospheric circulation is especially prominent in the North Pacific where fluctuations in the strength of the wintertime Aleutian Low pressure system, indicated by the North Pacific Index (NPI), co-vary with North

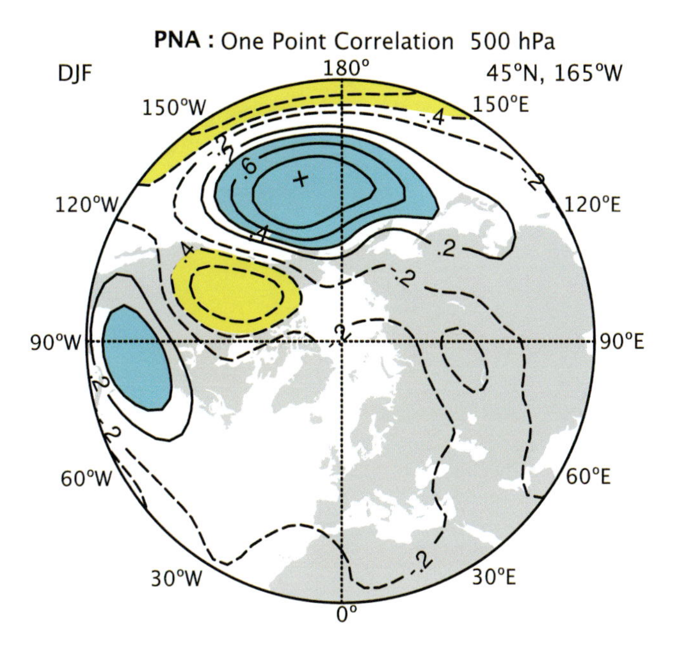

Fig. 11.6 The PNA teleconnection pattern, shown as one-point correlation maps of 500 hPa geopotential heights for boreal winter (DJF) over 1958–2005. The reference point is 45° N, 165° W, corresponding to the primary center of action of the PNA pattern, given by the + sign. Negative correlation coefficients are dashed, and the contour increment is 0.2. The main centers of action are colored. Adapted from Hurrell et al. (2003)

Pacific SST in what has been termed the Pacific Decadal Oscillation (PDO) or, its close cousin, the Inter-decadal Pacific Oscillation (IPO). The PDO and associated NPI occur on longer timescales and larger spatial scales than ENSO (Fig. 11.7).

The PDO/IPO pattern of SSTs is somewhat similar to that associated with ENSO, except that, by design, it is focused in the extratropics. They have been described as a long-lived El Niño–like pattern of Indo-Pacific climate variability or as a low-frequency residual of ENSO variability on multi-decadal timescales. Phase changes of the PDO/IPO are associated with pronounced changes in temperature and rainfall patterns across North and South America, Asia and Australia. Furthermore, ENSO teleconnections on interannual timescales around the Pacific basin are significantly modified by the PDO/IPO. Low PDO goes with high NPI values, indicative of a weakened circulation over the North Pacific (1900–24, 1945–76, 1999–2013), and predominantly high PDO values indicate a strengthened circulation (low NPI) (1925–44, 1977–98, and since 2014). Note, however, that the PDO time series (Fig. 11.7) includes all months of the year, while the NPI index is for a single value for the extended winter each year (and reversed in sign). The well-known decrease in Aleutian Low pressure from 1976 to 1977 is analogous to transitions that occurred from 1946 to 1947 and from 1924 to 1925, and these earlier changes were also associated with SST fluctuations in the tropical Indian and Pacific oceans.

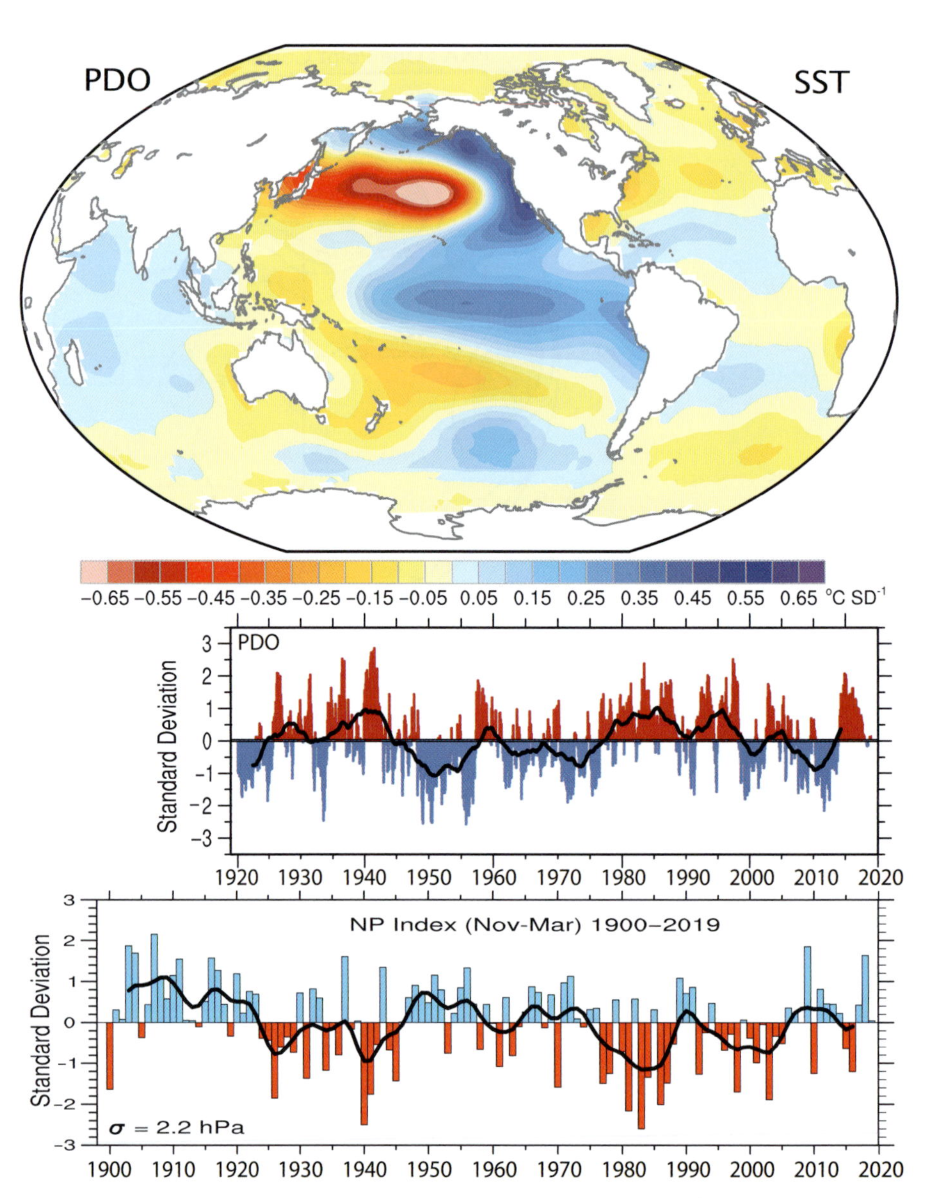

Fig. 11.7 The PDO. The pattern and time series of the first empirical orthogonal function (EOF)[1] of monthly SST anomalies over the North Pacific north of 20° N when the global mean SST is first removed (top and middle panels). Bottom panel is the NPI index for November to March (as an extended winter season), defined as the average MSLP anomaly in the Aleutian Low over the Gulf of Alaska (30–65° N, 160° E–140° W). When inverted, it is an index of the PDO, and hence the red values correspond to positive PDO. Both time series have the decadal variations called out in black by a decadal filter. Courtesy Adam Phillips

[1] EOF analysis sorts out the dominant covariability among all points that are coherent in time. Technically it is an eigenvector analysis of the covariance matrix.

The high PDO values relate to times of increases in the global mean surface temperature (GMST, Fig. 11.8) while the GMST no longer increases much for negative PDO values. From 1999 to 2013 this pause in the rise of GMST has also become known as a "hiatus" in warming; see Section 15.4. Although increases in GMST stalled, the ocean heat content and sea level continued to rise, showing that the heat from global warming was being redistributed within the ocean, both with depth and regionally in the West Pacific and Indian oceans. The main pacemaker of variability in rates of GMST increase appears to be the PDO, with aerosols likely playing a role in the earlier big hiatus from 1947 to 1976.

11.8 AMO

In the Atlantic sector, interannual variability is dominated by the NAO. However, Atlantic decadal variability has surprisingly large amplitude and has been termed the Atlantic Multi-decadal Oscillation, or AMO. North Atlantic SSTs show a 65- to 75-year variation ($\pm 0.2^{\circ}$C range), with a warm phase 1930–60 and after 1995, and cool phases during 1905–25 and 1970–95 (Fig. 11.9). The AMO has been linked to the Atlantic Meridional Overturning Circulation (AMOC) which extends from the Southern Ocean to the northern North Atlantic, transporting heat northwards throughout the basin, and sinking carbon and nutrients into the deep ocean. The decadal variability in the Atlantic arises mainly from changes in the AMOC volume transports. The AMOC is responsible for most of the meridional transport of heat and carbon by the mid-latitude northern hemisphere ocean and is associated with the production of about half of the global ocean's deep waters in the northern North Atlantic. The AMOC changes have been linked to abrupt climate change in the paleoclimate record, and freshwater flushing is especially suspected of contributing to rapid changes.

11.9 IOD

The Indian Ocean Dipole refers to the opposite changes in SST on each side of the tropical Indian Ocean which profoundly influence the tropical atmospheric circulation in and around the Indian Ocean. The IOD depicts an anomalous SST gradient between the western equatorial Indian Ocean (10° S–10° N, 50–70° E) and the southeastern equatorial Indian Ocean (10° S–0° N, 90–110° E). Events of one phase or the other usually start around May or June, peak between August and October and then rapidly decay when the monsoon arrives in the southern hemisphere around the end of spring. The IOD is one of the key drivers of Australia's climate and can have a significant impact on agriculture because events generally coincide with the winter crop-growing

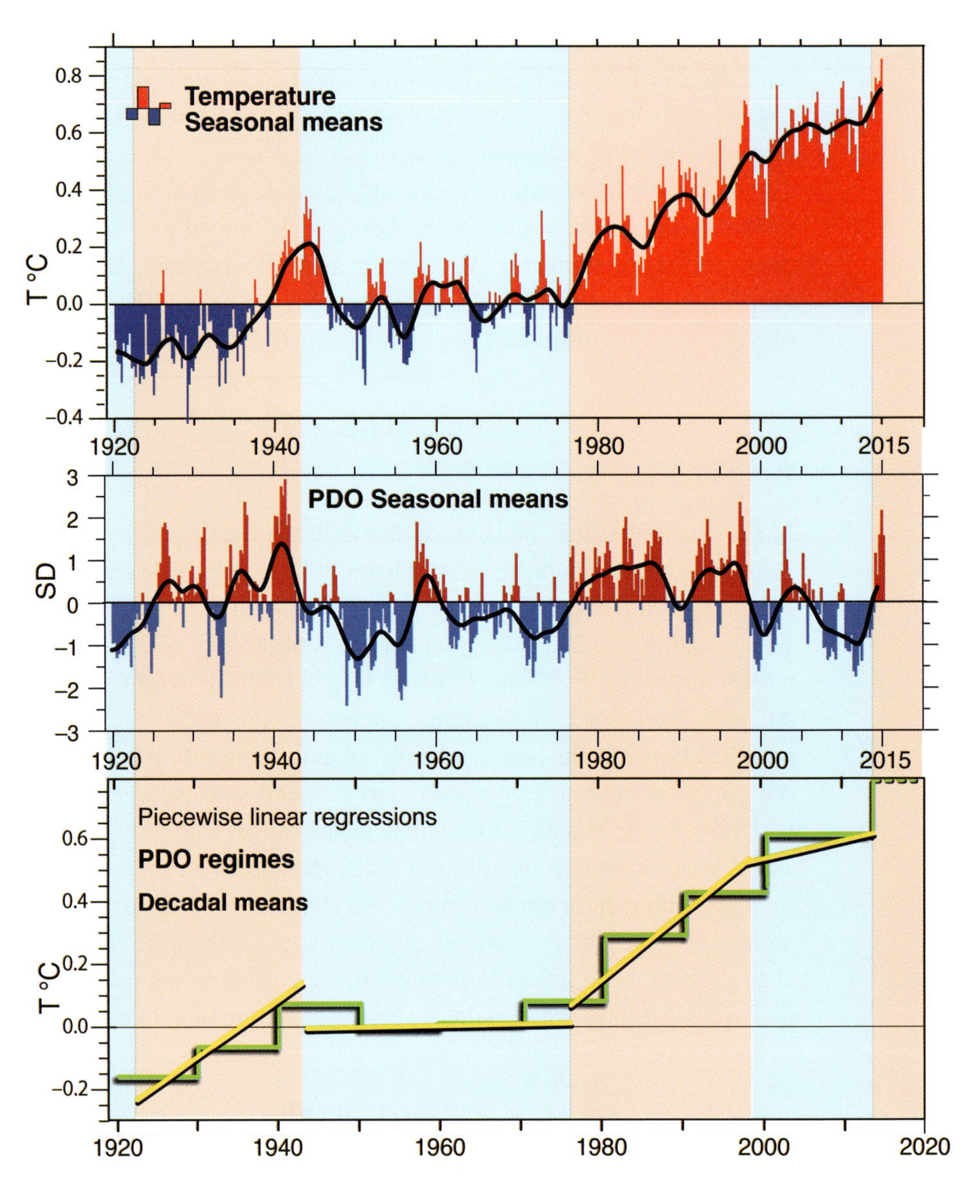

Fig. 11.8 A staircase of rising temperatures. (Top) Seasonal (December–January–February, etc.) global mean surface temperatures since 1920 (relative to the 20th-century mean) vary considerably on interannual and decadal timescales. (Middle) Seasonal mean PDO anomalies show decadal regimes (positive in pink; negative in blue) as well as short-term variability. A 20-term Gaussian filter is used in both to show decadal variations, with anomalies reflected about the end point of March to May 2015 (heavy black curves). (Bottom) Decadal average anomalies (starting 1921–30) of GMST (green) along with piecewise slopes of GMST for the phases of the PDO (yellow). Note how the rise in GMST (top) coincides with the positive (pink) phase (middle) of the PDO at the rate given in (bottom). Adapted from Trenberth (2015). Reprinted with permission from the American Association for the Advancement of Science

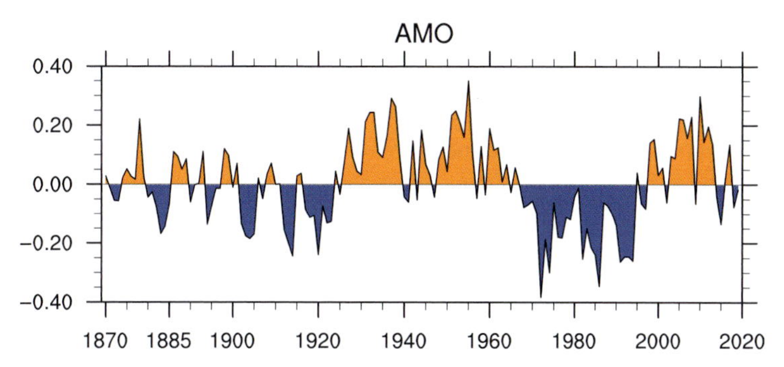

Fig. 11.9 The AMO computed as the time series of annual mean SST anomalies averaged over the North Atlantic (0–60° N, 0–80° W) as departures from the global mean SST. Courtesy Adam Phillips

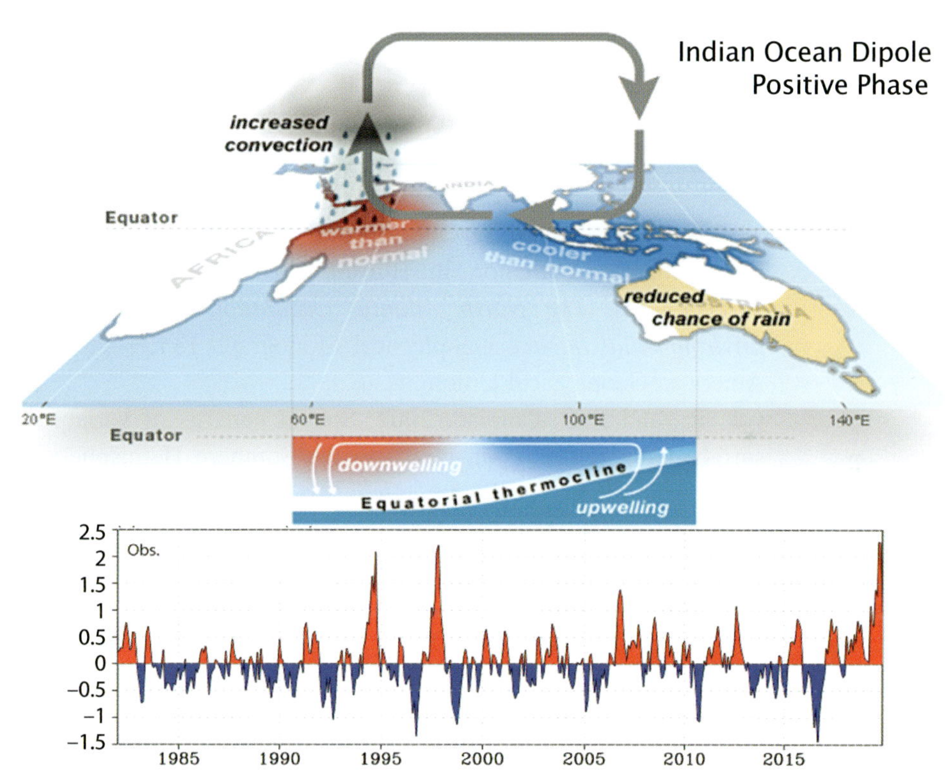

Fig. 11.10 Indian Ocean Dipole. (Top) schematic of the anomalous SSTs in the positive phase along with reduced rainfalls over Australia and the change in the local Walker Circulation in the atmosphere. (Middle) The subsurface changes in the thermocline and overturning in the tropical Indian Ocean. (Bottom) IOD given by time series of the Dipole Mode Index (DMI) of anomalous SST gradient between the western equatorial Indian Ocean (10°S–10° N, 50–70° E) and the southeastern equatorial Indian Ocean (10° S–0° N, 90–110° E). Adapted from CSIRO, Australia

season. During the positive phase (Fig. 11.10), westerly winds weaken along the equator, allowing warm waters to shift towards Africa and push down the thermocline. Changes in the winds also allow cool water to rise up from the deep ocean in the east, shoaling the thermocline there. Generally, this means there is less moisture than normal in the atmosphere to the northwest of Australia, but more around the horn of Africa. This changes the path of weather systems coming from Australia's west, often resulting in less rainfall and higher than normal temperatures over parts of Australia during winter and spring (Fig. 11.10). On the other hand, a negative IOD typically results in above-average winter–spring rainfall over parts of southern Australia as the warmer waters off northwest Australia provide more available moisture to weather systems crossing the country. Eastern Africa is then more at risk for drought.

In general, the IOD plays a secondary role to ENSO, but there have been occasions when the IOD has loomed especially large, such as in 2019 (Fig. 11.10), leading to an exceptionally strong drought, heatwaves, and widespread bushfires over southern parts of Australia.

References and Further Reading

Hurrell, J. W., Y. Kushnir, G. Ottersen, and M. Visbeck, 2003: An overview of the North Atlantic Oscillation. In: Hurrell, J. W., Y. Kushnir, G. Ottersen, and M. Visbeck, eds. *The North Atlantic Oscillation: Climatic Significance and Environmental Impact*. Geophysical Monograph **134**, 1–35. Washington, DC: American Geophysical Union.

Kwok, R., and J. C. Comiso, 2002: Spatial patterns of variability in Antarctic surface temperature: connections to the Southern Hemisphere Annular Mode and the Southern Oscillation. *Geophysical Research Letters*, **29**, 1705, doi: 10.1029/2002GL015415.

Marshall, G. J., 2003: Trends in the Southern Annular Mode from observations and reanalyses. *Journal of Climate*, **16**, 4134–4143. doi: 10.1175/1520-0442(2003)016<4134:TITSAM>2.0.CO;2.

Trenberth, K. E., 1990: Recent observed interdecadal climate changes in the Northern Hemisphere. *Bulletin of the American Meteorological Society*, **71**, 988–993.

Trenberth, K. E., 2015: Has there been a hiatus? *Science,* **349**(2649), 691–692. doi: 10.1126/science.aac9225.

Trenberth, K. E., and J. W. Hurrell, 1994: Decadal atmosphere–ocean variations in the Pacific. *Climate Dynamics*, **9**, 303–319.

Trenberth K. E., and J. W. Hurrell, 2019: Climate change. In: Dunn, P. O., and A. P. Møller, eds., *The Effects of Climate Change on Birds*, 2nd ed. Oxford: Oxford University Press, 5–25.

Trenberth, K. E., and D. J. Shea, 2006: Atlantic hurricanes and natural variability in 2005. *Geophysical Research Letters*, **33**, L12704. doi: 10.1029/2006GL026894.

Trenberth, K. E., G. W. Branstator, D. Karoly, A. Kumar, N-C. Lau, and C. Ropelewski, 1998: Progress during TOGA in understanding and modeling global teleconnections associated with tropical sea surface temperatures. *Journal of Geophysical Research*, **103,** 14291–14324.

Trenberth, K. E., P. D. Jones, P. Ambenje, et al., 2007: Observations: surface and atmospheric climate change. In: Solomon, S., D. Qin, M. Manning, et al., eds., *Climate Change 2007. The Physical Science Basis*. Fourth Assessment Report of the Intergovernmental Panel on Climate Change. Cambridge: Cambridge University Press, 235–336.

12 El Niño

The El Niño phenomenon causes the largest year-to-year perturbations and disruptions in weather and climate around the globe and especially throughout the tropics. It plays a major role in heat movements.

12.1 ENSO and the Mean Pacific Climate

The unique features of the tropical Pacific and its mean annual cycle set the stage for ENSO to occur. ENSO: El Niño–Southern Oscillation is the simultaneous occurrence of an El Niño and Southern Oscillation events. El Niño is the warm ocean component of ENSO and refers to an anomalous warming of the surface tropical Pacific Ocean east of the dateline to the South American coast. The Southern Oscillation corresponds to a global-scale pattern in mean sea-level pressure, and hence surface winds, that is the atmospheric component of ENSO. Historically 'El Niño' referred to the appearance of unusually warm water off the coast of Peru, where it was readily observed as an enhancement of the normal warming about Christmas (hence Niño, Spanish for "the boy Christ-child"), but in the last half-century the term came to be regarded as synonymous with the basin-wide phenomenon. The oceanic and atmospheric conditions in the tropical Pacific fluctuate somewhat irregularly between El Niño and the cold phase of ENSO: a basin-wide cooling of the tropical Pacific, named "La Niña" ("the girl" in Spanish). The most intense phase of each event typically lasts half of a year.

In the tropical Pacific, SSTs strongly influence the water vapor distribution and Total Column Water Vapor (TCWV), as well as precipitation (Chapter 5, see Fig. 5.4). While the major monsoons (Fig. 5.1) are strongly driven by the seasonal changes mainly because of the continental land influences, the ITCZs remain in the

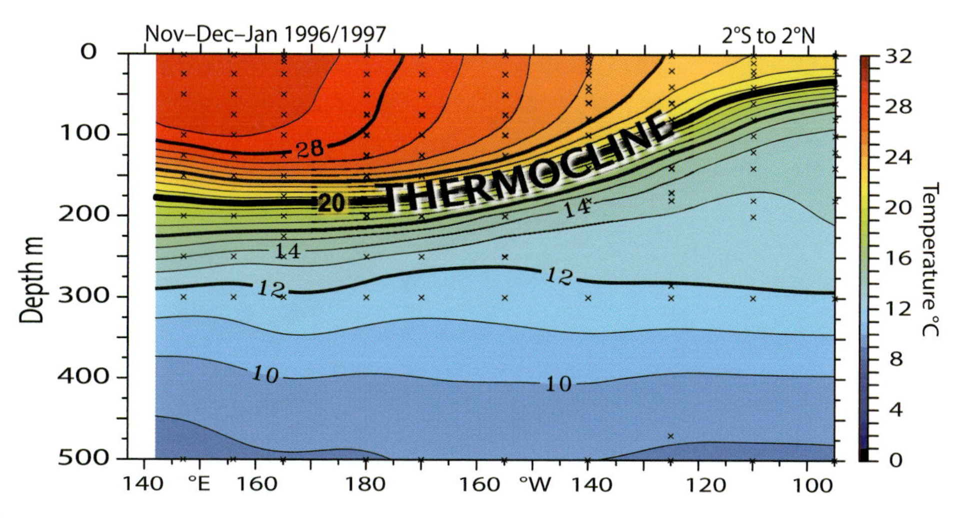

Fig. 12.1 Cross section along the equator of temperatures within the ocean with depth from the Tropical Atmosphere–Ocean (TAO) array (points given by crosses) for November–December–January 1996–97. The region from approximately 17 to 23°C, centered on the 20°C isotherm, is indicated as the main thermocline. Data downloaded August 15, 2018, courtesy GTMBA Project Office of NOAA/PMEL. www.pmel.noaa.gov/tao/drupal/disdel/

northern hemisphere year-round over the open ocean, away from land masses. In contrast, the Warm Pool of high SSTs in the western Pacific–Indonesian region (Fig. 12.1) migrates strongly across the equator accompanied by highest values in TCWV (Fig. 5.4). Surface trade winds, the southeast trades over and south of the equator and the northeast trades north of the ITCZ, converge moisture over the highest SSTs, leading to strong convection and upward motion overall, while the downward branch of the Walker Circulation over the cold tongue of SSTs in the upwelling region of the eastern equatorial Pacific suppresses rainfall (Fig. 5.4).

In the ocean, temperatures along the equator in the Pacific for November–December 1996 to January 1997, a fairly normal time, are shown in Fig. 12.1 with the 20°C isotherm highlighted as approximating the center of the thermocline: the strong temperature gradient in the vertical between the upper mixed layer and the deep cold abyss. In the west, in the Warm Pool, the thermocline is about 150–200 m deep, while in the east the thermocline is only 40 m deep on average. The Pacific sea surface slopes up by about 60 cm from east to west along the equator.

A schematic of the tropical Pacific Ocean for the top 200 m (Fig. 12.2) shows the relationships among the thermocline, cold tongue, the ITCZ, and the northeast and southeast trade winds. The South Equatorial Current (SEC) extends well into the southern hemisphere and is strong near the equator; the North Equatorial Current (NEC) lies north of about 10° N, and the North Equatorial Countercurrent (NECC) develops in between, north of the equator. The ITCZ typically lies between about 5 and

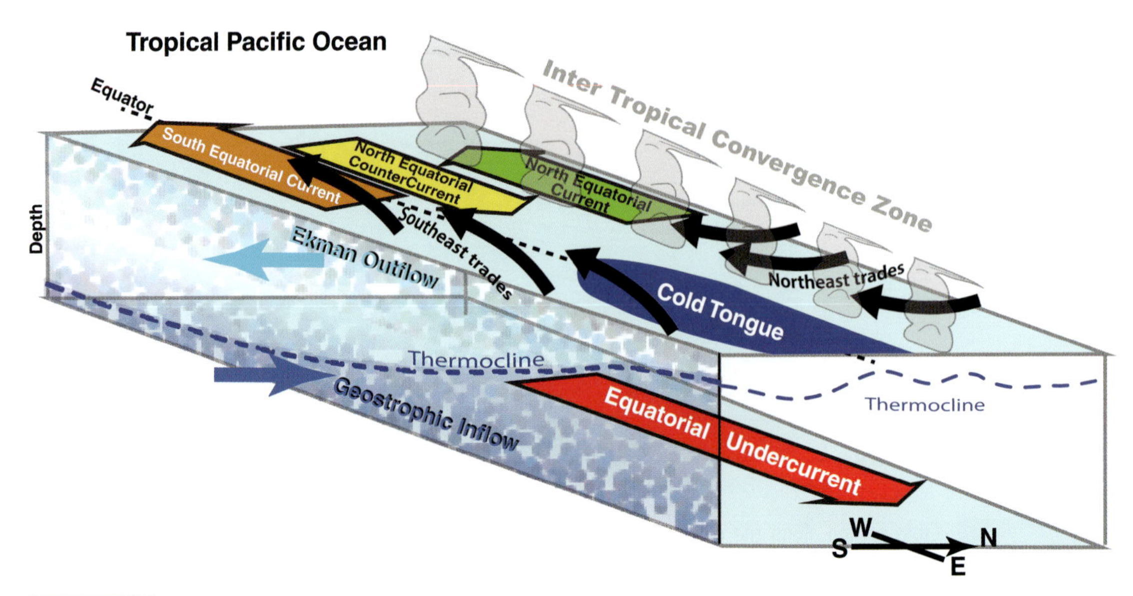

Fig. 12.2 Schematic of the tropical Pacific Ocean for the top 200 m. The equator is the dotted black line. Shown are the thermocline, cold tongue, and the northeast and southeast trade winds (black arrows). The latter extend to and across the equator, and the trade winds come together around 5–10° N in the ITCZ. For ocean currents, the South Equatorial Current extends into the southern hemisphere and is strong near the equator, the North Equatorial Current is north of about 10° N, and the North Equatorial Countercurrent develops in between, north of the equator. The equatorial undercurrent is strong along the equator, peaking at about 1 m s^{-1} at 100 m depth from 120 to 140° W. The trade winds create divergence at the surface owing to Earth's rotation, given here as Ekman outflow (speckled light blue), and compensated for by geostrophic inflow at greater depths (speckled blue) and upwelling. The surface winds drive the ocean currents, but the result depends mostly on the curl of the wind stress, and hence the countercurrent results from the minimum in easterlies.

10° N (Fig. 5.4). The equatorial undercurrent (Fig. 12.2) is strong along the equator below the surface, peaking at about 1 m s^{-1} at 100 m depth from 120 to 140° W. The trade winds create divergence at the surface owing to frictionally driven Ekman outflow (Section 8.2), and this is compensated for by geostrophic inflow at greater depths (Fig. 12.2). Surface winds drive the ocean currents, but the result depends mostly on the curl of the wind stress, and hence the countercurrent results from the minimum in easterlies (Chapter 8). Poleward of 25° N and 25° S, the prevailing westerly winds induce Ekman transports towards the equator, generating a convergence between 15–30° N and 15–30° S, with high sea levels around 10–15° N and 10–15° S, respectively.

In tropical ocean domains, the atmosphere and ocean are very strongly coupled. The surface winds drive surface ocean currents, which determine where the surface waters flow and diverge, and thus where cooler nutrient-rich waters upwell from below (Chapter 8). Thus, the winds largely determine the SST distribution, the differential sea levels and the heat content of the upper ocean. Then in turn the SSTs determine the winds. The result is high sea level in the western tropical Pacific and the huge Warm Pool, as waters driven by the easterly trade winds pile up. Waters from the Warm Pool flow into the tropical Indian Ocean as the Indonesian ThroughFlow, although the currents are complex owing to the small islands and varying depths of channels (Chapter 9, Fig. 9.5). In the eastern Pacific, upwelled cool waters result in a much shallower mixed layer (Fig. 12.1) and sea level is relatively low. The presence of nutrients in the upwelled cool surface waters along the equator and western coasts of the Americas in association with sunlight favors development of tiny plant species (phytoplankton) that are grazed on by microscopic sea animals (zooplankton) and in turn provide food for fish.

Although January and July represent extremes in temperature over land, following a few weeks after the solstice, extratropical SSTs rise until the end of summer – March in the south and September in the north. The equatorial cold tongue is strongest around September–October of each year, when the ITCZ is farthest north, nearly 10° N, and the southeast trades along the equator are strong. Hence this corresponds to the time when the northern oceans are warmest, and the southern oceans are coldest. On the northern side of the cold tongue, there are frequently tropical instability waves that feed on the resulting temperature gradients and act to transport ocean heat to reduce the cold tongue strength. In contrast, in March–April when SSTs are highest along the equator (Figs. 5.3 and 5.4), the trade winds are weaker, and the cold tongue is less pronounced. This is the time of year when modest SST anomalies can substantially alter the location of the highest SSTs, thereby shifting the ITCZ and SPCZ in ways that can lead to a change in phase of ENSO.

The average state of the tropical Pacific involves a fairly delicate balance between the atmospheric winds and SSTs; the surface winds are largely responsible for the tropical SST distribution which, in turn, determines the precipitation and associated latent heating, and thus the tropical atmospheric circulation and surface winds. However, average conditions never prevail, and instead the tropical Pacific sways back and forth between the El Niño and La Niña states. Although many of the same processes occur in the tropical Atlantic, the western part is dominated by the South American land mass instead of the Warm Pool, and hence the seasonal cycle plays a more prominent role there.

12.2 The Southern Oscillation

The Southern Oscillation (SO) was first discovered by Sir Gilbert Walker and named in the early 1930s, and is principally a global-scale seesaw in atmospheric sea-level pressure involving exchanges of air between eastern and western hemispheres (see Fig. 12.3: SLP) centered in tropical and subtropical latitudes with

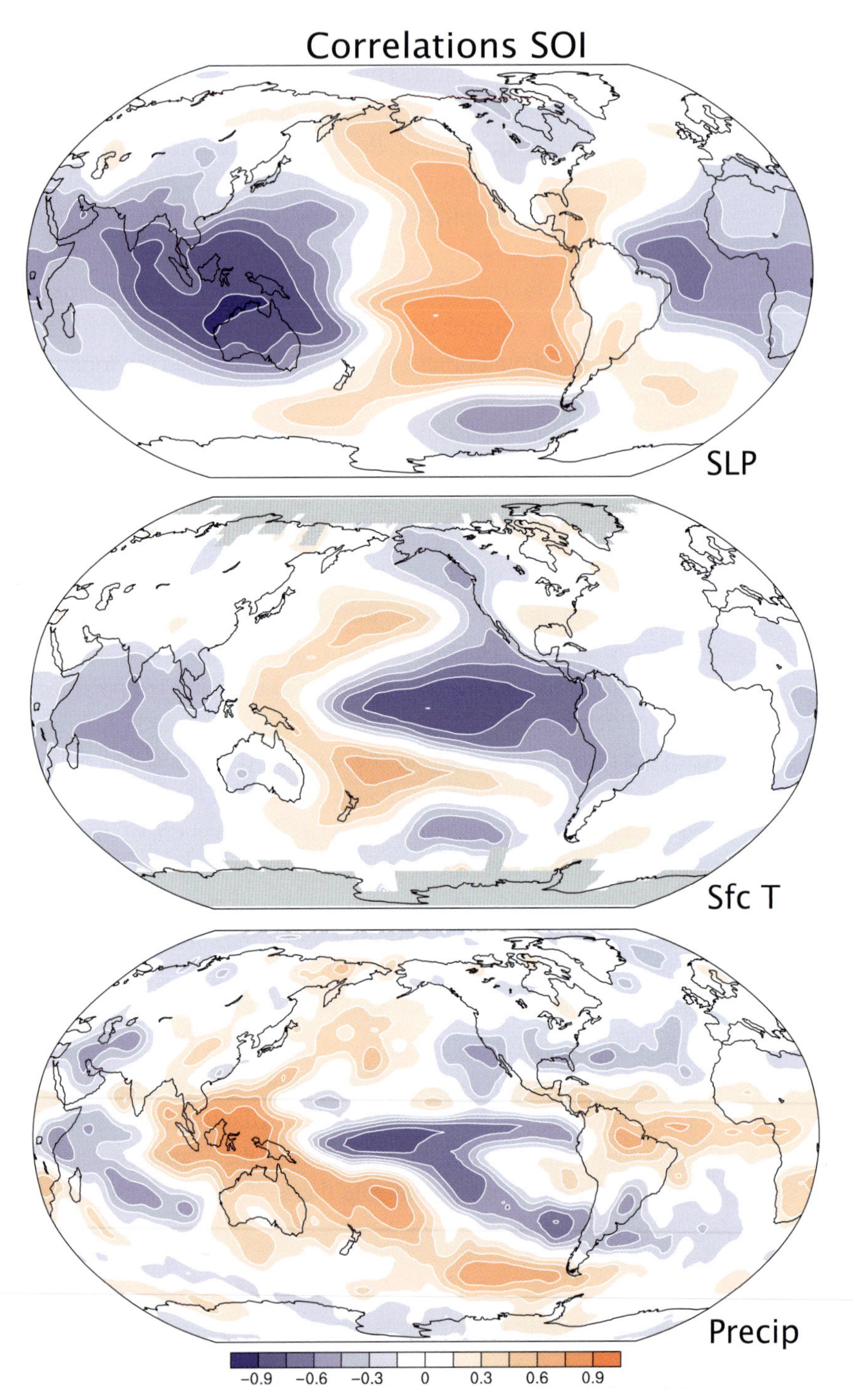

Figure 12.3 Correlations with the SO index, based on normalized Tahiti minus Darwin sea-level pressures, for annual (May–April) means for sea-level pressure (top) and surface temperature (center) for 1958–2004, and GPCP precipitation for 1979–2003 (bottom). Adapted from Trenberth et al., 2007

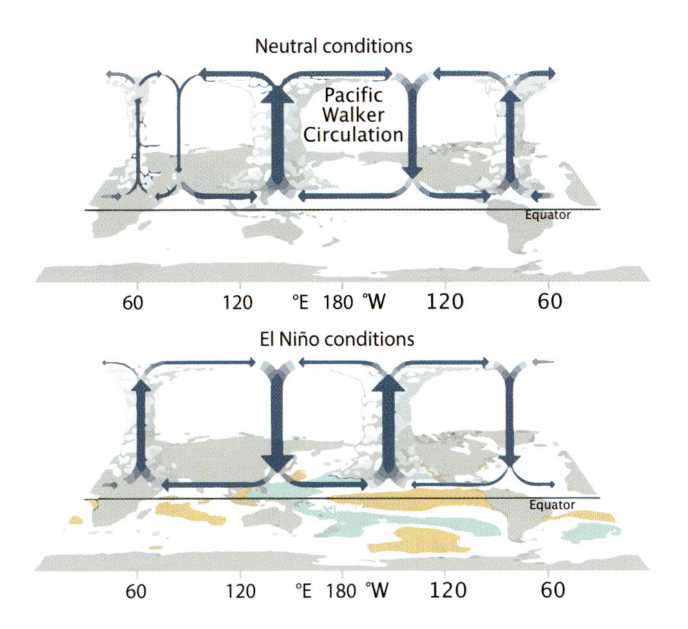

Fig. 12.4 The Walker Circulation and its changes with El Niño (December–February). (Top) During ENSO-neutral conditions, convection associated with rising branches of the Walker Circulation lies over the Maritime Continent, northern South America, and eastern Africa. (Bottom) During El Niño events, overlaid on a map of sea surface temperature anomalies, anomalous ocean warming in the central and eastern Pacific (orange) helps to shift a rising branch of the Walker Circulation to east of 180°, while sinking branches shift to over the Maritime Continent and northern South America. Based on NOAA Climate.gov drawings by Fiona Martin.

centers of action located near Indonesia and the tropical South Pacific Ocean (near Tahiti). Thus, there are the inverse variations in pressure anomalies (departures from average) at Darwin (12.4° S, 130.9° E) in northern Australia and at Tahiti (17.5° S, 149.6° W) in the South Pacific Ocean whose annual mean pressures are strongly and significantly oppositely correlated. During an El Niño event, the sea-level pressure tends to be higher than usual at Darwin and lower than usual at Tahiti. Consequently, the difference in pressure anomalies, Tahiti minus Darwin, appropriately weighted, is often used as a Southern Oscillation Index (SOI).

The correlations of SST and surface air temperature (Fig. 12.3: Sfc T) with the SOI then show the El Niño patterns. Warm ENSO events, therefore, are those in which both a negative SO extreme and an El Niño occur together. Higher than normal pressures are characteristic of settled fine weather with less rainfall, whereas lower than normal pressures are identified with ‘bad’ weather, more storminess and rainfall (Fig. 12.3: Precip). Thus, for El Niño conditions, higher than normal pressures over Australia, Indonesia, Southeast Asia, and the Philippines signal drier conditions or even droughts. Dry conditions also prevail in Hawaii, parts of Africa, and extend to the northeast part

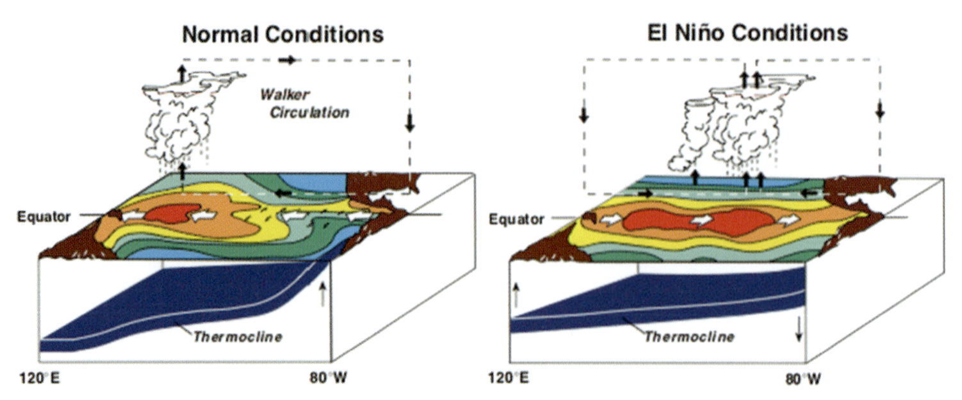

Fig. 12.5 Schematic of fairly normal conditions in the tropical Pacific, also called neutral (left), and El Niño conditions (right). Australia is left and the Americas are on the right, in brown. The thermocline is indicated (dark blue) along with the SST anomalies and the anomalous winds (white arrow) on the equator. Strongest convection occurs in the vicinity of the warmest water and is part of the east–west Walker Circulation, with large changes during El Niño. La Niña conditions are similar to, but a bit stronger than, the normal conditions.

of Brazil and Colombia. On the other end of the seesaw, excessive rains prevail over the central and eastern Pacific, along the west coast of South America, parts of South America near Uruguay, and southern parts of the United States in winter, often leading to flooding. When the pressure pattern in Fig. 12.3 reverses in sign, as for La Niña, the regions favored for drought in El Niño tend to become excessively wet, and vice versa.

Naturally, the changes at the surface in the atmosphere have consequences for the atmospheric circulation. In particular, in association with El Niño, there are substantial changes in the west–east overturning Walker Circulation near the equator (Figs. 12.4 and 12.5). During the warm phase of ENSO, the warming of the waters in the central and eastern tropical Pacific shifts the location of the heaviest tropical rainfall eastward toward or beyond the Date Line from its climatological position centered over Indonesia and the far western tropical Pacific, weakening the Walker Circulation. This shift in rainfall also alters the heating patterns that force large-scale waves in the atmosphere (Figs. 11.1 and 11.2). The waves in the airflow determine the preferred location of the extratropical storm tracks. Consequently, changes from one phase of the SO to another have profound impacts on regional temperatures (Fig. 12.3).

12.3 El Niño Events

El Niño events are quasi-periodic and occur every 3–7 years or so. ENSO is unique because it is the largest source of interannual variability of climate around the

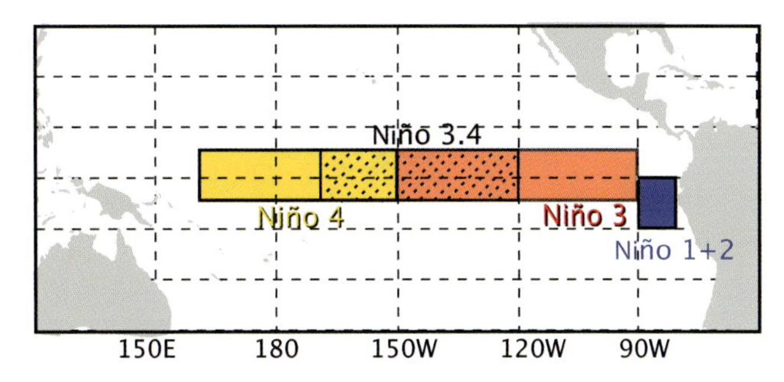

Fig. 12.6 The regions used for monitoring SSTs in the tropical Pacific: Niño 4 (yellow) is 5° N–5° S, 160° E–150° W, Niño 3 (red) is 5° N–5° S, 150–90° W, Niño 1+2 (blue) is 0°–10° S, 90–80° W. The Niño 3.4 region (dotted) is 5° N–5° S, 170–120° W.

world and it involves strong atmosphere–ocean interactions. That is to say, it is neither an atmospheric nor oceanic phenomenon, but rather it is a coupled phenomenon. It involves positive feedbacks, called the Bjerknes feedback (Section 13.2.6), and these can act as a form of instability designed to alleviate the buildup of heat in the tropical western Pacific. Variations in SSTs produce huge changes in tropical rainfall patterns and atmospheric circulation which alter the Walker and Hadley circulations (Figs. 12.4 and 12.5) and thus the jet streams and storm tracks in middle and high latitudes (Chapter 11). Some of the biggest changes are movements in the main climatological features, such as the ITCZ and SPCZ (Figs. 5.1 and 12.4), and thus their consequences cannot be anticipated unless their presence is first recognized. Hence ENSO effects play off the mean climate and are not simple linear additions or perturbations.

El Niño is often described by several indices which are used to determine whether an El Niño event is underway or not. The first is related to mean anomalies in SST east of the Date Line, in the so-called Niño 3.4 region as SST anomalies from 5° N to 5° S, 170 to 120° W (Fig. 12.6). This index is the best single indicator (Fig. 12.7), but is not sufficient by itself to describe the state of El Niño. Historically, the Niño 1, 2, 3, and 4 regions (Fig. 12.6) were defined first because of the availability of data from ship tracks. The two coastal regions Niño 1 and Niño 2 are usually combined, and Niño 1+2 is especially useful for the traditional coastal region. Niño 3.4 lies between the 3 and 4 regions and is often called the ONI: the Oceanic Niño Index. The second most important index involves SST gradients from west to east, summarized by the Trans Niño Index (TNI). The latter involves contrasting normalized values in the Niño 4 and Niño 1+2 regions (Fig. 12.6). There are intriguing lead–lag relationships. Because there are many "flavors" of El Niño, simply knowing an El Niño is underway is not sufficient to determine the outcomes.

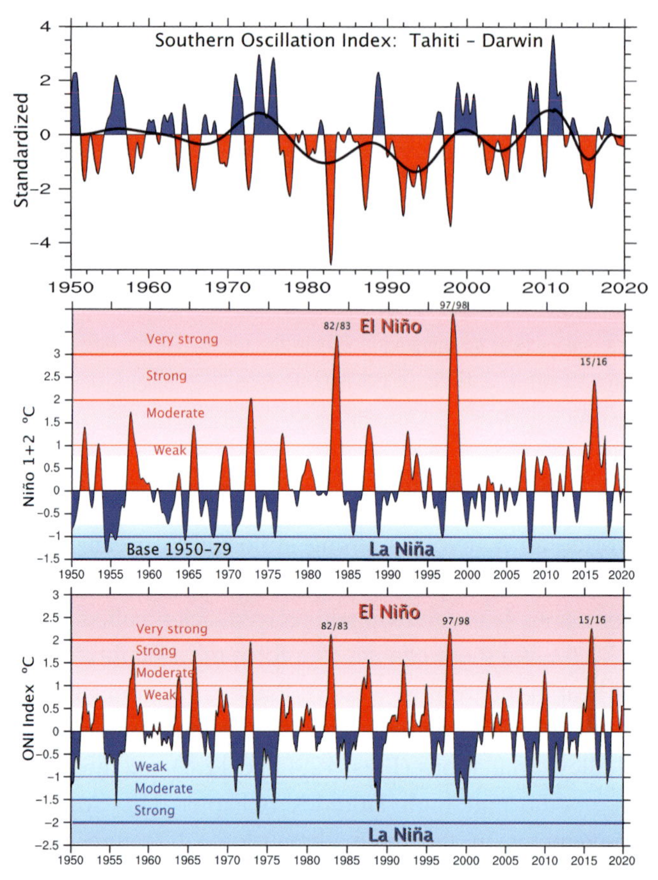

Fig. 12.7 Top: Southern Oscillation Index based on Trenberth (1984). Anomalies of monthly sea-level pressure at Tahiti normalized minus those at Darwin normalized, and then with the result normalized. An 11-term filter has been applied to the monthly data to show decadal variations. Red signifies El Niño conditions. Lower: Time series of SST anomalies from 1950 through 2020 relative to the means of 1950–79 for (middle) the traditional El Niño area Niño 1+2, and (bottom) the region most involved in ENSO Niño 3.4, labeled ONI (Oceanic Niño Index).

El Niño events clearly identifiable since 1950 in Fig. 12.7 occurred in 1951, 1953, 1957–58, 1963, 1965, 1969, 1972–73, 1976–77, 1982–83, 1986–87, 1990–95, 1997–98, 2002–03, 2004–05, 2006–07, 2009–10, and 2015–16. The 1990–95 event might also be considered three modest events after which conditions failed to return to below normal so that they merged together. Worldwide climate anomalies for several seasons have been identified with all of these events.

There have been three "super" El Niños, whereby the main index made it into the "very strong" category: 1982–83, 1997–98 and 2015–16 (Fig. 12.7). The 1997–98 event was the largest on record in terms of SST anomalies, and the

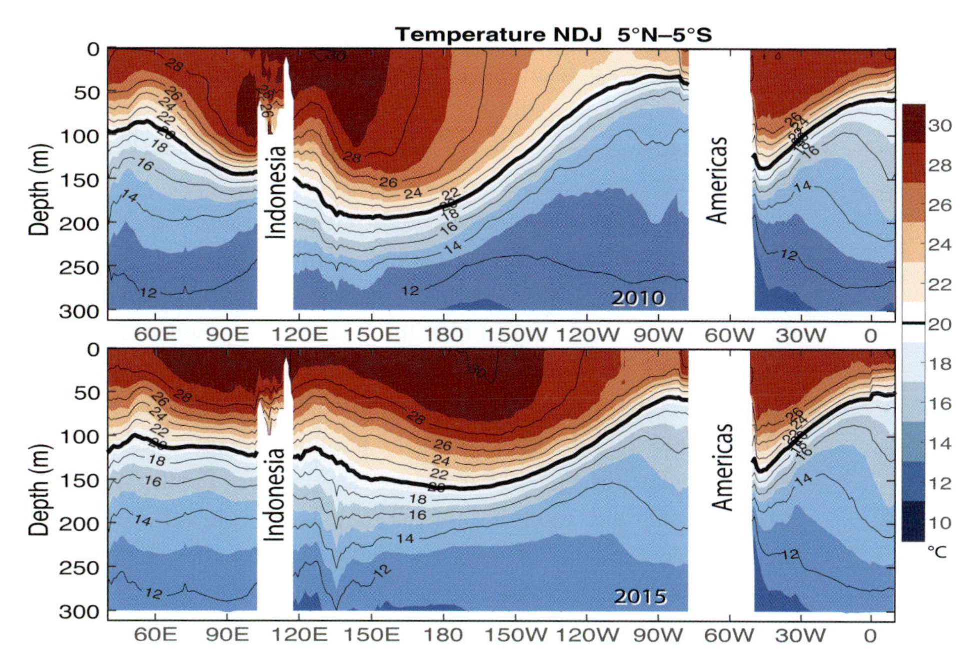

Fig. 12.8 A cross section from 5° N to 5° S along the equator illustrates the temperature structure in °C with depth for a La Niña in 2010 versus an El Niño in 2015, for November–December–January. The thermocline is the temperature gradient centered on about the 20°C isotherm, highlighted by the dark black line. The white spaces denote the land masses. Courtesy L. Cheng; http://159.226.119.60/cheng/

GMST in 1998 was the highest on record for the 20th century, but for the SOI, the El Niño event of 1982–83 still holds the record (Fig. 12.7). There are no "very strong" La Niña events, which also highlights some aspects of the asymmetry between the two phases: La Niña events tend to last longer or be double phased more often. After the 2015–16 event, a very unusual coastal El Niño occurred in the first few months of 2017 (Fig. 12.7, see Niño 1+2), causing devastating stormy weather over northern Chile, Peru, and Colombia.

An example for the tropical oceans (Fig. 12.8) shows how large the changes are with El Niño. Under normal conditions, and even more so with La Niña, strong trade winds pile up warm waters in the western tropical Pacific (Fig. 12.1), with profound effects on the thermocline. During El Niño, the trade winds weaken, which causes the thermocline to become shallower in the west and deeper in the eastern tropical Pacific (Figs. 12.8, 12.5), while sea level falls in the west and rises in the east by as much as 25 cm as warm waters surge eastward along the equator. Equatorial upwelling decreases or ceases, the cold tongue weakens or disappears, and the nutrients for the food chain are substantially reduced. The resulting increase in sea temperatures warms and moistens the overlying air so that convection breaks out, and the convergence zones and associated rainfall move to a new location with resulting changes in the atmospheric circulation (Figs. 12.4, 12.5, 11.1, and 11.2). A further weakening of the surface trade winds completes the positive feedback cycle, known as the Bjerknes feedback (Fig. 13.6), leading to an El Niño event.

Although ENSO has a clear preferred timescale, with a period of 3–7 years, every event is different. El Niño comes in many different "flavors," which has also been referred to as ENSO diversity or complexity. One form of variability arises in the tropical Pacific from just how the SSTs develop, where the warmest water lies, and what SST gradients emerge. The warmest water determines roughly where the main convection occurs while the gradients determine the anomalous trade wind strength, and thus the strength of the anomalous convection. The strong negative SST anomalies for 1997–98 in the tropical western Pacific were largely missing in 2015–16, with the result that the anomalous teleconnections were not as strong in the later event in spite of similarly high SST anomalies (Fig. 12.7). As well as the different flavors in the tropical Pacific, the teleconnections to higher latitudes occur through an ever-changing chaotic atmosphere involving vigorous baroclinic weather systems, jet streams, and frontal systems. Even with identical SST anomalies, the response can be quite different, just by chance, over a particular season, because El Niño events do not last long enough to sample all weather variability and thus provide stable statistics for each event.

The traditional ENSO indices (Fig. 12.7) identify with the ocean (e.g., ONI) or atmosphere (SOI), and accordingly these have many applications. In particular, SSTs and the ocean conditions can be specified to force an atmospheric model to determine the response; or the atmospheric winds can be used to drive an ocean model to determine the response there. There have been several discussions and proposals for more complete indices that vary with time of year and involve more variables. A Multivariate ENSO Index (MEI) initially included information from sealevel pressure, zonal and meridional surface wind components, SST, surface air temperature, and cloudiness, and varied with month of the year, but the availability, quality, and homogeneity of the records of several of these other variables are questionable as one goes back in time, and a reduced MEI was formulated using SLP and SST alone. Outgoing longwave radiation has also been used. As the nonlinearity of precipitation associated with SSTs is not well captured by anomalies alone, a new index based upon the longitude of tropical deep convection from 5° N to 5° S, the ENSO Longitude Index, has been proposed and allows the diversity of ENSO to be captured with a single index that accounts for nonlinearity of convective response.

The character of recent ENSO events seems to be different than before 1976. Several events have had a lot more activity in the central Pacific, while the

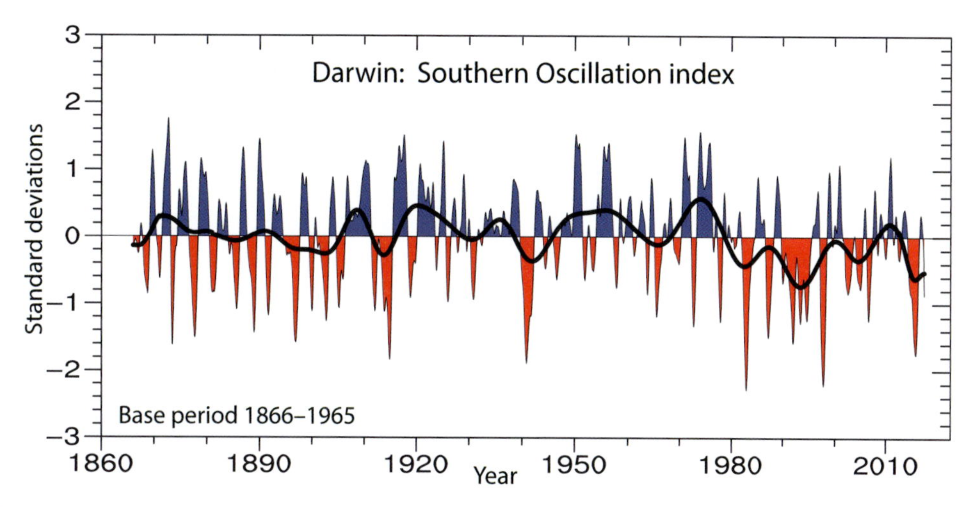

Fig. 12.9 The Southern Oscillation Index from the 1860s to illustrate El Niño events in red and La Niña events in blue along with a decadal filter (black) to show decadal variability and the trend for bigger and stronger El Niños after 1976.

traditional South American coastal temperature anomaly is not as affected, and instead an anomaly arises in the central Pacific (Nino 3.4), called El Niño Modoki. Modoki is Japanese for "similar, but different" and is related to the role of gradients across the Pacific described by the TNI index, and the so-called Central Pacific (CP) El Niño events as compared with the more traditional East Pacific (EP) events. However, this pattern is also part of the evolution of ENSO, and it is unclear how much this is simply part of the continual variety: the different flavors of El Niño, or whether it might be part of a climate change signal. Longer time series of the main indices (Fig. 12.9) show that although the SO has a typical period of 3–7 years, the strength of the oscillation has varied considerably. There were strong variations from the 1880s to the 1920s and after about 1950, but weaker variations in between, with the exception of the major 1939–41 event. A remarkable feature of the SOI is the decadal and longer-term variations in recent years, which is lacking from earlier periods. Again, this raises the specter of climate change.

12.4 ENSO and Hurricanes

The higher SSTs and change in SST patterns have profound effects on the distribution of tropical storms and hurricanes, typhoons, and cyclones (Fig. 7.12). The latter three terms are synonymous but apply to different regions.

Fig. 12.10 In El Niño events, the warming in the central Pacific creates more favorable conditions for tropical storms. The southern summer of 2015–16 was especially active, here showing one example for January 19, 2016, of Victor, which was at peak intensity as a Category 3 severe tropical cyclone just east of Tonga, but was followed by the strongest storm on record for the southern hemisphere, a Category 5 severe tropical cyclone (hurricane), Winston, that barreled into the Fijian islands on February 20, 2016, with 44 fatalities. IR image from GOES West, https://www.star.nesdis.noaa.gov/GOES/fulldisk.php?sat=G17#, courtesy NOAA

As detailed in Section 10.7, with higher SSTs and ocean heat content, "activity" increases but can be manifested as increases in numbers, size, duration, and intensity, as well as heavy rains and risk of flooding.

In El Niño, activity increases in the central and eastern Pacific Ocean, including the central Pacific in the southern hemisphere, but decreases in the Pacific Northwest, Australian region, and Atlantic. In La Niña, the activity increases in the Pacific Northwest and Atlantic. In 2015, as part of the 2015–16 El Niño, severe tropical cyclone Pam ripped through Vanuatu in the South Pacific in March, causing enormous damage, enabled by warm waters from the El Niño. Less than a year later, as part of extra activity in the southern hemisphere (Fig. 12.10), the strongest tropical cyclone on record in the southern hemisphere (Winston) severely damaged Fiji. The 2015 northern hurricane season featured by far the greatest number of category 4 and 5 hurricanes/typhoons on record (25 versus the previous record of 18).

Because of the enhanced activity in the Pacific and the changes in atmospheric circulation throughout the tropics (Fig. 12.4), there is a decrease in the number of tropical storms and hurricanes in the tropical Atlantic during El Niño (Fig. 7.12). Good examples are 1997 and 2015, with 1997 one of the quietest Atlantic hurricane seasons on record. In contrast, the El Niño events of 1990–95, 1997–98, and 2015–16 terminated before the 1995, 1998, and 2017 hurricane seasons, which unleashed storms and placed those seasons among the most active on record in the Atlantic.

12.5 Movement of Heat and Energy with ENSO

There are many theories about the ENSO phenomenon, but one fundamental question is whether ENSO plays a vital role in the climate system beyond the obvious weather regime changes. Increasingly, it appears that the answer is yes. As noted above, heat builds up in the tropical western Pacific during normal and La Niña conditions, and during El Niño events heat in the ocean first spreads across the Pacific, and then along the Americas outside of the tropics, all the while causing major changes in convection such that there is a major loss of ocean heat through evaporative cooling to the atmosphere. The increased atmospheric water vapor fuels storms and has a strong greenhouse effect. In turn this leads to heating of the atmosphere by latent heat from rainfall, and there is a mini-global warming. The latter stages of an El Niño event produce the highest GMSTs. The result is a recharge and then a discharge of energy during the ENSO cycle, and ENSO effectively acts as a relief valve for the Pacific Ocean, keeping it cooler than it otherwise would be. Consequently, there has been some speculation about whether ENSO events could become more frequent and/or bigger.

The actual energy budget is quite involved. While there is a strong decrease in ocean heat content in the tropical Pacific Ocean during El Niño, mainly through enhanced air–sea heat fluxes into the atmosphere driven by high SSTs, there is also a lot of redistribution of heat both laterally and vertically. The exchange of heat into the atmosphere is largely from evaporation that cools the ocean and provides moisture for increased rainfalls in the central Pacific. There is also a considerable discharge of heat from the equatorial region of the tropical Pacific (5° N–20° S) during and after El Niño into the region north of about 5° N (north of the ITCZ) through ocean heat transport. There are strong zonal transports, with cooling in the west and warming in the east Pacific in the lead-up to El Niño via eastward-propagating waves in the ocean, as well as slower westward propagation into the Indian Ocean. In this process, there is a huge exchange of heat in the vertical across about 100 m depth, so that temperature changes from 0 to 100 m are opposite to those from 100 to 300 m. Again, this is related to ocean currents and internal waves that change the depth of the thermocline. The east tropical

Pacific warming is mainly in the upper 100 m, while in the 100–300 m layer the cooling is mainly in the Warm Pool and tropical west Pacific, as the thermocline shoals there (e.g., Fig. 12.8).

Energy compensations with the atmosphere also play a major role, as moisture and heat losses in the tropical Pacific Ocean are partly compensated by warming in the tropical Atlantic and Indian Oceans via teleconnections. For the Atlantic this comes about through changes in the Walker Circulation (Fig. 12.4), sunny skies and lighter winds (less evaporative losses) during El Niño, and this also occurs into the tropical Indian Ocean, but a change in the Indonesian ThroughFlow also contributes there. An atmospheric bridge also exists to the extratropics to the North and South Pacific, although the signal is strongest into the South Pacific. Globally, all of these exchanges add up to a net loss of ocean heat during and after El Niño because of air–sea exchanges of 0.3 W m^{-2} (0.15 PW) globally. This is what leads to the atmosphere becoming warmer and the GMST higher following El Niño (Fig. 2.4).

Climate models have great difficulty in correctly simulating the ITCZ, SPCZ, and cold tongue, and accordingly ENSO is often not simulated very well. As ENSO plays a key role in redistributing heat and thereby provides a mechanism for abating global warming, at least regionally, the shortcomings of climate models imply that they may overestimate warming.

12.6 Impacts

Changes associated with ENSO produce large variations in weather and climate around the world from year to year and often these have a profound impact on humanity because of droughts, floods, heatwaves and other changes which can severely disrupt agriculture, fisheries, the environment, health, energy demand, and air quality, and also change the risks of fire. The normal upwelling of cold nutrient-rich and CO_2-rich waters in the tropical Pacific is suppressed during El Niño. Normally, the presence of nutrients and sunlight fosters development of phytoplankton and zooplankton to the benefit of many fish species. Therefore, El Niño–induced changes in oceanic conditions can have disastrous consequences for fish and seabirds and thus for the fishing and guano industries, for example, along the South American coast. Around the top end of Australia in the 2015–16 El Niño event, prawns were adversely affected, and the catch of red-leg banana prawns was the lowest ever. Other marine creatures may benefit so that unexpected harvests of shrimp or scallops occur in some places. Rainfall over Peru and Ecuador can transform barren desert into lush growth and benefit some crops, but can also be accompanied by swarms of grasshoppers, and increases in the populations of toads and insects. Human health is affected by mosquito-borne diseases such as malaria, dengue, and viral encephalitis, and by water-borne diseases such as cholera. In Africa, Rift Valley fever outbreaks may occur. Economic impacts can be large, with losses typically overshadowing gains.

ENSO in both phases is the largest cause of drought around the world, causing loss of agricultural production, widespread human suffering and loss of life. El Niño also influences the incidence of fires, especially in Australia, Indonesia, and Brazil. With the fires come air-quality and respiratory problems in adjacent areas up to 1000 km distant. Meanwhile, flooding occurs in Peru and Ecuador, and also in Chile, and coastal fisheries are disrupted. Very wet conditions can occur in California and the Southeast USA, but not always.

Extremes of the hydrological cycle such as floods and droughts are common with ENSO and are apt to be enhanced with global warming. For example, the modest 2002–03 El Niño was associated with a drought in Australia, made much worse by record-breaking heat. A strong La Niña event took place in 2007–08, contributing to 2008 being the coolest year since the turn of the 21st century. 2016 is by far the warmest year on record to date, in part because of the El Niño event (Fig. 2.4). All of the impacts of El Niño are exacerbated by global warming.

References and Further Reading

Cheng, L., K. E. Trenberth, J. Fasullo, M. Mayer, M. Balmaseda, and J. Zhu, 2019: Evolution of ocean heat content related to ENSO. *Journal of Climate,* **32,** 3529–3556, doi: 10.1175/JCLI-D-18-0607.1.

Trenberth, K. E., 1984: Signal versus noise in the Southern Oscillation. *Monthly Weather Review*, **112**, 326–332

Trenberth, K. E., 1994: The different flavors of El Niño. 18th Annual Climate Diagnostics Workshop, November 1–5, 1993, Boulder, CO, 50–53.

Trenberth, K. E., 1997: The definition of El Niño. *Bulletin of the American Meteorological Society*, **78**, 2771–2777.

Trenberth, K. E., G. W. Branstator, D. Karoly, A. Kumar, N-C. Lau, and C. Ropelewski, 1998: Progress during TOGA in understanding and modeling global teleconnections associated with tropical sea surface temperatures. *Journal of Geophysical Research*, **103,** 14291–14324.

Trenberth, K. E., J. M. Caron, D. P. Stepaniak, and S. Worley, 2002: Evolution of El Niño Southern Oscillation and global atmospheric surface temperatures. *Journal of Geophysical Research*, **107**(D8), 4065. doi: 10.1029/2000JD000298.

Trenberth, K. E., P. D. Jones, P. Ambenje, et al., 2007: *Observations: Surface and Atmospheric Climate Change*. In: Solomon, S., D. Qin, M. Manning, et al., eds., *Climate Change 2007. The Physical Science Basis*. Fourth Assessment Report of the Intergovernmental Panel on Climate Change. Cambridge: Cambridge University Press, 235–336.

Wolter, K., and M. S. Timlin, 2011: El Niño/Southern Oscillation behaviour since 1871 as diagnosed in an extended multivariate ENSO index (MEI.ext). *International Journal of Climatology*, **31**, 1074–1087. doi: 10.1002/joc.2336.

13 Feedbacks and Climate Sensitivity

Many important feedback processes that occur in the climate system are described. The concepts of climate sensitivity are outlined and related to model projections of climate change.

13.1 Amplifying Factors in Climate

The climate is changing. The main reason is because of human-induced changes in atmospheric composition which produce warming from increased greenhouse gases. This is referred to as a *forcing* of the climate system. There are many other forcings, both natural and anthropogenic. The issue then is to determine the consequences in terms of the change in climate and its impacts. There is a direct response to just about any forcing, and in some cases that is the answer we seek. But in many or most cases, it is not so simple. Rather, the initial change provokes other responses, especially in the atmospheric and ocean circulation, that in turn cause other changes to occur. If the response amplifies the original change, then it is referred to as a *positive feedback.* Whereas if the response offsets and reduces the outcome, then it is a *negative feedback*. The size of some effects is quantified in Section 13.5.

13.2 Feedbacks

13.2.1 Planck Radiation Feedback

The best example of negative feedback is from radiation. If warming occurs, then temperatures rise. The ultimate rise in temperature in the absence of feedbacks is determined by the amount of heat, and the mass and specific heat of the body. But all bodies radiate, as seen in Chapters 2– 4, and a pure black body radiates with the fourth power of absolute temperature. Therefore, as the body heats up, it radiates more and loses some of the heat. This is referred to as the Planck radiation feedback and it is quite strongly negative.

13.2.2 Ice–Albedo Feedback

The simplest idealized example of a positive feedback is that involving snow and ice (Fig. 13.1). Surfaces covered in snow or ice are bright white and highly reflective of solar radiation. This is to say, they have a high albedo. As the temperatures increase, and snow and ice melt, leaving behind much darker ocean or land, more radiation is absorbed instead of reflected, and the temperature rises even more. Yet another feedback with snow is from dust and aerosols, which may be deposited on snow and make it darker, and hence more likely to melt as greater sunshine is absorbed.

This example is very useful and certainly applies in nature, but it is nevertheless oversimplified, because of other consequences. For instance, increased open water from sea-ice melt may lead to more evaporation and atmospheric water vapor, thereby increasing fog and low cloud amount, offsetting the change in surface albedo. Another example is that as snow and ice are removed on land, permafrost

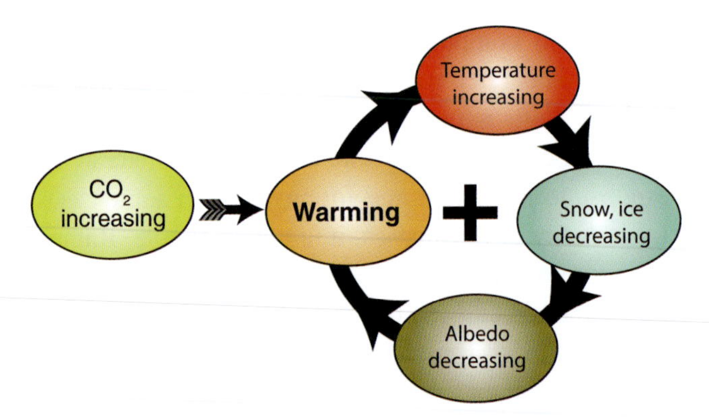

Fig. 13.1 Example of ice and snow–albedo feedback, which is positive. The decrease in albedo means more solar radiation is absorbed, increasing heating.

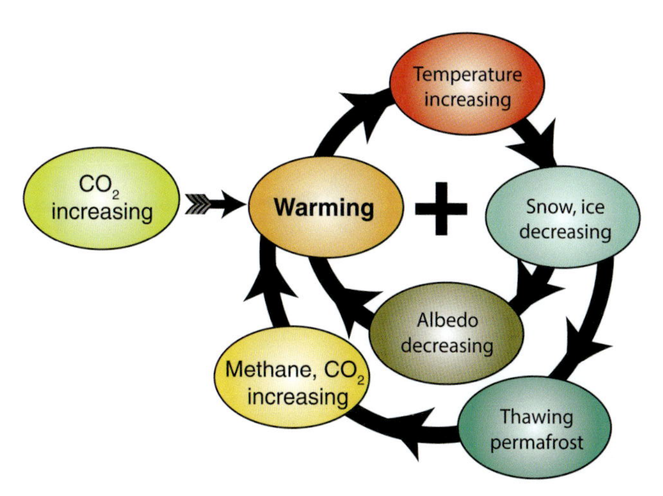

Fig. 13.2 Land–biogeochemical feedback. An increasing positive feedback results on longer timescales on land as permafrost thaws and vegetation decay in the soils is accelerated, releasing methane or carbon dioxide into the atmosphere.

thaws, opening up extensive regions with decaying vegetation embedded in frozen soils. This greatly increases methane emissions from the soils in wet areas, and carbon dioxide emissions in dry areas (Fig. 13.2). This is, again, just an example, because in the ocean, as the warming is promoted, it may affect methane hydrates (clathrates) deposited on the sea floor, and further accelerate methane emissions. Many of these are absorbed in the sea water and do not make it into the atmosphere, but the potential is certainly there for more positive feedbacks.

Both land and ocean biogeochemistry are complex and involve numerous possible feedbacks related to changes in land use, vegetation extent and type, all of which affect albedo and evapotranspiration. Surface roughness may also be affected. On land, heatwaves and drought may lead to wildfires and demise of vegetation, which in turn may affect atmospheric circulation and climate. Moreover, the carbon cycles of atmosphere–ocean or atmosphere–land exchanges of carbon, carbon dioxide or methane, and so forth, can play a role.

13.2.3 Water Vapor Feedback

Water vapor feedback relates to the strong tendency for water vapor to increase as temperatures increase because the relative humidity tends to remain about the same. This is because the water-holding capacity of the atmosphere increases with temperatures, as given by the Clausius–Clapeyron equation (Chapter 10), and therefore water vapor feedback is distinctly positive (Fig. 13.3). Some of the global warming goes into evaporation, and more moisture cycles through the

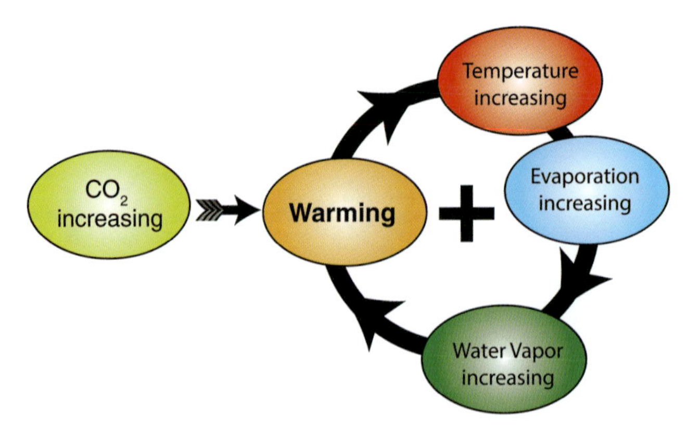

Fig. 13.3 Water vapor feedback, whereby increased heating promotes increased evaporation and higher atmospheric temperatures that lead to increased atmospheric water vapor, which is a powerful greenhouse gas. As a result, the warming is amplified and this constitutes a strong positive feedback.

atmosphere as part of the hydrological cycle. The result is an increase in water vapor in the atmosphere, which is observed to be happening, and because water vapor is a powerful greenhouse gas, it amplifies the original warming. However, the average lifetime of a water vapor molecule in the atmosphere is just 9 days as it is quickly involved in the hydrological cycle. This is why water vapor is primarily a feedback, rather than a forcing of climate variability and change. Irrigation and many other uses of water by humans can certainly act as a forcing, but that is not the primary role.

This process has been strongly evident in climate models and is also clearly observed and documented in nature. Empirical relationships automatically fold in other dependencies. Water vapor increases by 7% per degree Celsius warming of the atmosphere as long as moisture is available. In reality these processes appear to apply quite well to the total column water vapor, and, although the main increases may be at lower altitudes in the atmosphere, while the upper tropospheric water vapor is more critically important for the net greenhouse effect, the feedback clearly arises.

13.2.4 Cloud Feedback

Relationships with clouds are complex. Fig. 13.3 is somewhat simplified, because increased water vapor can also lead to increased cloud, although it is the relative humidity rather than the total moisture amount that matters. Moreover, the extra moisture can promote instability in the vertical in the atmosphere and thereby increase convection. The result can bring in other feedbacks (Fig. 13.4). This figure is designed to illustrate possible cloud feedbacks. As noted in Chapter 2, clouds have dual and conflicting effects on climate.

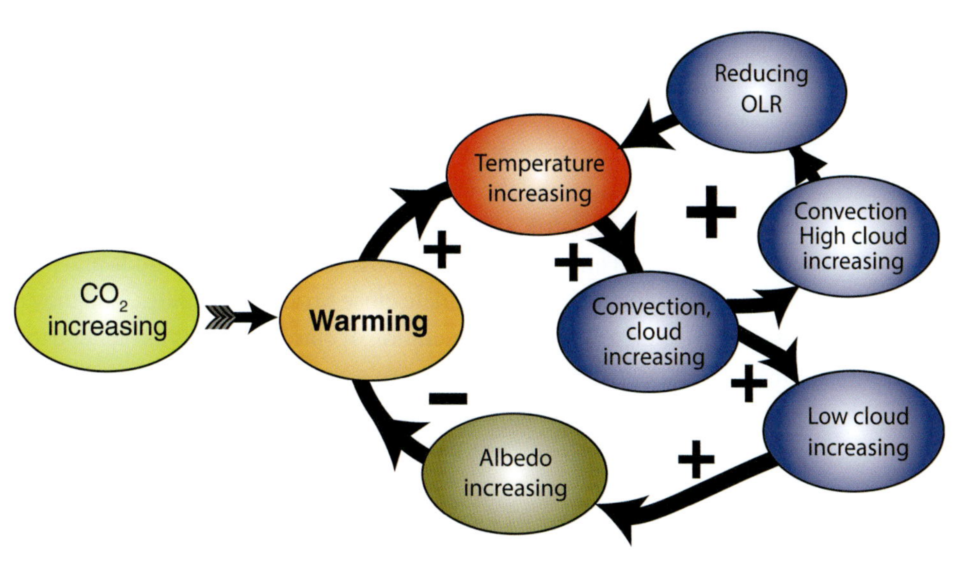

Fig. 13.4 Cloud feedback. Clouds have a greenhouse effect that can result in a positive feedback (top right loop) but they also reflect solar radiation and can have a negative feedback (as for the bottom loop). OLR is outgoing longwave radiation. Here the increase in albedo reflects solar radiation and reduces warming. The net outcome depends on how clouds change in amount and height.

Clouds affect the climate by reflecting incoming solar radiation back to space, which tends to cool the climate, and by trapping outgoing infrared radiation, which tends to warm the climate. Clouds cool the planet on average at present, but the issue is how they change as the climate changes. Increases in low cloud cool the planet because they are highly reflective and yet they are relatively warm in their emissions to space. In contrast, high clouds have very cold cloud tops that radiate much less to space. Increasing high cloud warms the planet, but increasing low cloud cools the planet, provided that sunshine exists (not applicable to the high latitudes in winter or at night). In the tropics, there is a remarkable cancellation between these two effects for high clouds so that they have very little signature in the net radiation (see Fig. 5.9).

The short observational record is dominated by ENSO variations. During El Niño (Chapter 12), the warming of the tropical Pacific Ocean occurs in the east, where there is normally upwelling on the equator and along the coast of the Americas, that in turn promotes the development of stratocumulus clouds. The clouds are extensive, bright but shallow and act to cool the planet. In El Niño, when these clouds burn off as the SSTs rise, more sunshine reaches the surface and further increases SSTs, leading to a positive feedback. This process is, however, linked to the changes in the Walker Circulation across the Pacific (Chapter 12), and accordingly there is also a dynamic atmospheric circulation component, because the subsidence in the atmosphere above the surface is weakened and so is the temperature inversion layer near the 850 hPa level that promotes the existence of the stratocumulus clouds. This aspect has often not been considered in some studies, which then have concluded that clouds produce a positive feedback.

Depending on how the complexities of climate change alter the mix of low and high clouds, clouds can have either a positive or negative feedback effect, depending on their extent, altitude, and the size of the water droplets, and their diurnal and seasonal variability. Atmospheric aerosols compound the issues of how clouds behave, both from natural and anthropogenic aerosols and pollution. Clouds and precipitation have been difficult for models to simulate close to observed values, and they differ from one model to the next. The latest generation of climate models has tended toward a positive feedback in part because of the way aerosols are parameterized. Observations are confounded by atmospheric circulation changes, such as subsiding air. For net cloud feedback, the uncertainties remain large.

13.2.5 Soil Moisture Feedback

An extra complication to the changes associated with clouds is the effects of rainfall on the land surface. While cloud feedback can go either way, any rainfall that is associated with clouds, especially convection, moistens the soil, thereby triggering evaporative cooling. This is a negative feedback as heat that might have gone into raising temperature instead re-evaporates the surface moisture (Fig. 13.5). The net result depends a lot on the nature of surface vegetation and the degree to which the surface moisture infiltrates the soil and flows into streams. Nevertheless, the hydrological cycle can play a major role in the net outcome. Clearly the complexity is a challenge in order to correctly model the result.

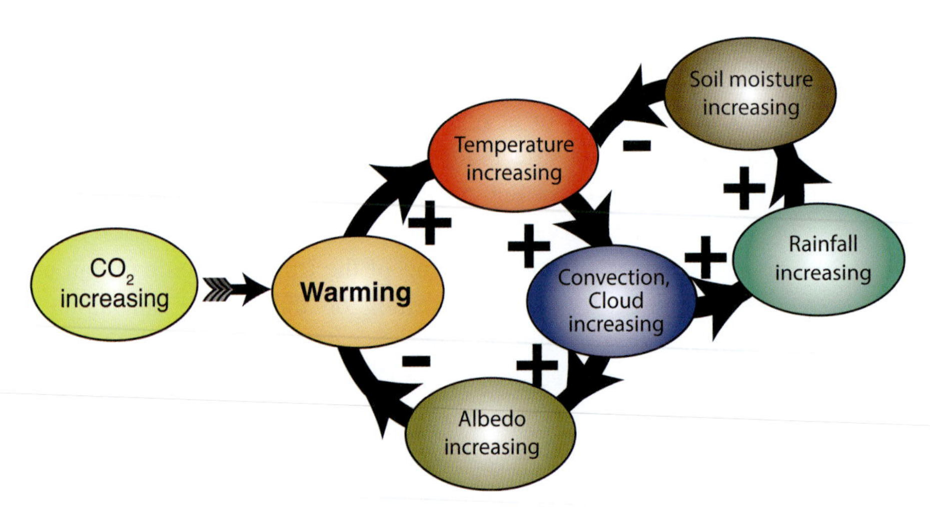

Fig. 13.5 Soil moisture feedback can be negative in terms of the effects on surface temperatures on land. The result is that dry spells and less soil moisture promote heatwaves, while wet spells are cool.

13.2.6 Bjerknes Feedback

In Chapter 12 dealing with ENSO, the Bjerknes feedback (Fig. 13.6) was described. As an El Niño develops, perhaps initiated by a chance disturbance in the atmosphere over the tropical Pacific Ocean, SST warming occurs in the eastern and central Pacific, reducing the temperature gradient at the surface and the associated atmospheric pressure gradient that drives the easterly trade winds. The Walker Circulation weakens and, as the trade winds relax, more warm water surges eastwards in the upper ocean and upwelling of cold waters in the equatorial region of the eastern Pacific is abated, increasing the warming, as a strong positive feedback.

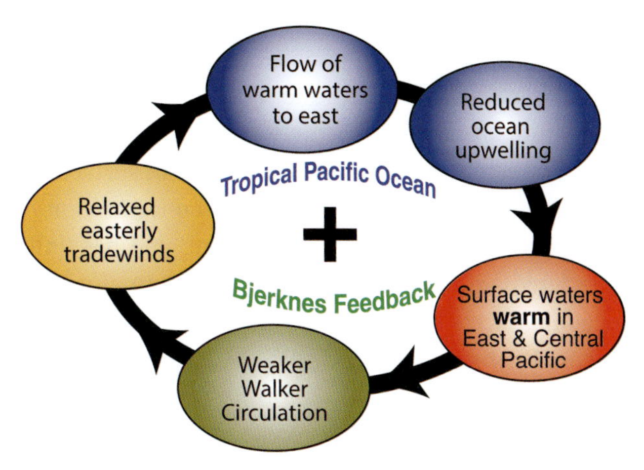

Fig. 13.6 In the tropical Pacific Ocean, the Bjerknes feedback is a positive feedback of warming in the central and eastern Pacific, that weakens the trade winds and upwelling, further increasing the movement of waters in the top 150 m eastwards.

There are many other interactions within the climate system that produce feedback effects. The above examples are intended to be illustrative, and ideally the full set of complex interactions is put into a climate model as processes, and in this way the net result may be determined. However, it then becomes very important to include all related feedbacks, and this has been difficult to do for both aerosols and the carbon cycle. Some processes are well enough understood to be included, but others, although known to be in play, may not be known well enough. For example, aerosols block the Sun and most have a cooling effect (see Section 10.4). However, they also have indirect effects on clouds and precipitation. With more aerosols present, the cloud liquid water is redistributed over more particles that serve as cloud condensation nuclei, making more, smaller, cloud droplets and a brighter cloud. But what does this do to the lifetime and thickness of the cloud, and the onset of precipitation? Because all of these processes involve microphysics, they have to be parameterized in climate models, and although progress has been made, the processes are not yet completely represented or fully understood.

13.3 Climate Sensitivity Defined

Climate sensitivity is the term used by climate scientists to express the relationship between the human-caused emissions that add to Earth's greenhouse effect – carbon dioxide and a variety of other greenhouse gases – and the temperature changes that result from these emissions. It is an idealized value but widely used to depict the sensitivity of individual climate models to a climate perturbation. In that regard, it has proven to be quite useful, but in other ways it is rather flawed as a metric, because it is an abstract quantity and cannot really be observed.

Traditionally the most widely used value is the *Equilibrium Climate Sensitivity* (ECS), which is that of the net global mean surface temperature (GMST) change in response to doubling carbon dioxide concentrations from pre-industrial values of 280 ppmv to 560 ppmv at equilibrium, and has units of °C. Usually the changes considered are those of the physical system alone, and do not include changes in vegetation, ice sheets (Antarctica and Greenland), and other similar factors that are specified as fixed. With no feedbacks at all, the increase in GMST to doubling CO_2 would be about 1°C, and the primary well-understood amplifications come from water vapor feedback and ice–albedo feedback. For many years, the assessment of ECS has been that it lies between 1.5 and 4.5°C, and the central value of 3°C is regarded as reasonable.

ECS can be calculated from observations or models as

$$\mathrm{ECS} = -F_{2\times \mathrm{CO}_2}/\lambda \tag{13.1}$$

where $F_{2\times \mathrm{CO}_2}$ is the radiative forcing at doubling of carbon dioxide concentrations, and λ, the feedback parameter in W m^{-2}/°C, represents the top-of-atmosphere (TOA) radiative flux change per degree of surface temperature, as further explained below; see Eq. (13.2). Some models have been run for thousands of years to determine the equilibrium change in GMST, but more commonly the value is approximated based upon model runs of a few hundred years. The deep ocean takes millennia to reach true equilibrium; see Fig. 13.7 for an example based on a step function increase in carbon dioxide.

In the short run, for so-called transient model runs, the actual warming is less than suggested by the climate sensitivity due to the thermal inertia of the ocean, and it may take some time after a doubling of the concentration is reached, presuming no further forcing change occurs, before the climate reaches a new equilibrium. A "*transient climate response*" (TCR) has also been defined as the

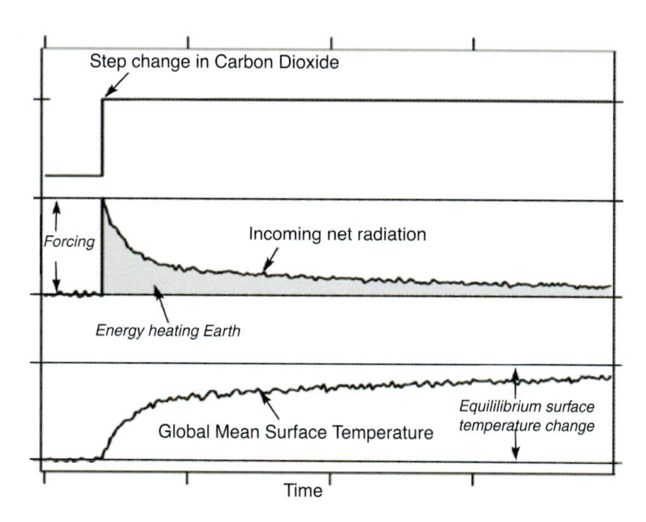

Fig. 13.7 An illustration of a climate model time series to determine the ECS. Following a step doubling of carbon dioxide concentrations, there is an instantaneous increase in the net radiation forcing, but this energy imbalance declines over time as Earth warms up, as given by the GMST value at the bottom. Over time the latter approaches an equilibrium value, the ECS, as the energy imbalance decays to zero.

actual value for GMST increase at the time of doubling CO_2 with increases of 1% per year, which occurs after 70 years. TCR is less than ECS, but the values in models are correlated.

13.4 Determining Climate Sensitivity

The global mean surface temperature (GMST) response to perturbations to Earth's energy budget is traditionally approximated by the following linear equation:

$$\Delta R_T = \Delta F + \lambda \Delta T + \varepsilon \tag{13.2}$$

where Δ refers to a change, R_T is the TOA net radiation downward, ΔF is the effective net change in radiative forcing perturbation, $\lambda \Delta T$ is the change in net radiation due to a temperature change ΔT, and ε is the leftover noise and internal variability. ΔR_T is Earth's Energy Imbalance. λ is the feedback parameter, and, in equilibrium, when $\Delta R_T = 0$, then $\Delta T = -\Delta F/\lambda$. User beware, sometimes λ is defined with the opposite sign. Here T is commonly the GMST, and hence this simple model does not explicitly include the vertical structure of the atmosphere. λ can be decomposed into the sum of individual feedback values.

The effective radiative forcing (ΔF; units: W m^{-2}) quantifies the change in the net TOA radiative budget of Earth system due to an imposed perturbation (e.g., change in carbon dioxide concentration, change in incoming solar radiation). It is expressed as a change in net downward radiative flux at the TOA, R_T, following the adjustments in both tropospheric and stratospheric temperatures, water vapor, clouds, and some surface properties, such as surface albedo, prior to any GMST change. These adjustments affect the energy budget both at the TOA and at the surface. Accounting for such processes gives an estimate of radiative forcing that is more representative of the climate change response associated with forcing agents than simple stratospheric temperature–adjusted radiative forcing, commonly used in the past.

The forcings that should be considered in determining λ in Eq. (13.2) include the small but steady increase in greenhouse gases, the 11-year solar cycle which has an amplitude of order 0.1%, volcanic aerosols from multiple eruptions albeit small in recent times, and tropospheric aerosols. Tropospheric aerosols do not appear to greatly affect sensitivity results. For the time period after 2000, in the absence of the 1991 Mt. Pinatubo eruption effects, which had decayed by then, the impacts of the forcing terms should be modest.

For doubled carbon dioxide levels, $\Delta F = 3.7$ W m^{-2}. As noted above, the main negative feedback is the Planck response from outgoing longwave radiation (see Chapter 3), and for an Earth without climate feedbacks other than the Planck response, λ is expected to have a value of about -3.2 W m^{-2} K^{-1}; hence values of λ less than (greater than) this correspond to a climate with a globally negative (positive) feedback.

Greenhouse gases from industrial activities are estimated to have trapped roughly 3 watts per square meter so far. However, the actual changes in radiative forcing involve much more than those associated with changes in greenhouse gases. Changes in low-lying clouds from all industrial pollution sources are thought to have blocked about 1 watt per square meter of energy globally, and thus cloud changes caused by industrial pollution have produced a global cooling effect that is about one-third as strong as the warming from increased greenhouse gases.

13.5 Estimates of Climate Sensitivity

It is difficult to determine feedbacks from observations, and even more difficult to determine the climate sensitivity. Complications arise from regional forcings such as from aerosols and ozone changes, from nonlinearities, and lack of uniform responses on various timescales. Regional forcings or responses necessarily drive temperature and pressure gradients and thus change the atmospheric circulation that compounds the regional responses through disparate changes in cloud and precipitation, for example. Timescales involved in the various responses and adjustments to a change can vary a lot.

To try to make estimates of climate sensitivity using the observational record, key variables needed are the direct observations of surface temperatures and TOA radiation, as well as estimates of radiative forcings. Greenhouse gas concentrations, led by carbon dioxide, continue to rise, and increase radiative forcing at a rate of about 0.3 W m^{-2} per decade. Tropospheric aerosols arising from human activities also play a role that is mainly important regionally, and may compensate for some warming of greenhouse gases.

Warming since the pre-industrial period is measured to be around 1°C with small uncertainty. Together with estimates of Earth's Energy Imbalance (Chapter 14) and the global radiative forcing that has driven the observed warming, the instrumental temperature record enables global energy budget estimates to be used to make estimates of ECS and TCR. Observation-based studies have confirmed that water vapor increases in amount in the atmosphere as it warms, with correlations of 0.7 between monthly anomalies of total column water vapor and average tropospheric temperature. In turn, observations show that global R_T is negatively correlated with water vapor, although the correlation is stronger with OLR alone, −0.5, than for the total R_T, highlighting the positive feedback effects of water vapor. Accounting for clouds and vertical structure of water vapor can increase the relationship somewhat.

A major issue with the simple form of the energy balance expressed in Eq. (13.2) is that it does not explicitly deal with the three-dimensional structure of the atmosphere. Any influences of vertical structure, such as through changes in the vertical temperature profile with height, occur only through the feedback parameter. Yet OLR emanates from various levels in the atmosphere. As seen in Chapter 3, the main emissions to space occur from the middle of the atmosphere around the 500 hPa level (5.5 km altitude), not directly from the surface. Accordingly, it is quite possible to include aspects of the full vertical structure of temperatures in the atmosphere and examine relationships with the TOA radiation, for instance by replacing the GMST in Eq. (13.2) with the 500 hPa value or the tropospheric average value of temperature. This thereby includes the effects of the intervening atmosphere above the surface and the radiative fluxes at TOA. Of course, progression to higher in the atmosphere means that there are fewer greenhouse gases and especially water vapor to provide feedbacks, and thus it is expected that values of the feedback parameter are more negative than at the surface and somewhat closer to the black-body value.

Several recent attempts have been made to estimate λ using observation-based estimated values of ΔF and ΔT using linear regression, which is a straight-line fit to the scatter of points of one variable plotted against the other, and values range from about −1.15 to −2.3 W m^{-2} K^{-1}. The strength of these values depends on how high the correlation coefficient is between the two variables. Unfortunately, the correlation is not very high, and for the GMST it is about −0.2. However, vertically averaged tropospheric temperatures correlate significantly with net incoming radiation R_T at −0.6 globally. This all suggests that the simple model given by Eq. 13.2 is not a very good one.

Moreover, the externally forced signal is quite small for the short period that spans 2000–20, the period of high-quality TOA observations (see Fig. 14.1), and the dominant variations seen in the energy balance arise from internal natural variability. In the latter case, the radiation changes at TOA are largely a response to temperature and other variations rather than a cause of them, thereby limiting their relevance to forced variability.

A few studies have examined how well the ECS can actually be determined from observations, and results using models suggest that the natural variability is a substantial obstacle to estimating the long-term response to forcings. Given a 20-year record, one factor is simply that any 20-year period can give quite different results just by chance because of natural variability, with values 0.3–0.5°C in error. Differences in sea ice and the albedo feedback play a big role. Important, subtler differences arise from the TCR vs. ECS, because some feedbacks only kick in when the changes are already big enough to matter. That effect leads to underestimates of ECS by about 0.2°C.

Overall, the estimates of the feedback parameter in W m^{-2} K^{-1} are about −3.2 for Planck black-body radiation, 1.3 for water vapor (including changes in vertical temperature profile), and 0.4 for surface albedo. Clouds may have a slight positive feedback (positive feedback parameter) but not far from zero, and quite uncertain. The overall value is likely about −1 to −1.5; giving an ECS of about 2.5–3.7°C. Taking all factors and modeling into account, the traditional range of ECS values stands firm except it is now deemed very unlikely for ECS to be below 2°C, so that the assessed range is 2–4.5°C. Indeed, a new World Climate Research Programme assessment in mid-2020 suggests 2.3–4.5°C. The TCR is estimated to lie between 1.5 and 2.4°C, with a central value of 1.9°C.

Yet a number of the latest generation of climate system models have climate sensitivities over 5°C, and these relate to cloud and aerosol effects in models with prognostic aerosol schemes. Climate models show systematic biases in cloud droplet numbers over the Southern Ocean, where human influences are tiny, suggesting that models have a way to go on this issue. In particular, over high latitudes, where mixed-phase (ice and rain) clouds are important, models decrease ice fraction and amplify warming in questionable ways. Based upon paleoclimate simulations, that can be used to test the response of a model to a fairly large change in forcing, such high values do not appear to be viable.

In summary, ECS has proven to have numerous computational difficulties related to timescales of response, along with how different feedbacks are modeled in various components of the climate system. ECS has proven very useful for comparing model results. The relationship of ECS to historical climate change likely depends a great deal on processes connected to ocean heat uptake. This requires a better quantification of OHC changes, as taken up in Chapter 14. Climate models appear to fall short in not having the heat penetrate deep enough quickly enough. Therefore, in models, it also requires much improved ocean uptake of heat and better fidelity of temperature changes at depth in the model ocean versus the observations available. It further requires improved temperature observations through the full depth of the global ocean.

References and Further Reading

Charney, J. G., B. Stevens, I. H. Held, et al., 1979: *Carbon Dioxide and Climate: A Scientific Assessment*. Washington, DC: US National Academy of Sciences.

Dessler, A. E., 2020: Potential problems measuring climate sensitivity from the historical record. *Journal of Climate*, **33**, 2237–2248.

Dessler, A. E., and P. M. Forster, 2018: An estimate of equilibrium climate sensitivity from interannual variability. *Journal of Geophysical Research*, **123**, 8634–8645. doi: 10.1029/2018JD028481.

Dessler, A. E., T. Mauritsen, and B. Stevens, 2018: The influence of internal variability on Earth's energy balance framework and implications for estimating climate sensitivity. *Atmospheric Chemistry and Physics*, **18**, 5147–5155. doi: 10.5194/acp-18-5147-2018.

Trenberth, K. E., Y. Zhang, J. T. Fasullo, and S. Taguchi, 2015: Climate variability and relationships between top-of-atmosphere radiation and temperatures on Earth. *Journal of Geophysical Research*, **120**, 3642–3659. doi: 10.1002/2014JD022887.

Trenberth, K. E., Y. Zhang, and J. T. Fasullo, 2015: Relationships among top-of-atmosphere radiation and atmospheric state variables in observations and CESM. *Journal of Geophysical Research*, **120**, 10,074–10,090. doi: 10.1002/2015JD023381.

Zhu, J., C. J. Poulsen, and B. L. Otto-Bliesner, 2020: High climate sensitivity in CMIP6 model not supported by paleoclimate. *Geophysical Research Letters*, **47**. doi: 10.1038/s41558–020-0764-6.

14 Earth's Energy Imbalance Estimates

The chapter focuses on estimates of the components of Earth's Energy Imbalance (EEI) and how the values have changed in the past two decades. Reservoirs of extra heat, including sea-level rise, are the memory of climate and directly relevant to how the climate is changing.

14.1 Inventory of Increasing Energy post-2000

Global Earth's Energy Imbalance (EEI) is a fundamental metric of climate change, and the local distribution of the imbalance has implications for regional climate variations. It has been a major challenge to rein in the uncertainties and reasonably establish the EEI. Previous chapters have exploited the local energy imbalance estimates to examine resulting heat transports and anomalies, and teleconnections. The atmosphere and oceans are dynamically active, and many phenomena attempt to move heat to where it can best be either lost in some sense, such as by radiation to space, or perhaps sequestered by being buried deep in the ocean. Although there is somewhat constrained effectiveness in many phenomena, such as hurricanes or ENSO, in redistributing heat and keeping regions cooler than they otherwise would be, these aspects are often not replicated well in climate models. Hence it is vital to understand the net heat gain, and how much and where heat is distributed within the Earth system. How much heat might be readily purged and serve as a negative feedback to warming?

14.2 How to Measure EEI

To measure EEI, options include:

(1) *Direct measurements from space* of ASR, OLR, and net R_T are not accurate enough. They appear to be good for relative changes after 2000, as the errors seem to be systematic and relate to geometry of the Sun, Earth and satellite holding the instrument, and under-sampling of the radiation and clouds. The diurnal cycle in clouds and precipitation suggests the potential for biases if not fully sampled. Instead it is necessary to adjust the net TOA radiation in its global mean to match an assumed EEI based on changes in ocean heat content (OHC) and other components, as given below. Nevertheless, these measurements provide reliable changes over time.

(2) *Take an inventory of where all the energy has gone.* Direct measurements of the energy in all components turn out to be the only viable option, but this requires derivatives to get the changes over time. The observations have not been consistent as the observing system has evolved. It is vital to assess the effects of changing instrumentation and sampling in both space and time.

(3) *Use climate models with specified forcings.* The resulting EEI depends on how good the model and the forcings are. As discussed in Chapter 1, climate models have improved enormously, and latest Earth System Models include many processes formerly left out or parameterized (see also Chapter 16). Nevertheless, approximations and assumptions remain, and the reproduction of accurate clouds remains a challenge.

(4) *Use atmospheric reanalyses.* Atmospheric analyses are produced in numerical weather prediction (NWP), and the reanalyses of all data with a state-of-the-art atmospheric model that avoids operational changes are very useful for climate studies. Thus far, the analysis increments introduced to adjust the predicted model climate state to the observations mean that reanalyses do not conserve energy or mass. Nor do they have accurate enough forcings – the specifications of the Sun's irradiance and composition of the atmosphere, including aerosols. These aspects are improving but are not yet adequate.

(5) *Use surface fluxes* (assumes no atmospheric heat capacity). Bulk flux estimates from surface observations and analyses contain large systematic errors. These can be useful locally.

Sidebar 14.1: Time Filters

Many time series are quite noisy, and it is desirable to apply a time filter to damp out the noise and reveal the underlying tendencies. A filter that removes the high frequencies is called a low-pass filter, since the low frequencies and trends pass through unscathed. The desirable characteristics of such filters are (1) they should be easily understood and transparent; (2) they should avoid introducing spurious effects such as ripples and ringing; (3) they should remove the high frequencies; and (4) they should involve as few weighting coefficients as possible, in order to minimize end effects.

Classic low-pass filters widely used have been the binomial set of coefficients that remove $2\Delta t$ fluctuations, where Δt is the sampling interval. The 1–2–1 filter has weights ¼[1, 2, 1] and completely removes $2\Delta t$ fluctuations. Hence it means adding $x(t-1) + 2x(t=0) + x(t+1)$ and dividing by 4, where the x's are numbers in the time series of values involving the time before $(t-1)$ and after $(t+1)$ the value in question $(t=0)$. However, combinations of binomial filters are usually more efficient.

A very effective filter with only five weights 1/12 [1,3,4,3,1] has as its response function (ratio of amplitude after to before) of 0.0 at 2 and $3\Delta t$, 50% at $6\Delta t$, 69% at $8\Delta t$, 79% at $10\Delta t$, 91% at $16\Delta t$, and 100% for zero frequency, and so for yearly data ($\Delta t = 1$) the half-amplitude point is for a 6-year period, and the half-power point (half the squared amplitude) is near 8.4 years. Another useful filter, often used with annual values ($\Delta t = 1$) designed to remove fluctuations on less than decadal timescales has 13 weights 1/576 [1,6,19,42,71,96,106,96,71,42,19,6,1]. Its response function is 0.0 at 2, 3 and $4\Delta t$, 6% at $6\Delta t$, 24% at $8\Delta t$, 41% at $10\Delta t$, 54% at $12\Delta t$, 71% at $16\Delta t$, 81% at $20\Delta t$, and 100% for zero frequency, and so for yearly data the half-amplitude and half-power points are about 12 and 16 years period.

Even if the original series were anomalies as departures from the monthly mean value, there is often a seasonal cycle in variance. For instance, temperatures in the extratropics typically have a lot more variability in winter than in summer. Hence, another commonly used filter is the 12-month running mean, also called a moving average, applied to monthly data. It has uniform weights over 12 points 1/12 [1,1,1,1,1,1,1,1,1,1,1,1]. This has a response function at 24 months period with the filtered series containing 64% of its original value. The filtered amplitude is 0 for a 12-month period and it removes any seasonality in the time series (6, 4, 3, and 2 months). However, it lets through a significant amount of the signal shorter than the window length, and worse, it actually inverts it. At 8 months period the amplitude is about 20% with inverse phase. Hence it has quite large ringing effects. A further disadvantage is that it is not a symmetric filter and displaces values by half a month.

To accommodate both of these points, a 13-point filter is commonly used with weights 1/12 [0.5,1,1,1,1,1,1,1,1,1,1,1,0.5]. This can be thought of as taking two 12-month moving averages shifted by a month and then averaging the results. The latter is equivalent to applying a 2-month moving average. This filter also effectively removes 12-month fluctuations but it is now centered, and at 8 months period the inverse response is now only 7%, so that the ringing is much reduced. This is what is used in this book as a "12-month running mean filter."

Finally, another low-pass filter, widely used and easily understood, is to fit a linear trend to the time series although there is often no physical reason why trends should be linear, especially over long periods. The overall change in the time series is often inferred from the linear trend over the given time period. LOWESS (Locally Weighted Scatterplot Smoothing) is a weighted running mean trend over a certain specified interval and is a useful low-pass filter.

Therefore, it is necessary to determine where all of the extra energy goes in order to complete the inventory. The extra heat or energy:

(1) Warms the land and atmosphere;
(2) Warms the ocean, and heat storage in the ocean causes expansion and raises sea level;
(3) Melts land ice, which raises sea level by adding mass to the ocean;
(4) Melts sea ice and warms melted water;
(5) Evaporates moisture, which feeds rainstorms and clouds, possibly altering albedo and subsequent EEI.

Careful studies have determined limits for these values, and by far the bulk of EEI (over 90%) goes into the ocean as ocean heat storage. Atmospheric energy and its changes are quite well established. Ocean heat content has considerable uncertainty in terms of variability, ice volumes are loosely known although time series exist, but land changes have only ballpark values. Improvements are needed.

Box 14.1: Heat Units

The heat needed to raise sea level through expansion by 1 mm is about 0.12 ZJ (10^{21} J), with 1 mm of sea level equivalent to 362 km^3 volume. 1 ZJ per year is equivalent to 32 TW. Melting one gigatonne (Gt) of ice consumes about 3.34×10^{17} J, and 1 Gt yr^{-1} of ice melt takes about 0.0106 TW; 1 W m^{-2} (globally) is 510 TW.

14.3 TOA Radiation

Earth's energy budget encompasses the major energy flows of relevance for the climate system (Chapter 3). All energy that enters or leaves the climate system does so in the form of radiation at the top-of-atmosphere (TOA). The TOA energy budget is determined by the amount of incoming net absorbed shortwave (solar) radiation (allowing for the reflection) (ASR) and the outgoing longwave (thermal) radiation (OLR). To the extent that the TOA solar irradiance is a function of season and latitude, anomalies in ASR are simply opposite to those in reflected shortwave radiation, although changes in irradiance such as with the 11-year solar cycle alter this. The net TOA radiation downwards R_T = ASR − OLR. For a steady-state climate, the outgoing and incoming radiative components are essentially in balance in the long-term global mean, although significant fluctuations around this balanced state arise through internal climate variability.

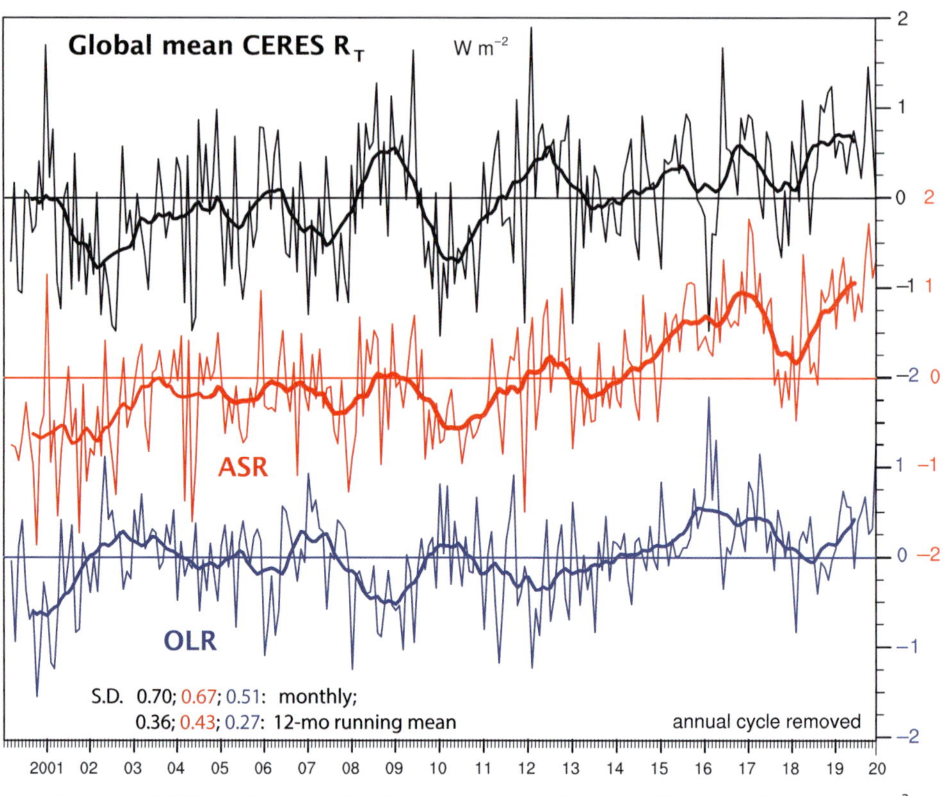

Fig. 14.1 Satellite-based (CERES) radiation at TOA, for the net R_T (which is also EEI), ASR and OLR in W m^{-2} as departures from monthly means. Shown are the monthly anomalies, and the heavy curve is the 12-month running mean. Data courtesy Yongxin Zhang

Reliable TOA radiation measurements begin in March 2000 from CERES: Clouds and Earth's Radiant Energy System (Fig. 14.1), although less accurate values exist from Earth Radiation Budget (ERB) satellite measurements in the 1980s. The global mean net TOA radiation downwards is too small to measure directly from satellite, and the mean (zero) EEI value on Fig. 14.1 assigned by CERES is 0.7 W m^{-2}. However, instruments are far more stable than they are absolutely accurate, with calibration stability <0.3 W m^{-2} per decade (95% confidence), and hence there is considerable confidence in the changes from year to year. Accordingly, these should represent the changes in EEI (Fig. 14.1).

The TOA radiation has a strong ENSO signal but it has varied in form. The cold La Niña in 2008–09 reduced OLR and increased the net radiation received, thereby recharging the ocean heat, while a discharge of heat occurred in the 2002–03 and 2009–10 El Niños (Fig. 14.1). The increase in GMST with the 2016 El Niño resulted in part from heat coming out of the ocean (Section 14.5), and low clouds were reduced, increasing both ASR and OLR; OLR was further increased by high temperatures.

There is clearly large month-to-month variability in global TOA radiation (Fig. 14.1). Standard deviations of monthly anomalies are 0.70 W m^{-2} (354 TW), and hence fluctuations of ± 1 W m^{-2} are common in the net downward radiation. Yet small signals associated with climate change of just a few tenths of a W m^{-2} over a decade are sought. Most of the monthly variability has very short timescales and is associated with weather systems, such as the MJO in the tropics, or weather noise in the extratropics. Some noise arises from partial sampling of weather systems, such as a low-pressure system one month and the associated high-pressure perturbation the next. More generally, outside of the tropical oceans, increased cloud, precipitation, and water vapor reduces ASR and thus cools the surface, which in turn reduces OLR a week or two later, and often the reverse happens the next month. Over the ocean this reduces surface temperature and thus SSTs, while over land surface moisture and evapotranspiration likely also play a role.

The strongest relationship overall is that increased temperature leads to more OLR. In particular, land areas, especially in the extratropics, provide a strong window for radiation to escape to space because the atmosphere tends to be dryer than over the oceans, so that there is less greenhouse effect from water vapor. The biggest exception is in the deep tropics where high SSTs are accompanied by deep convection and the extensive cloud reduces ASR while the high cloud tops reduce OLR, so that these effects mostly cancel for R_T. Consequently, there is a lot more high-frequency variability in the radiation fields than there is in temperature fields, highlighting the role of clouds and transient weather systems in the radiation statistics.

Hence the EEI undergoes large monthly variability associated mostly with weather systems, and the standard deviation of the 12-month running mean for EEI is 0.36 W m^{-2} (184 TW) with an interannual range of about 1.2 W m^{-2}. Note the increasing trend (Fig. 14.1), with distinctly higher values in the past five years in EEI to over 1.2 W m^{-2}; an important factor is reduced albedo from a combination of loss of Arctic sea ice and other ice mass (see Fig. 14.7), plus reduced cloud, which have increased ASR.

14.4 Atmosphere

Energy in the atmosphere includes internal energy, potential energy, latent energy, and kinetic energy. The net contribution of the atmosphere to the increased energy storage is fairly small overall, and from 1979 to 2010 the warming and changes in

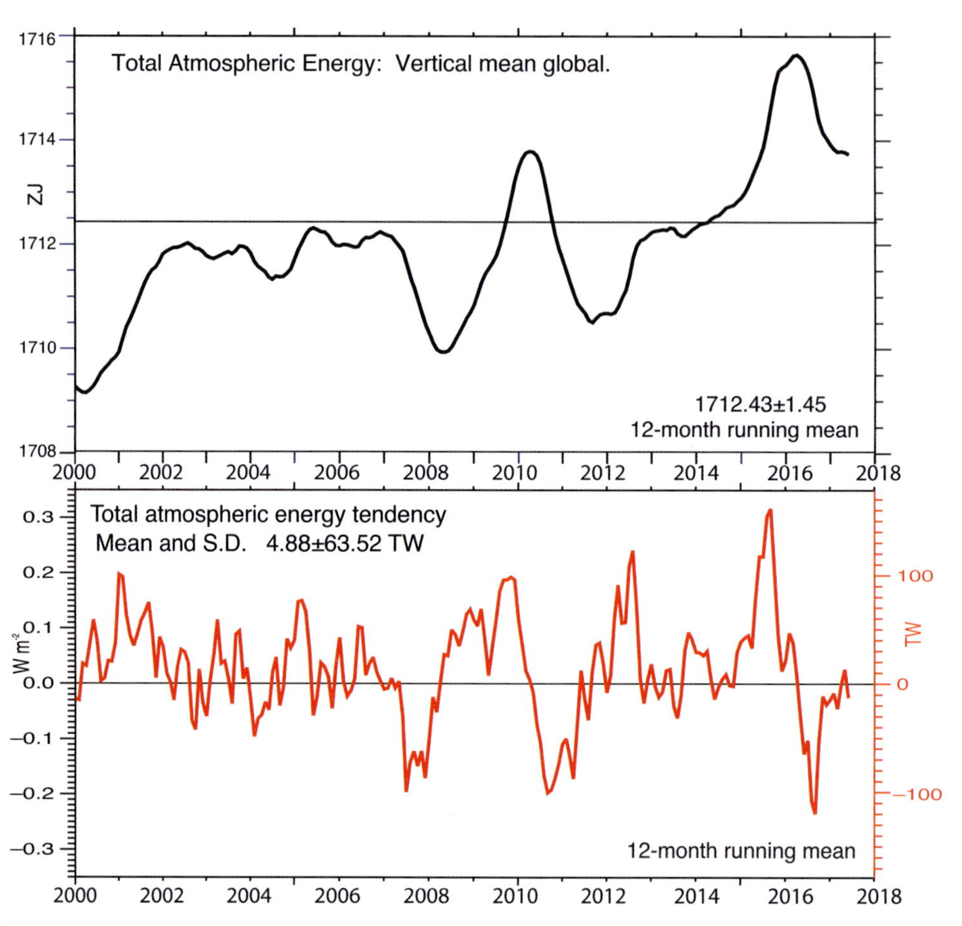

Fig. 14.2 Total global vertically integrated atmospheric energy, based on ERA-Interim reanalyses as 12-month running means, differenced to give the tendency (change) in both TW (right axis) and W m^{-2} (left axis).

water vapor combined were previously estimated by IPCC in 2013 to average about 2 TW. Because the warmest years on record are all after 2014, the rate has increased (Fig. 14.2), and the mean annual rate of increase in energy from 2000 to 2018 is 4.9 $\pm$ 34 TW (0.01 W m^{-2}). Here the 34 is a 95% error bar expressing considerable uncertainty in the mean value. The huge interannual variability is mostly associated with ENSO (see Fig. 2.4 for the GMST variations). The highest atmospheric energy value was in 2016, augmented by energy out of the ocean from the major El Niño. This peak value built up over 4 years. The lowest value in this interval was in 2008, when a strong La Niña occurred. However, big changes also occurred late last century (not shown) in association with the 1997/98 El Niño event in which the atmospheric energy peaked slightly higher than in 2016 but dropped like a stone in 1999.

Fig. 14.2 also gives the energy changes, and fluctuations exceeding 0.1 W m^{-2} are not uncommon. If the tendency time series (Fig. 14.2 lower) seems unduly noisy for a 12-month running mean, it is. The monthly variability of atmospheric energy is an order of magnitude larger, depicting huge variability from one month to the next. The standard deviation of 12-month running mean energy atmospheric changes is 63.5 TW, which accounts for over a third of the annual standard deviation of R_T of 184 TW (Fig. 14.1), highlighting that it is not atmospheric energy or temperatures so much as clouds that cause the TOA variability.

14.5 Ocean Heat Content (OHC)

Sea-level rise consists of two major components, one from the melt of land ice that adds more volume to the ocean, and the second from the expansion of the ocean through increases in temperature. The former can be estimated after 2002 via gravity measurements from space (Section 14.10). In 2009, crude estimates of the accumulated heat in the various components of the climate system compared with sea-level rise observations revealed a substantial shortfall, and blame was assigned to the considerable uncertainty in the ocean component at that time. It has been difficult to pin down the OHC changes, because prior to about 2005 when the Argo program ramped up, the analyses of OHC suffered from inadequate numbers of observations. Values of OHC also suffered from poor mapping methods and biases in some data used.

Prior to Argo, the main temperature soundings were from expendable bathythermographs (XBTs) which were dropped from ships of opportunity. As the rate of fall was not measured, the depth of measurements depended on the assumed fall rate and design of the XBT. All of the earlier data from different brands have now been recalibrated both in terms of the instrumental readings and their depths using experimental studies of fall rates in different conditions to narrow the uncertainties.

The Argo era after about 2005, when something like 3800 observations became available at any time (Fig. 14.3), however, has provided detailed monthly mean maps that have now been analyzed in several ways, although several products suffer from serious shortcomings (user beware). These analyses in turn provided a means to greatly improve analyses of earlier periods. Fully using observations in terms of how representative they are in both space and time was the key, along with appropriate adjustments to the data. Dr. Lijing Cheng (Institute for Atmospheric Physics [IAP], China) introduced innovative analyses of the past. Evaluation of how good past analyses were, say in the 1960s when a sparse distribution of observations existed, was assessed by subsampling the rich

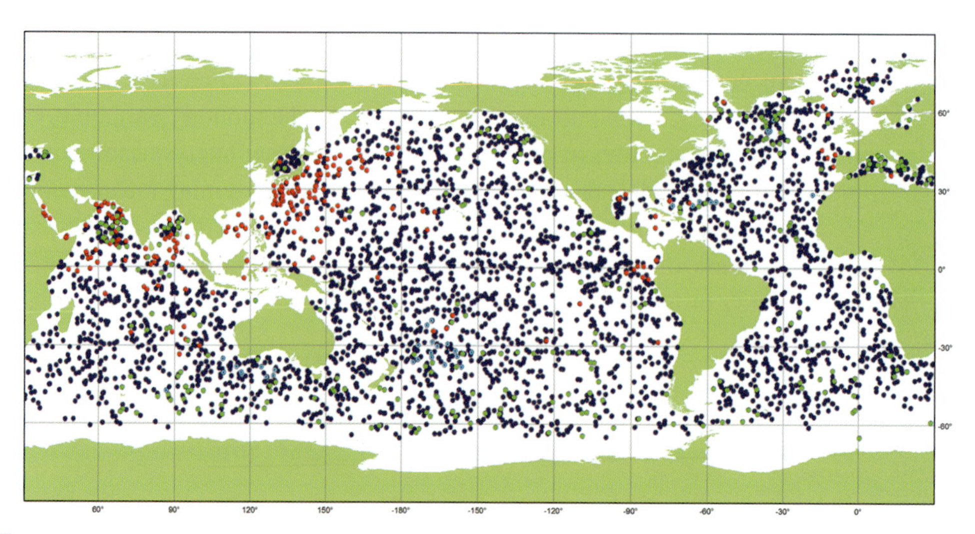

Fig. 14.3 Argo observations in September 2017 (from jcommops.org), showing the base Argo and related soundings in dark blue and red, biogeochemical soundings in light green, and deep Argo soundings (extending below 2000 m depth) in light blue.

Argo period to the locations of the old observations and a full analysis performed. This could then be evaluated from the result using all observations to establish the truth and thus the error bars. Tests showed that results were credible and within acceptable limits after 1958, the International Geophysical Year, but not before then. The IAP analysis provides a long-term perspective of ocean change but suffers from some noise.

Ocean reanalyses of past data are also available, and use is also made of ORAS5 (see Chapter 9). The reanalysis system works very well for the post-2005 Argo period, but not for earlier times. Adjusted ORAS5 reanalyses were used to produce the fields in Chapter 9.

A cross section of IAP zonal mean temperatures as a function of depth (Fig. 14.4) shows warming in the upper oceans everywhere. In the tropics the main warming occurs in the top 100 m or so, and the largest and deepest warming is in the southern oceans south of 30° S. The cooling in the subtropics of the southern hemisphere from about 300 to 1000 m depth is linked to the warming at higher latitudes by the ocean circulation, with cooling where upwelling is taking place, bringing up colder waters from below.

The warming of the oceans is better seen from the cross sections of each ocean (Fig. 14.5), and while the southern oceans warm at all longitudes, the subsurface subtropical cooling is evident only in the Pacific and Indian oceans. Warming does not penetrate as deep in those oceans as in the Atlantic, where warming is revealed everywhere south of about 40° N.

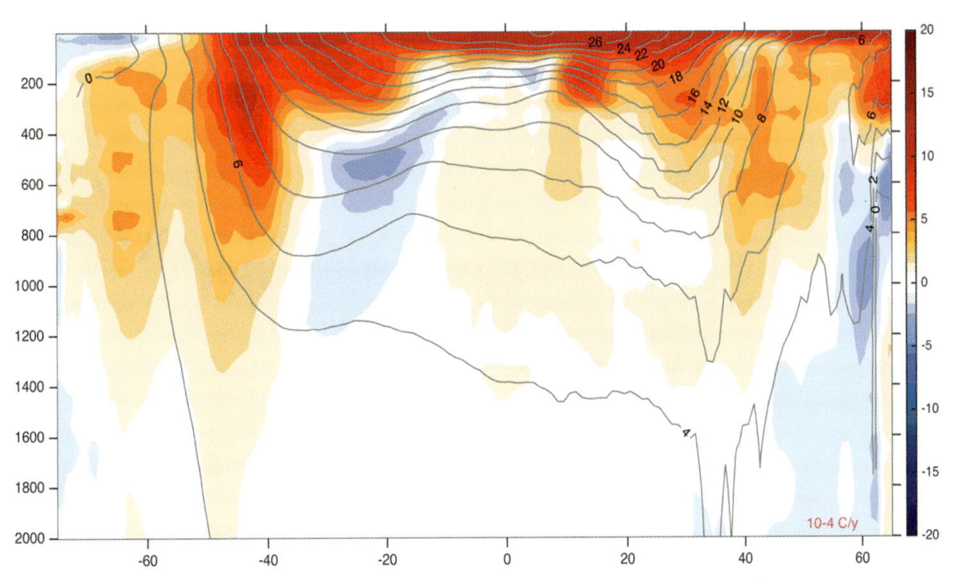

Fig. 14.4 The linear trend of zonal mean ocean heat changes for 1960–2016 (colors) in 10^{-2} °C/century, superposed on the average temperatures (°C) as black lines. The axes are depth in m at left and latitude from 60° S to 60° N. Courtesy Lijing Cheng (2017)

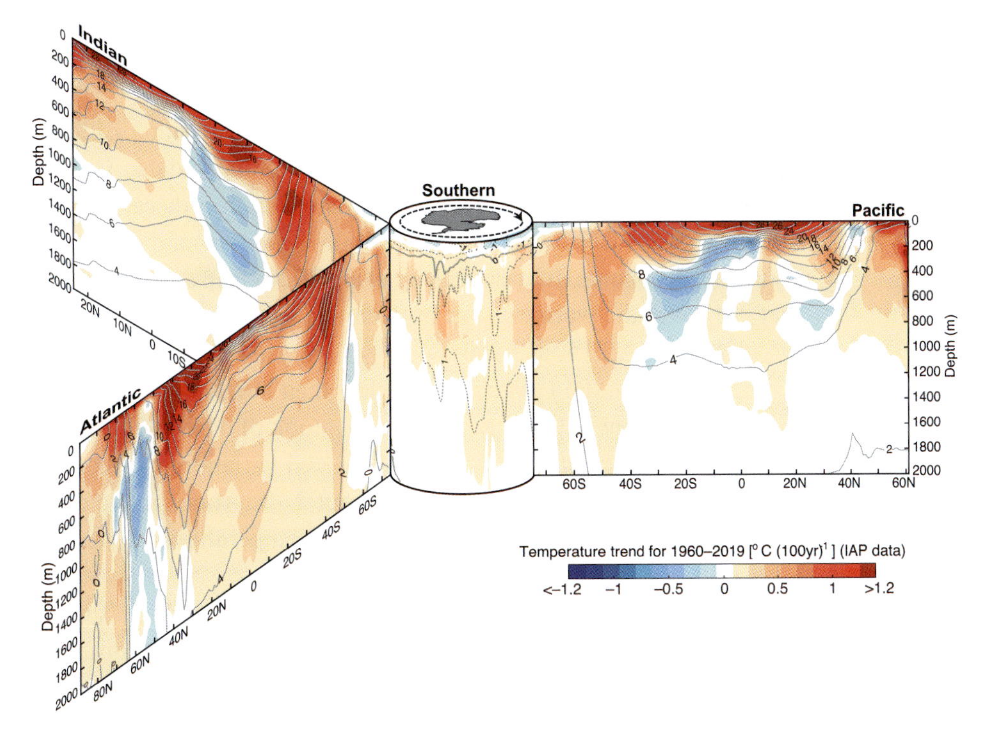

Fig. 14.5 An update of Fig. 14.4 to 2019 and divided into the three oceans as transects for the Atlantic, Pacific and Indian oceans extending from the northern limits to the Southern Ocean at 75° S. The latter joins all oceans. From Cheng et al. (2020). © American Meteorological Society. Used with permission

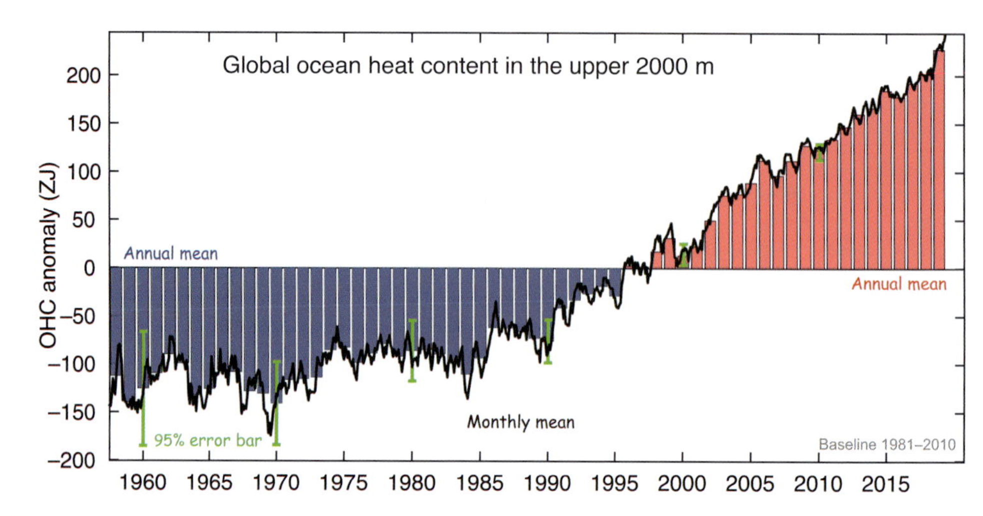

Fig. 14.6 Global ocean heat content for the top 2000 m after 1958 through 2019, showing the monthly anomalies and annual bars relative to 1981–2010 mean, in zettajoules (ZJ). The green error bars are 95% uncertainties. Adapted from Cheng et al. (2020)

Values of global annual mean OHC and by layer (Figs. 14.6–14.8) reveal that rates of change are small prior to 1970, when the upper 300 m layer begins to warm, and after about 1990 the heat penetrates to below 300 m depths. The rate of increase of IAP OHC is about 220 zettajoules (1 ZJ = 10^{21} J) over the past 20 years, from 2000 through 2019 (Fig. 14.8), compared with 5 ZJ for the atmosphere. Even higher rates are present in 2020 in preliminary data. The OHC trend is so large that the variability about the trend is difficult to discern accurately; the 95% error bar is ±9 ZJ after 2000 (Fig. 14.6). Figs. 14.6 and 14.8 show the monthly anomalies for the IAP dataset Cheng et al., 2020), and there is no doubt whatsoever that 2019 was the warmest year on record (2020 has likely surpassed it), followed by 2018, 2017, 2015, and 2016.

The OHC for IAP in 2016 was 5 ZJ less than in 2015, and 13 ZJ less than in 2017 (Fig. 14.8), owing to the huge El Niño that took of order 8 ZJ out of the ocean. The ORAS5 reanalyses have a smaller but related variation and featured lower values from July 2015 to February 2016 by about 2 ZJ. Atmospheric energy went up by about 2 ZJ (Fig. 14.2) and the higher surface temperatures led to an upward spike in OLR (Fig. 14.1) in early 2016.

Comparison with the CERES EEI estimate from Fig 14.1 (see Fig. 14.8) shows that the ORAS5 changes are quite plausible, and contributions from land and ice help make up differences. The rates of change of 12-month running mean OHC have a standard deviation for IAP of 0.9 W m^{-2} after 2000, which is much larger than the 0.37 W m^{-2} for ORAS5 (Fig. 14.8) and the 0.36 W m^{-2} for the net EEI (Fig. 14.1), and even though there is limited cancellation from atmospheric energy

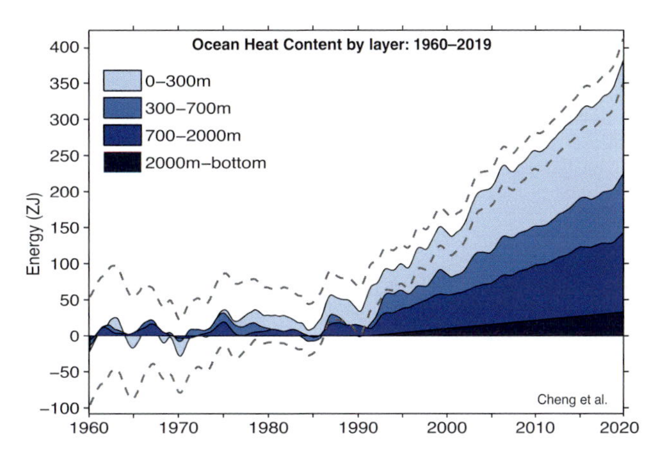

Fig. 14.7 Ocean heat content for various layers, in ZJ. An error bar is also given for the total as gray dashed lines. These values have been smoothed using LOWESS to remove fluctuations less than about 20 years period. Adapted from Cheng et al., 2020

(Fig. 14.2), some of the IAP variability is spurious. Substantial discrepancies remain among different OHC products, and combining the OHC changes with surface fluxes (Chapters 8 and 9) implies nonphysical ocean heat transports across the southern ice boundary around Antarctica, when integrated from the north. However, the results in Chapter 9 corrected for this and the consistency among three independent products during the 2015–16 El Niño demonstrates the power and potential behind this kind of analysis.

The increase of OHC of 220 zettajoules over 20 years for IAP (Fig. 14.8) expressed as a rate is equivalent to 350 TW (terawatts 10^{12} W), or 0.7 W m^{-2}, where the global area, rather than that of the oceans, is assumed. ORAS5 features slightly higher values of 0.8 W m^{-2} (Fig. 14.8). Adding in a contribution from below 2000 m depth (Fig. 14.7) for the past 15 years (2005–19) gives an estimated best value of 430 $\pm$ 70 TW.

Another important effect of climate change in the oceans is that warm, light surface waters are warming faster than the cold deeper waters (Fig. 14.4). Global warming is consequently tending to make the oceans more stable. Sea-water density depends not just on temperature but salinity too (Chapter 10, Fig. 10.3), because fresh water is lighter than salty water. Further, the melting of ice is leading to the accumulation of fresh, light water at the surface in subpolar regions and the presence of ice means that extra heat is taken up by melting rather than raising temperature. Both heating and freshening lead to more stably stratified oceans, and there has been a 6% increase in the average stratification of the upper 200 m of the world oceans over the past half century. Consequently, this inhibits the ability of the ocean to sequester heat and carbon dioxide, with implications for acidification (Chapter 6) and marine productivity (Chapter 8).

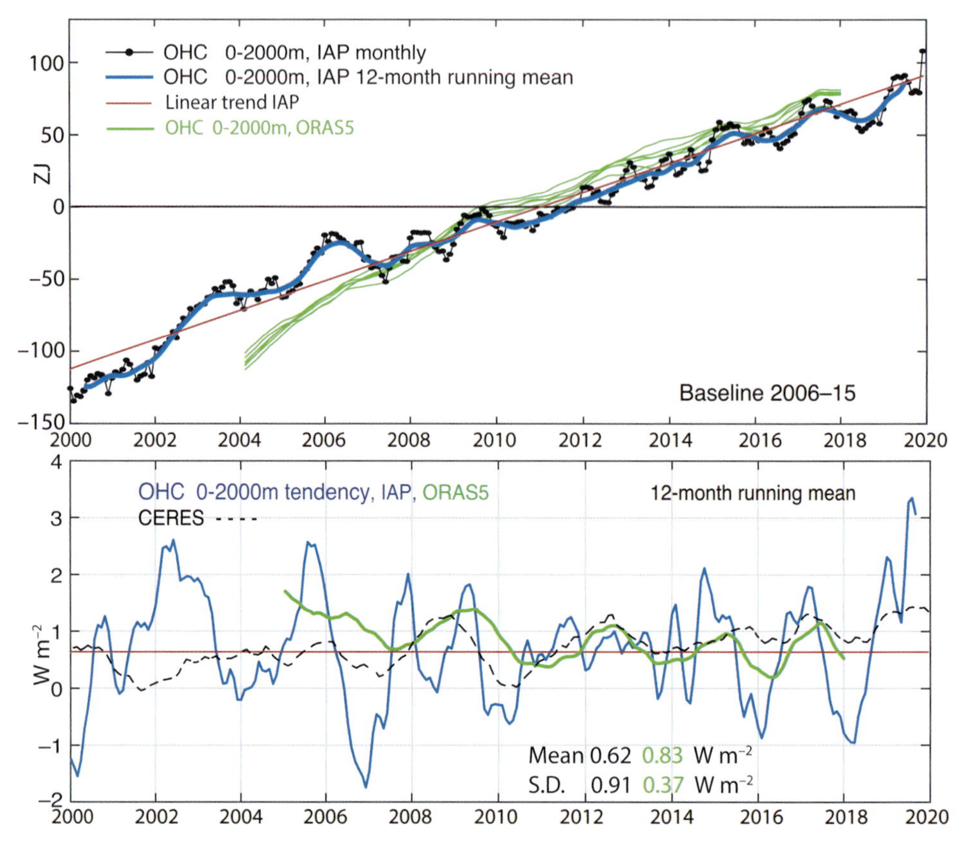

Fig. 14.8 Detailed view of the IAP OHC from 2000 through 2019 showing monthly and 12-month running mean values of OHC in ZJ and the rates of change for 12-month running means (below) in W m^{-2}. Also in green for 2004–16 are the OHC from ORAS5 for five ensemble members, and the rates of change of the ensemble mean (below). For comparison, the CERES 12-month running mean (from Fig 14.1) is added as a dashed black curve. For the top 2000 m the linear rate of increase is 0.62 W m^{-2} for IAP for 2000—19, and 0.8 W m^{-2} for ORAS5 for 2005—16. The standard deviations of the rates of change in W m^{-2} are 0.91 (IAP), 0.37 (ORAS5), and 0.36 (CERES). These do not include the region below 2000 m. IAP data courtesy Lijing Cheng

14.6 Ice

Melting of ice contributes a significant amount to where EEI is deposited. The best ice-related record is probably that of sea ice extent in the Arctic (Fig. 14.9) which has decreased by about 45% since 1979 in September, when the annual minimum occurs. The extent is readily determined from satellite observations because microwave measurements can see through cloud, although they are confounded

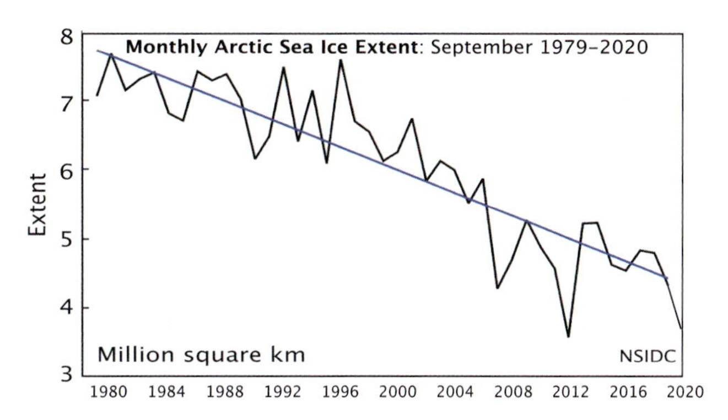

Fig. 14.9 Arctic sea ice extent in million square km for the month of September from 1979 to 2020. Based on data from the National Snow and Ice Data Center. The linear trend through 2019 is also indicated in blue.

by surface melt ponds of water. However, the volume is what is important for EEI. Pan-Arctic Ice Ocean Modeling and Assimilation System (PIOMAS) estimates of volume ice changes have trends in volume of $4 \pm 1 \times 10^3$ km^3/decade for 2000–16, and hence the net trend in latent heat of fusion from melting Arctic sea ice is about 3.8 TW on an annual basis.

Antarctic sea ice is briefly discussed in Section 11.4 where it is noted that the atmospheric circulation plays a dominant role. The satellite record reveals that a gradual, overall increase since 1979 in Antarctic sea ice extent reversed abruptly in 2014, and values since 2015 have been the lowest on record. Changes in volume are not known but overall are not significant.

There are considerable disparities in estimated volumes of land ice melt in different studies, some of which arise because they include contributions from different sources, such as whether frozen tundra is included. Ice sheet melt or gain is a major concern and is fairly well measured by satellite laser altimetry since 2003. Ice sheets gain mass through snow accumulation and lose it through (i) surface melt runoff (Greenland, 50–65%); (ii) iceberg calving (Antarctica, ~50%, and Greenland, 15–25%); and (iii) basal melting of floating ice shelves (Antarctica, ~50%) and tidewater glaciers (Greenland, 15–25%). Greenland has a sea-level potential of about 7 m if it all melted, and there has been widespread thinning of the ablation zone and thinning and retreat of tidewater glaciers. Antarctic ice has a sea-level rise potential of about 58 m. Ice sheet gains have occurred in East Antarctica from increased snowfall accumulation while in West Antarctica basal melting of ice shelves is occurring owing to warmer waters in the surrounding ocean, reducing their buttressing capability and increasing ice discharge into the ocean. It has been difficult to obtain a combined estimate of floating and grounded ice changes.

A recent assessment puts the overall ice sheet mass change for Greenland between 2003 and 2019 at -200 ± 12 Gt yr^{-1}, with accumulation in the high interior, and melt around the coast. The sea-level equivalent change is 8.9 mm. This value, based on Smith et al. (2020), is somewhat less than another recent estimate (-268 ± 14 Gt yr^{-1}), and the error bars do not overlap. What is clear is that melting of Greenland has accelerated somewhat since 2006. A startling record event of surface melting for Greenland occurred in early August 2019, so that total ice mass lost in 2019 through the summer melt season was as high as 2012, the previous record, with ice loss of about 300 Gt and thus 0.1 ZJ of energy or 3.2 TW. The five highest melt mass loss years have occurred since 2010. Hence the estimates of heat uptake range from 2.1 to 2.9 TW; we use 2.6 TW.

In Antarctica, the ice sheet has gained mass from snowfall, especially over East Antarctica, but has lost mass through melting ice shelves and calving. The pattern consists of thickening along the steep slopes of the Antarctic Peninsula and around the coast, while East Antarctica gained less mass as distances from the coast increase, highlighting the snow accumulation aspect. However, these gains were more than offset by dramatic, ongoing mass loss around the margins, especially West Antarctica, since 2003 in response to rapidly shrinking ice shelves. It is important to distinguish between floating and grounded ice loss because only the latter is reflected in sea-level rise. Hence from 2003 to 2019, while East Antarctica is estimated to have gained 196 Gt yr^{-1} (106 floating, 90 grounded), West Antarctica lost about 245 Gt yr^{-1} (-76, -169), and the Antarctic Peninsula lost a further 53 Gt yr^{-1} (-14, -39), for a net loss of Antarctic ice of about 103 Gt yr^{-1} (15, -118), but with quite a large uncertainty from the floating ice component (15 ± 65 Gt yr^{-1}). (Values based on Smith et al., 2020. Note values may not add up exactly because of the fractions attached to each estimate and round off.) Indeed, other estimates place the ice shelf gain over East Antarctic much higher, but with much greater losses for West Antarctica (by 25 Gt yr^{-1}). Another estimate has much larger losses from basal melting of ice shelves, but with large uncertainty. Taking the net loss as 113 Gt yr^{-1} for this period amounts to 1.2 TW.

Together, the ice sheets contributed ~14 mm sea-level equivalent to the global oceans from 2003 to 2019 (8.9 mm from Greenland and 5.2 mm from Antarctica). Melting ice shelves do not contribute to sea-level rise. In addition, glacier and small ice cap loss of ice is about 1 mm sea-level equivalent per year, or 362 km^3 yr^{-1} since about the 1970s. Hence this element contributes about 333 Gt/yr^{-1} or 3.4 TW. Thawing of permafrost might also be considered under the category of ice melt, but here we prefer to keep that category under the heading of "Land." Accordingly, the heat uptake to melt ice is about 10–12 TW (0.02 W m^{-2} globally).

14.7 Land

Historical land-use and land-cover change (LULCC) from clearing for agriculture and pasture, and from wood harvest, represent about a third of all human CO_2 emissions. Land-use and land-cover change also have large impacts on the surface energy budget through changing surface albedo, and on hydrology and surface roughness through changing vegetation and land use practices such as irrigation. Surface changes are widespread. Forest loss in the tropics is outweighed by increases in the extratropics to give a 7% increase globally from 1982 to 2016. Bare ground has decreased by 3%, mostly because of human activities.

There have been relatively few investigations of how ground and soils warm in comparison to the atmosphere. The continental heat uptake occurs through a heat flux into the solid surface of the lithosphere, although water flows can be important where they occur. It appears that land warming for the top meter of soil may keep pace with air warming except mainly where snow and ice occur, in northern latitudes. A surface signal takes about 50 years to penetrate to a depth of 50 m. A portion of anthropogenic warming is being stored in deep soils.

The main measurements used for assessing land warming are borehole temperatures, and these are profiles of temperature with depth down quite deep boreholes. A borehole may be constructed for the extraction of water, oil, or natural gas, as part of a geotechnical investigation to assess ground properties perhaps for construction purposes, environmental site assessment, mineral exploration, as a pilot hole for installing piers or underground utilities, for geothermal installations, or for underground storage of unwanted substances. Hence most boreholes are drilled for other purposes, and therefore are heavily biased as to where they are located. The assessments to date have not adequately considered the heterogeneity of the land, such as whether there are various kinds of rock, soils (which may or may not have vegetation and root systems embedded), fractures, and water flows.

Based upon borehole temperature changes, in the late 20th century the rate of warming of the land after 1950 was estimated to be of order 6 TW. This has no doubt increased since the turn of the century. Several estimates of heat uptake into land, based partly on models, put the values at 6–7 TW, increasing to perhaps 10–12 TW after 2000.

A recent assessment of surface waters in lakes, rivers, and reservoirs finds them all warming, but by far the biggest effect on energy, by an order of magnitude, is the increase in water storage on land especially in reservoirs. By constructing reservoirs, humans are not only redistributing mass from the oceans to the land, but also the thermal energy carried within this water. Inland waters cover about 2.6% of the global continental area. The artificial reservoirs are estimated to have increased global lake volume by 3.2% overall, but only modestly since 2005. Since 2005, the mean trend in global lake, river, and reservoir heat uptake is about 0.4 TW and the energy in increased mass of waters in reservoirs is about 0.9 TW.

Consequently, if it is assumed that these are somewhat independent complementary estimates, the total land heat uptake is 13–15 TW, and 14 TW (0.03 W m^{-2}) is used as an overall estimate for land warming. Only recently has there been an assessment of permafrost thaw. The average temperature increase of the permafrost is 0.3°C, and if it is assumed that this applies to the top 20 m and fades to zero by 30 m, then the warming of permafrost is 2.2 TW.

14.8 Variability of EEI

Earlier, the large variabilities in EEI (Fig. 14.1) and atmospheric energy content (Fig. 14.2) were noted, and the latter is often offset by changes in OHC, in particular during ENSO events (Chapter 12). Another way to examine the total EEI is through comprehensive climate system model simulations (Fig. 14.10). As shown by the red curves for individual runs and the gray shaded region overall (Fig. 14.10) the interannual natural variability can be ±0.1 W m^{-2}; indeed, the net ENSO natural variability is order ±0.1 W m^{-2}, even though there is compensation between atmosphere and ocean. Model results suggest large short-term perturbations of several W m^{-2} from volcanoes (Fig. 14.10). The model ensemble mean

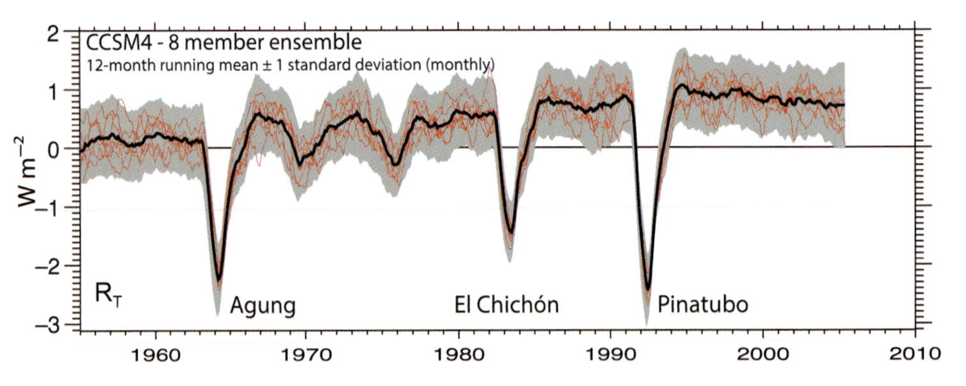

Fig. 14.10 The net TOA radiation (down; W m^{-2}) from CCSM4 run at 1° resolution for eight ensemble members. Shown are the ensemble 12-month running mean (black line) with ±1 standard deviation of the individual monthly values (gray shaded region) and the eight individual ensemble values as 12-month running means. The last 50 years includes the Mt. Agung, El Chichón, and Mt. Pinatubo volcanic eruptions. Adapted from Trenberth et al. (2014). © American Meteorological Society. Used with permission

values in the late 1990s for EEI were 0.9 W m^{-2}, dropping to 0.8 W m^{-2} in the 2000s, as the solar irradiance declined with the sunspot cycle.

The measured solar irradiance (Figs. 4.2 and 4.3) was quite variable but ran about 0.2 W m^{-2} above the average in 2005 (end of Fig. 14.10) and dropped to 0.4 W m^{-2} below average by the sunspot minimum in 2008. Values were again higher by 0.5 W m^{-2} from 2012 to 2016 in the more recent sunspot maximum. The effects on the net energy imbalance EEI are about ¼ of these solar irradiance changes: close to $\pm$0.1 W m^{-2}. (The factor of 4 comes from the ratio of the cross section of Earth's area πa^2 that intercepts the Sun's radiation to the surface area of the sphere $4\pi a^2$.) The model EEI (Fig. 14.10) agrees very well with the observational assessment, both in terms of the values and their variations (cf. Fig. 14.1), although it stops in 2005.

14.9 Total EEI Inventory

The ocean was assessed as taking up slightly more than 0.8 W m^{-2} for 2005–2019 and a best value is estimated to be 430 $\pm$ 70 TW. For the atmosphere the value is 3.4 TW but with an extremely large uncertainty owing to large variability in time. The total non-ocean component takes up about 30 TW of energy compared with about 430 TW for the ocean and is 7% of the total (460 TW) in this accounting, or about 0.06 W m^{-2} globally. The total EEI is therefore close to 0.90 $\pm$ 0.15 W m^{-2} since 2005. Here the error bar is a 95% value based upon 15-year trends and estimates of uncertainty, which is smaller than earlier estimates owing to growing confidence in the continued Argo data and improved analyses. These total EEI values are much larger than IPCC (2013) assessed, but they do vary over time, with the sunspot cycle, volcanic eruptions and El Niño events. Note also from Fig 14.1 that values of EEI are distinctly higher after 2012 and are about 1.2 W m^{-2} after 2018.

Table 14.1 Contributions of various components for the destination of the EEI in TW after 2005. Values related to ice are shaded

TW	Atmosphere	Land	Permafrost	Arctic sea ice	Glaciers	Greenland	Antarctica
Ice				3.8	3.4	2.6	1.2
Non-Ocean	3.4	14	2		11		
Non-Ocean				30			
Ocean				430±70			

14.10 Sea-Level Rise

Although the EEI is very small compared to the natural flows of energy and cannot be discerned at any given time, it accumulates in the oceans which are now warmer than they used to be, and the air above the oceans is warmer by order 1°C and moister as a result. On land, as long as the surface is wet or there are lakes and rivers nearby, much of the heat goes into evaporating moisture and driving the hydrological cycle. Clearly this happens also over the oceans. However, in dry conditions the heat effects accumulate (Chapter 10). The EEI also accumulates by melting glaciers and ice on land to add water to the ocean's volume, and sea-ice melts in the oceans. The result is a steady increase in global sea level of over 3 mm yr^{-1} since satellite altimeter measurements of sea level were initiated in late 1992.

Traditionally, sea level has been measured at coastal stations and islands in tide gauges (Fig. 14.11), but relatively few such measurements extend back to 1900. Moreover, local sea level is subject to local gravity, tides, wind, and atmospheric pressure effects, and is typically only somewhat representative of larger areas when averaged over time. Mean sea level at a location is typically computed from 19 years of hourly measurements.

On land, it is not just the sea-level rise, but also the movement up (uplift) and down (subsidence) of coastal land that matters. In many coastal areas, the rate of

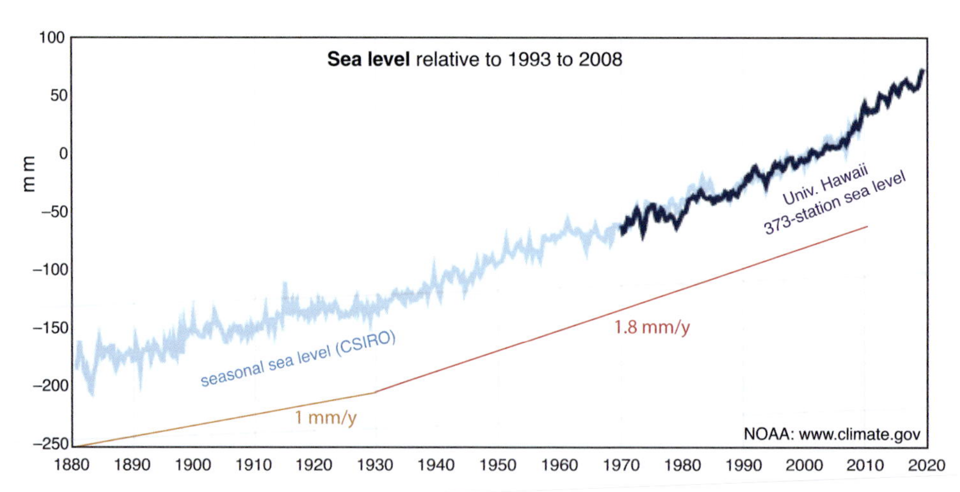

Fig. 14.11 Estimates of global sea level based on tide gauges from Church and White (2011) (pale blue) and, the more recent part of the time series, from the University of Hawaii Sea Level Center based on a weighted average of 373 global tide gauge records collected by the US National Ocean Service and partner agencies worldwide. Weights are adjusted to prevent overemphasizing regions where there are many tide gauges located in close proximity. The values are change in sea level in millimeters compared to the 1993–2008 average. Adapted from NOAA

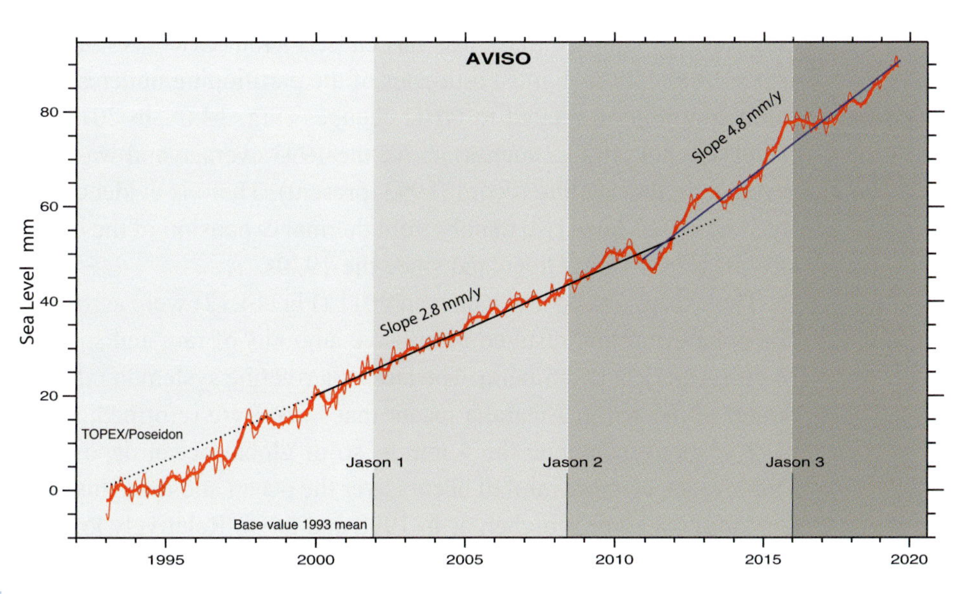

Fig. 14.12 The global sea-level record based on altimetry from space in mm, relative to the annual mean for 1993, based on data from AVISO. The series of satellite missions is indicated along with some slopes. www.aviso.altimetry.fr/en/home.html.

relative sea-level rise is much higher as a result of subsidence resulting from human causes, such as groundwater pumping and fossil fuel extraction. Coastal wetlands normally accumulate sediments and act as natural levees against storms to provide protection against rising sea levels and local subsidence. Experience shows that mangrove ecosystems did not develop unless relative sea-level rise was less than 6–7 mm yr^{-1}. Sediment supply to the coast has declined as a result of damming rivers and mining and export of sediment, increasing mangrove vulnerability to sea-level rise and reducing their ability to reduce unwanted water flow and wave turbulence.

Since late 1992, altimeters have been launched on satellites to provide global sea-level coverage about every 1–3 weeks, with revisit times to the same location varying from 10 to 35 days. Since then a continuous record exists to millimeter accuracy of the true global sea-level value (Fig. 14.12). The global rate of sea-level rise has doubled from 1.8 mm yr^{-1} over the 20th century to approximately 3.2 mm yr^{-1} since 1992, but with even higher values after 2012 (Figs. 14.11 and 14.12).

Global mean sea level has risen about 210–240 mm (8–9 inches) since 1880, with about a third of that coming in just the last two and a half decades (Fig. 14.11). The rising water level is mostly due to a combination of meltwater from glaciers and ice sheets and thermal expansion of seawater as it warms. Changes in terrestrial storage of water also contribute, especially at times of major dam construction and reservoir development, such as in the 1970s. Some estimates suggest that land-ice mass loss has caused about twice as much rise in sea level since 1900, although most estimates of the partitioning underestimate the thermosteric contribution related to OHC changes (Fig. 14.6). In 2019, global mean sea level was 8.4 cm (3.2 inches) above the 1993 average and was the highest annual average in the satellite record (1993–present). There is evidence of acceleration in the rate of sea-level rise mainly from thermal expansion of the ocean and increased ice-mass loss from Greenland since the 1970s.

The exceptionally low values in 2011 (Fig. 14.12) were associated with a major La Niña event that resulted in massive amounts of rain and snowfall in Australia, North America, and Siberia. The lack of a riverine system to return the precipitated water to the sea in Australia meant that Lake Eyre reformed, and all told, it took 1–2 years to evaporate the 5 mm or so of global ocean deposited on land. In El Niño years, as more rainfall occurs over the ocean and droughts are more prevalent on land, sea level is higher, as in 1998 and 2016. Relatively low values in the first 2 years of the record likely resulted from the major volcanic eruption, Mt. Pinatubo, in June 1991 that cooled the climate system.

Sea-level rise is expected to cause massive upheavals to civilization in coming decades:

- forcing millions of people to abandon the coast as rising seas flood populated areas and major cities;
- opening the way for climate change–amplified hurricanes to drive higher storm surges farther inland;
- knocking out transportation systems and sewage treatment plants;
- inundating prime agricultural land and barrier islands; and
- infiltrating aquifers with salt water.

As global sea level rises so will the frequency and depth of high-tide flooding. "Nuisance" flooding, also called "sunny-day flooding," is already a growing problem in places like Miami Beach, Norfolk, and San Francisco in the United States. It is associated with high tides but occurs in benign weather conditions, except that the wind helps pile up waters along the coast or in bays. It causes short-term public inconveniences such as flooded streets and closed roads. Chronic flooding can strain city budgets and compromise infrastructure. Along much of the US coastline, high-tide flooding is three to nine times more frequent than 50 years ago. Coastal water levels considered as floods today will become the new normal for high tides in coming decades.

Major flooding already occurs with hurricanes, and many houses were destroyed by hurricane Harvey in the Houston area in 2017, yet they were outside the so-called "100 year" flood zone. A recent comprehensive flood assessment for the United States by the First Street Foundation (2020) suggested that there are at least

6 million households that are unaware they are living in homes that have a 1% chance of flooding in each year, putting them within a 100-year flood zone. At least 14.6 million properties have a substantial level of risk now, and that could grow to 16.2 million by 2050, although there are considerable uncertainties.

14.11 The Way Forward

This chapter has presented detailed time series where available of EEI (R_T), atmospheric energy (AE) and its tendency, and the ocean heat content (OHC) and its tendency. The remaining components come under "land" and "ice" and neither is close to being available in a form that can examine interannual variability. However, they make up only about 5% of the total. Unfortunately, the other three do not add up that well. Previous attempts to match the R_T from atmospheric reanalyses with CERES values revealed both substantial biases and unrealistic changes over time in the reanalyses. These should be improved in the latest reanalyses, which may eventually provide closure of the energy budget interannually from the standpoint of the TOA radiation and the atmosphere. Discrepancies among OHC products reveal that they too require substantial improvement.

Progress is needed before the surface fluxes of energy from atmospheric reanalyses become realistic enough to be used with OHC to close the ocean heat budget. In Chapters 8 and 9, by creating constraints and adjustments, the net surface energy flux was deduced and used along with OHC to compute the ocean heat transport as a residual. This should be the objective, and it is a reasonable goal on an annual basis. It should then be possible to refine the exchanges of energy among the various climate system components during ENSO. On the other hand, shortening the timescale to seasonal would be a major challenge owing to the huge natural variability and inadequate temporal sampling of TOA radiation – one satellite is not enough – and extra complications from seasonal sea ice. Nevertheless, it is a challenge worth taking on as it could lead to much better forecasts on seasonal to interannual timescales.

Better tracking of the energy and its flows through the climate system provides potential for much improved analyses of anomalous conditions, such as major marine heatwaves or sea ice perturbations, and how well they are supported by the rest of the system or whether they are shallow and transient. These concepts are extremely important for storms over the ocean and especially tropical storms and hurricanes. High SSTs are important in these cases, but they cannot be sustained without support from the upper ocean heat content. Similarly, over land, the extent and depth of soil moisture anomalies relate to drought and wildfire risk. In the high latitudes and Arctic, the warming and thawing of permafrost matters not just for the physical climate but also for the biogeochemical processes that relate to how much methane and carbon dioxide may be emitted as a powerful feedback to the warming.

Once the distribution of energy and how it moves are better pinned down, the information also provides solid data for testing, verifying and improving climate and weather models. The forefront of the research and needs relates to coupled processes and exchanges. Those prospects are what help build much better projections of the future.

References and Further Reading

Biskaborn B. K., S. L. Smith, J. Noetzli, et al., 2019: Permafrost is warming at a global scale. *Nature Communications*, **10**, 264. doi: 10.1038/s41467-018-08240-4.

Cheng, L., K. Trenberth, J. Fasullo, T. Boyer, J. Abraham, and J. Zhu, 2017: Improved estimates of ocean heat content from 1960–2015. *Science Advances*, **3** (3), e1601545. doi:10.1126/sciadv.1601545. http://advances.sciencemag.org/content/3/3/e1601545.

Cheng, L., J. Abraham, Z. Hausfather, and K. E. Trenberth, 2019: How fast are the oceans warming? Observational records of ocean heat content show that ocean warming is accelerating. *Science*, **363**, 128–129. doi:10.1126/science.aav7619.

Cheng, L., J. P. Abraham, J. Zhu, et al., 2020: Record-setting ocean warmth continued in 2019. *Advances in Atmospheric Sciences*, **37**, 137–142. doi: 10.1007/s00376-020-9283-7.

Church, J. A., and N. J. White, 2011: Sea-level rise from the late 19th to the early 21st Century. *Surveys in Geophysics*, **32**(4–5), 585–602. doi: 10.1007/s10712–011-9119-1.

First Street Foundation, 2020: First national flood risk assessment. https://assets.firststreet.org/uploads/2020/06/first_street_foundation__first_national_flood_risk_assessment.pdf.

IPCC: Intergovernmental Panel on Climate Change, 2013: *Climate Change 2013. The Physical Science Basis*, ed. Stocker, T. F., et al. Cambridge: Cambridge University Press.

Kwok, R., 2018: Arctic sea ice thickness, volume, and multiyear ice coverage: losses and coupled variability (1958–2018). *Environmental Research Letters*, **13**, 105005. doi: 10.1088/1748-9326/aae3ec.

Nerem, R. S., B. D. Beckley, J. T. Fasullo, et al., 2018: Climate-change–driven accelerated sea-level rise detected in the altimeter era. *Proceedings of the National Academy of Sciences USA*, **115**, 2022–2025.

NSIDC: National Snow and Ice Data Center, 2019: http://nsidc.org/greenland-today/.

Schweiger, A., R. Lindsay, J. Zhang, M. Steele, H. Stern, and R. Kwok, 2011: Uncertainty in modeled Arctic sea ice volume. *Journal of Geophysical Research*, **116**, C00D06. doi: 10.1029/2011JC007084.

Smith, B., H. A. Fricker, A. S. Gardner, et al., 2020: Pervasive ice sheet mass loss reflects competing ocean and atmosphere processes. *Science*, **368**, 1239–1242. doi: 10.1126/science.aaz5845.

Song. X.-P., M. C. Hansen, S. V. Stehman, et al., 2018: Global land change from 1982 to 2016. *Nature*, **560**, 639–643. doi: 10.1038/s41586-018-0411-9.

Trenberth, K. E., J. T. Fasullo, and M. A. Balmaseda, 2014: Earth's energy imbalance. *Journal of Climate*, **27**, 3129–3144. doi: 10.1175/JCLI-D-13-00294.

Trenberth, K. E., J. T. Fasullo, K. von Schuckmann, and L. Cheng, 2016: Insights into Earth's energy imbalance from multiple sources. *Journal of Climate*, **29**, 7495–7505. doi: 10.1175/JCLI-D-16-0339.

15 Attribution and the Hiatus

Detection and attribution of climate variations and extremes are essential first steps in building confidence in projections. Enlightened approaches are described. A major pause in warming from 1999 to 2013 called "the hiatus" is used as an illustration.

15.1 Attribution of Warming

Temperatures on Earth are rising at the surface and especially for the oceans. But how is it known that this is due to human activities? The main way this question has been addressed in most places is to use climate modeling, where a climate model can be run with and without various forcings and the results compared with observations. In these models, the increase in carbon dioxide is generally assumed. But how is it known that the increased carbon dioxide comes from human activities?

As detailed in Chapters 2 and 6, carbon dioxide and other greenhouse gases are increasing in the atmosphere. Chapter 2 briefly examined the carbon cycle to understand the natural and perturbed flows of carbon through the climate system. Carbon dioxide is emitted from decaying plants and trees as a natural part of the annual cycle. Beginning in spring, photosynthesis in plants takes up carbon dioxide from the air and converts it into the wood, twigs and leaves of plants, enabling the plant to grow; but in the fall, many plants have completed their productive cycle, and leaves and dead branches fall to the ground and decay.

The sources and sinks of carbon dioxide can be ascertained by considering the different kinds of carbon: the various carbon isotopes. Normal carbon is C_{12}, has an atomic weight of 12, and is common in plants. C_{13} is a somewhat heavier carbon that can be quite common in volcanic gas. C_{14} is the heavy carbon produced naturally in the atmosphere by cosmic rays and is radioactive, with a half-life decay time of 5730 years, which is why it is not found in fossil fuels like coal and oil deposits. Observations show that the rise in carbon dioxide comes from the light carbon, C_{12}, and not C_{13} or C_{14}, and hence it is not coming from volcanoes, but instead comes from burning of fossil fuels that are missing the heavy carbon isotopes.

15.2 Attribution of Climate Events

One of the most prevalent straw man arguments is that scientists claim climate change "causes" extreme events, when in fact, climate scientists make careful distinctions between "causality" and "influence" – two very different things. Weather events and climate variability continue, regardless of climate change, and often with a much bigger signal. However, that does not mean that there is not an influence of climate change on the outcomes.

A related issue is that scientists use uncertainty to represent the diverse possible different outcomes because of the natural variability but this should not be mistaken to mean that scientists do not know. Those who believe that the evidence shows our current path crosses dangerous climate limits and may lead to severe environmental and social disruption cannot prove that a catastrophic future will happen, but the argument is still that humanity must do what we can to avoid such stresses.

Climate model simulations that account for changes in forcings have now reliably shown that global surface warming of recent decades is a response to the increased concentrations of greenhouse gases and sulfate aerosols in the atmosphere. When the models are run without these forcing changes, the remaining natural forcings and intrinsic natural variability fail to capture the increase in GMSTs since the mid-1970s. But when the anthropogenic forcings are included, the models simulate the observed GMST record with impressive fidelity. Moreover, the models can be used to show that the greenhouse gas increases alone overdo the observed warming and that the aerosols act to offset some of the warming.

The ability of coupled climate models to simulate the temperature evolution on continental scales, and the detection of anthropogenic effects on each continent except Antarctica, has also improved. No climate model that has used natural forcing alone has reproduced either the observed global mean warming trend or the continental mean warming trends. Attribution of temperature change on smaller than continental scales and over timescales of less than 50 years or so is more difficult because of the much larger natural variability on smaller space and timescales. This also applies to assigning a cause to extreme events.

The conventional approach to attribution of climate events is to characterize the event and ask (1) whether the likelihood or strength of such events has changed in the observational record, referred to as "detection"; and (2) whether this change is consistent with the anthropogenic influence, as found in one or more climate models. This traditional approach is statistical and based on data analysis and a fingerprint from climate model runs. For instance, given a major heatwave in Russia, an analysis is first performed on the temperature records in the region, hopefully long reliable records, and a signal of the event is determined. Then climate model simulations are analyzed for the same quantities using models both with and without human-forcings. The odds of the event can be estimated – how likely it was to have occurred by chance, with and without these forcings.

This approach has had considerable success with extremes that are strongly governed by thermodynamic aspects of climate change, especially those related to temperature, but dealing with all aspects of the problem has often confounded results. It is computationally expensive and takes time because it requires many climate model runs with and without climate change present to sort out how unusual the weather event was and how the odds were changed by climate change. Because of the infinite natural chaotic variety of weather and the often uncertain nature of the human influences, such changes are mostly very small and get lost in the weather-noise. Indeed, weather conditions present in the actual event are never simulated by chance. Moreover, the huge computational demand limits the near-real-time commentary required by the media.

The conventional approach works well for large anomalies of temperature, such as in heatwaves. It does not work well or at all for storms or precipitation. The conventional approach is severely challenged when evaluating climate extremes that are strongly governed by aspects of the atmospheric circulation, including local precipitation. Moreover, the conclusions drawn are nearly always that the event was dominated by natural variability. This is the correct conclusion from that approach, but it is meaningless. It resulted in the misleading mantra that *"We can't attribute a single event to climate change."*

> *The result of bad and misleading statements about attribution, of which there have been many, is to grossly underestimate the role of humans in climate events of note in recent times to the detriment of perceptions about climate change and subsequent policy debates.* (Trenberth, 2011)

The fundamental problem with the conventional approach is that scientists lean over backwards to prevent what are called in statistics "Type I" errors, in which one might conclude that there is a human influence when there is not. The bias comes from using 95% confidence levels and the way questions are posed. The approach is inherently conservative and prone to false negatives, referred to as "Type II" errors, which underestimate the true likelihood of the human influence.

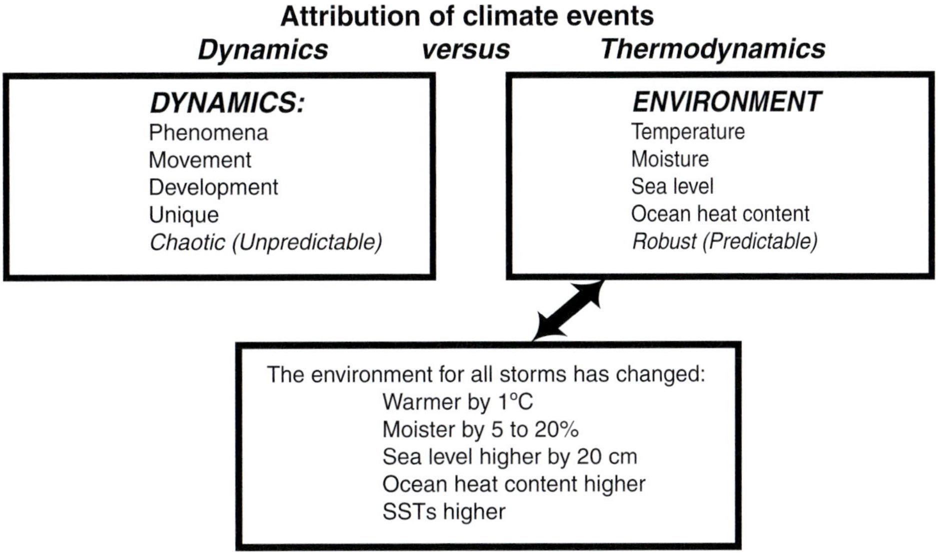

Attribution of climate events
Dynamics* *versus* *Thermodynamics

DYNAMICS:
Phenomena
Movement
Development
Unique
Chaotic (Unpredictable)

ENVIRONMENT
Temperature
Moisture
Sea level
Ocean heat content
Robust (Predictable)

The environment for all storms has changed:
Warmer by 1°C
Moister by 5 to 20%
Sea level higher by 20 cm
Ocean heat content higher
SSTs higher

Fig. 15.1 In attribution of events, such as storms, the dynamics (left) are unpredictable except for a very short time, while the thermodynamic aspects are pervasive and create a different environment for all events.

In addition, every event has been treated as if it were the only such event and it does not recognize the multitude of other similar events around the world. A new approach asks a different question and properly accounts for previous experience, that is known in statistics as "priors," and makes use of Bayesian statistics.

Instead it is recommended to adopt a conditional approach and address a different question. Assume that the phenomena and weather event were natural and would have happened anyway (this is consistent with expectations from models); and ask instead "*what was the impact of the change in the environment from climate change on the outcome?*" Of particular note is that the following changes in the environment are pervasive and affect the weather outcomes:

- Increases in SSTs by 0.6–1°C supported by much higher and increasing ocean heat content;
- Higher sea level by 20 cm;
- Warmer and moister air by 5–20% over and near oceans;
- Higher temperatures on land.

Figure 15.1 illustrates these aspects by emphasizing the thermodynamic facets of the problem versus the dynamic aspects. The latter are unique to every event and chaotic in character. The thermodynamic aspects relate to the changes in the environment in which the storm or event is occurring.

Using atmospheric models, initialized with real weather conditions, as in numerical weather prediction (NWP), numerical experiments can be run with environmental conditions from, say, 1900, versus those for the current atmospheric state. It is important that the model is able to replicate and forecast reasonably well the evolution of the storm or event for 2 weeks or so. Then the differences can be taken. Moreover, this can be done dozens of times to assess the reproducibility and statistics compiled.

A particular example is the extensive floods in and around Boulder, Colorado, in September 2013, explored by Pall and colleagues (2017). Boulder experienced over 230 mm on September 12, 2013, and 436 mm over 8 days. A high-resolution NWP model was used to simulate the events given the observed synoptic-scale meteorology in mid-September 2013 and assumed these conditions would have been similar in the absence of anthropogenic forcing. But the prescribed lower boundary SST conditions, greenhouse gas concentrations and radiative forcing were either current-day values or those estimated for 1855 using a climate model simulation. Using this "conditional event attribution" approach they found that anthropogenic drivers increased the magnitude of heavy northeast Colorado rainfall for the wet week in September 2013 by 30%, with the occurrence probability of a week at least that wet increasing by more than a factor of 1.3. By comparing the convective and large-scale components of rainfall, they found that this increase resulted in part from the additional moisture-carrying capacity of a warmer atmosphere – allowing more intense local convective rainfall that induced a dynamical positive feedback in the existing larger-scale moisture flow – and also in part from additional moisture transport associated with larger-scale circulation change.

This process has now been applied to a number of examples of extreme precipitation, flooding, and hurricanes and the results clearly show how the changed environment has altered the outcome. As expected, a 10% increase in atmospheric moisture immediately contributes to 10% heavier rains (or snows), but the results have often suggested a change of order 30% resulting from the longer-lasting, more intense and bigger storms; see also Section 10.4.

As noted by Trenberth (2012):

> *Scientists are frequently asked about an event "Is it caused by climate change?" The answer is that no events are "caused by climate change" or global warming, but all events have a contribution. Moreover, a small shift in the mean can still lead to very large percentage changes in extremes.*
>
> *The answer to the oft-asked question of whether an event is caused by climate change is that it is the wrong question. All weather events are affected by climate change because the environment in which they occur is warmer and moister than it used to be.*

Trenberth et al. (2015) went on:

> *Here, we suggest that a different framing is desirable, which asks why such extremes unfold the way they do. Specifically, we suggest that it is more useful to regard the extreme circulation regime or weather event as being largely unaffected by climate*

> *change, and question whether known changes in the climate system's thermodynamic state affected the impact of the particular event.*

In attribution studies, changing the null hypothesis from "there is no anthropogenic global warming effect" to one that recognizes the changed environment can completely change the outcome. Because the questions are posed differently, their answers have a different meaning and the conditional approach focuses more on impacts. However, the resistance of the mainstream attribution community to adopting the ideas put forward was picked up on and formed part of the topic of a special study by the National Research Council (2016), *Attribution of Extreme Weather Events in the Context of Climate Change*; which found that the strongly conditioned approach is completely acceptable and that the traditional approach is limited by the adequacy of the modeling tools available.

In particular, as the consideration turns from extremes such as heatwaves to extreme precipitation and storms, the strongly conditioned approach has much to recommend it. The result is that *"While we cannot say that these events were due to global warming (poorly posed question), it is **highly likely** that they would not have had such extreme impacts without global warming!"*

Hence, human-induced climate change is real, but it intersects with natural variability. It is when the natural variability is going in the same direction as climate change that records are broken, and measurements go outside of previous bounds. The result is that thresholds are crossed and things break, or burn, or critters die! Moreover, it has substantial costs – which are typically underestimated by economists.

15.3 Storylines

In the light of increasing demands for attribution in near real time, and other actionable information that enables adaptation and the building of resilience to expected threats, it is important to recognize the physical processes going on that cause changes. A key aspect of this requirement is the representation of uncertainties. The conventional approach to representing uncertainty in physical aspects of climate change is probabilistic, based on ensembles of climate model simulations. Yet the models are not perfect and involve assumptions and approximations (see Chapter 16). In contrast to weather forecasting, where many forecasts are issued and verified each day, the shortness of the observational record from a climate change perspective makes it very difficult to directly assess the reliability of future climate projections.

As developed in a somewhat ad hoc manner for attribution, as given above, a more formal alternative approach called a "storyline" is emerging. This approach does not seek to quantify probabilities, but instead develops descriptive "storylines," "narratives" or "tales" of plausible future climates. A storyline is a physically self-consistent unfolding of past events, or of plausible future events or pathways. No a priori probability of the storyline is assessed; emphasis is placed instead on understanding the driving factors involved, and the plausibility of those factors.

Storylines also offer a powerful way of linking physical with human aspects of climate change. However, storylines can be perceived as anecdotal and thus unscientific. It is therefore important to understand their basis and how they contribute to representing and communicating uncertainty in the physical aspects of climate change.

For example, a storyline of a past event could reveal the actions taken in response, whether useful or not, and thus which mitigating actions against possible future events of a given type might work, and which did not. Preparation for future events could be built on a storyline based on this event (or ones of a similar type, real or imagined) combined with climate change information. This may also be regarded as an analog approach for the impacts and adaptation aspects of an event. Regardless of what caused a flooding, how best to respond may benefit enormously from past experience.

For example, the increased moisture content in a warmer atmosphere would lead to increased moisture transport and increased risk of flooding. Climate change can also be expected to decrease the probability of rain-on-snow events in the autumn, but to increase it in the winter. This change in seasonality could matter for impacts since the ground is potentially frozen during winter, which would result in increased direct surface runoff rather than infiltration of water into the soil, and hence more flooding. On the other hand, the snowpack is likely colder during winter and will thus require more energy input to be heated to the melting point, which could reduce the amount of flooding.

Behavioral psychology shows that humans have difficulty responding rationally to risks from events from outside of their experience, even when accurate quantitative information on these risks, and the benefits of rational mitigation actions, is available. The tendency is to act as though the probability of a bad outcome is less than it really is if an event of that type has not happened recently (or ever), and more than it really is, if it has. For instance, building new infrastructure to prevent the next disaster is much more likely immediately after such a disaster took place. In contrast, studies of such an event before it has happened are much less likely to get support for mitigation actions. Examples are the flooding of the subway system with super storm Sandy in New York in 2012 and the flooding in Houston with hurricane Harvey in 2017. Despite many warnings ahead of time, flood control and abatement measures were not implemented. Often this applies also to individuals in failing to take out flood insurance or continuing to build in vulnerable coastal locations. Many new examples abound from the COVID-19 pandemic responses.

The uncertainty in future global temperature increase is substantial and relates to models and climate sensitivity uncertainties. The observed warming provides only a very weak constraint on model performance, and its direct applicability is a matter of considerable current debate. A storyline approach attempts to articulate the much richer physical understanding of how climate processes change as climate warms to provide a more mechanistic way of constraining the problem. In all of these cases, it is also important to note that the risks come not just from climate change but all aspects of sustainable development. This is readily incorporated into the storyline approach. Storylines can raise risk awareness by framing risk in an event-oriented rather than a probabilistic manner.

An approach led by Ted Shepherd (University of Reading, UK) has a typology of four reasons for using storylines to represent uncertainty in physical aspects of climate change:

(i) improving risk awareness by framing risk in an event-oriented rather than a probabilistic manner, which corresponds more directly to how people perceive and respond to risk;
(ii) strengthening decision-making by allowing one to work backward from a particular vulnerability or decision point, combining climate change information with other relevant factors to address compound risk and develop appropriate stress tests;
(iii) providing a physical basis for partitioning uncertainty, thereby allowing the use of more credible regional models in a conditioned manner; and
(iv) exploring the boundaries of plausibility, thereby guarding against false precision and surprise.

15.4 The Hiatus

Global warming took off in terms of the increases in GMST in the mid-1970s (Fig. 2.4), and the increases in temperature went well above the noise of natural variability: every decade after the 1960s was warmer than the one before, and the decade of the 2010s has been the warmest on record by far. The warmest year in the 20th century was 1998, a major El Niño year. However, the warming stalled, or paused after that, and by about 2010 until 2015 deniers of climate change were becoming vociferous and proclaiming that global warming was a myth. The depiction was as captured in Fig. 15.2. From 1998 to 2013, the global surface warming rate (Chapter 2; Fig. 2.4) was significantly lower than in the previous

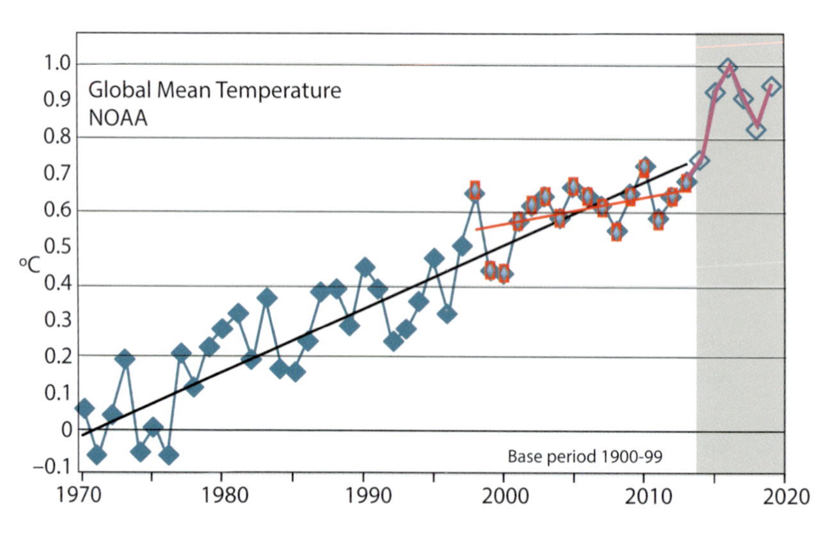

Fig.15.2 Global mean surface temperature from NOAA, as anomalies relative to 1900–99 plotted with linear trends for 1970–2013 (blue) and 1998–2013 (red). The more recent values, unavailable at the time, are plotted in violet with open square points (shaded gray). Updated and extended from Trenberth et al. (2014)

25 years. Some have termed this a global warming hiatus, although it is also referred to as a "pause."

Of course, with hindsight of the GMST values since then, where the GMST value for every single year is higher than any that went before 2014, it is readily apparent that this was just an unduly long pause. The hiatus ended in 2013. So, what was going on to bring this about?

It is readily apparent that there is variability in GMST from year to year (Figs. 2.4 and 15.2), and decade to decade; this is expected and thought to arise largely from internal natural variability. In fact, there are two hiatus intervals featuring much lower rates of increase. The first was from about 1943 to 1975, and the second was from 1999 to 2013.

Increased warming from increases in heat-trapping greenhouse gases can be offset by visible pollution and atmospheric aerosols, which are mostly also a product of fossil fuel combustion. For 1945 to 1970 there were increases in pollution in the atmosphere arising from post–World War II industrialization in Europe and North America, especially over the Atlantic, and some volcanic activity that increased aerosols in the stratosphere. The clean air acts of the 1970s in developed countries then brought that era to an end. Almost certainly, this forcing was a factor in the first hiatus.

After the signal of human-induced climate change emerged from the noise of natural climate variability in the 1970s, expected rates of change were very much in step with the rate observed from 1975 to 1998, but not the slower rate from 1999 on. The year 1998 featured a mini–global warming associated with the biggest El Niño event on record – the 1997–98 event. Prior to that event, ocean heat that had built up in the tropical western Pacific and, during the event, it spread across the Pacific and into the atmosphere, invigorating storms and warming the surface especially through latent heat release, while the ocean cooled from evaporative cooling. It was not until 2015 that another strong El Niño took place, signaling the end of the hiatus.

For 1999–2013, the hiatus in surface warming was mainly evident in the central and eastern Pacific. There is strong decadal variability in the Pacific, known in part as the Pacific Decadal Oscillation (PDO) or Inter-decadal Pacific Oscillation (IPO) – the former is northern hemisphere focused but the two are closely related (see Chapter 11). The PDO was a major player in these hiatus periods, as has been well established by observations and models. There were major changes in Pacific trade winds, sea-level pressure, sea level, rainfall, and storm locations throughout the Pacific and Pacific-rim countries, but the changes also extended into the southern oceans and across the Arctic into the Atlantic. The effects were greatest in winter in each hemisphere. There is good but incomplete evidence that these changes in winds alter ocean currents, ocean convection and overturning, resulting in changes in the amount of heat being sequestered at greater depths in the ocean during the negative phase of the PDO. The result (Fig. 11.8) is that during the positive phase of the PDO the GMST increases, while during the negative phase it stagnates. The latter effect has been successfully modeled. Results suggest that total Earth's Energy Imbalance is largely unchanged with the PDO, but during the positive phase more heat is deposited in the upper 300 m, where it can influence the GMST, while in the negative phase, more heat is dumped below 300 m, contributing to the overall warming of the oceans, but likely irreversibly mixed and lost to the surface.

Although the Atlantic also features decadal variability, it appears that the Pacific has dominated recently. The Arctic has also featured large changes in recent years, somewhat out of step with the hiatus. While Arctic amplification is important, the evidence suggests that it is mainly responding to influences from elsewhere, and especially the Pacific, although feedbacks from the snow/ice–albedo mechanisms no doubt help amplify the changes.

The conclusion is that although human-induced climate change is relentless and largely predictable, at any time and especially locally it can be masked by natural variability whether on interannual (El Niño) or decadal timescales. The predominant driver of the slowdowns in GMST is the PDO which redistributes heat within the ocean.

A simple illustration is given of how decadal variability plus a trend can lead to pauses in GMST (Fig. 15.3). Indeed, the combination of decadal variability plus a

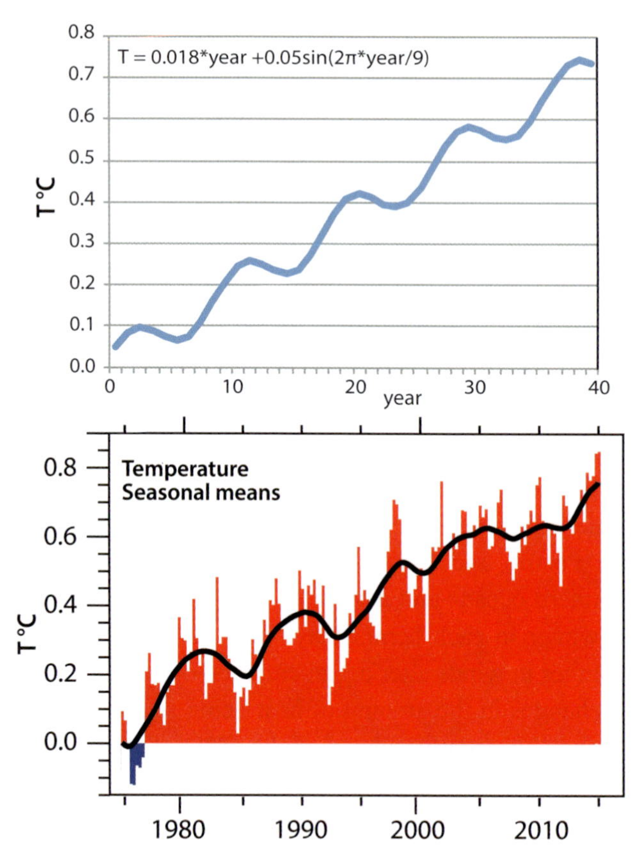

Fig. 15.3 The top panel is the superposition of a simple sine wave with amplitude 0.05°C and a period of 9 years combined with a linear trend in GMST of 0.18°C/decade, to be compared roughly with the 40-year period of GMST in the bottom panel, where the black line is a decadal smoothing. Why 9 years? In the observations, the dips come from two volcanic eruptions, El Chichón in April 1982 and Mt. Pinatubo in June 1991, and then the switch to negative PDO following the 1997/98 El Niño event.

trend from increasing greenhouse gases makes the GMST record more like a rising staircase than a monotonic climb (Fig. 11.8). Figure 15.3 shows how variability can indeed lead to a pause in the climb in GMST but with consequences for more abrupt changes at other times.

Some conclusions are:

1. While there was a hiatus in the rise of GMSTs from 1999 to 2013, there was no slowdown in sea-level rise and only up to about 20% of the hiatus can be linked to changes in the energy imbalance and changes in the Sun and volcano pollution.

2. Global warming continues but can be manifested in different ways than simple increases in surface temperature, as seen through rising sea levels, increasing ocean heat content, and melting Arctic sea ice and land ice.
3. Natural variability played the major role in the hiatus, especially through the PDO, which changes winds and ocean currents in such a way as to affect how much heat gets deposited into the deeper ocean, below 300 m depth.
4. The positive phase of the PDO from 1976 to 1998 enhanced surface warming somewhat by reducing the amount of heat sequestered by the deep ocean, while during the 1999–2013 negative phase of the PDO more heat was deposited at greater depths, contributing to the overall warming of the oceans but cooling the surface somewhat relatively.
5. The hiatus is long over, and the six years since 2014 are the warmest on record.

References and Further Reading

National Research Council, 2016: *Attribution of Extreme Weather Events in the Context of Climate Change*. Washington, DC: The National Academies Press, 165pp. doi: 10.17226/21852.

Pall, P., C. M. Patricola, M. Wehner, et al., 2017: Diagnosing conditional anthropogenic contributions to heavy Colorado rainfall in September 2013. *Weather Climate Extremes*, **17**, 1–6. doi: 10.1016/j.wace.2017.03.004.

Shepherd, T. G., E. Boyd, R. A. Calel, et al., 2018: Storylines: An alternative approach to representing uncertainty in climate change. *Climatic Change*, **151**, 555–571. doi: 10.1007/s10584-018-2317-9

Trenberth, K. E., 2011: Attribution of climate variations and trends to human influences and natural variability. *WIREs Climate Change*, **2**, 925–930. doi: 10.1002/wcc.142.

Trenberth, K. E., 2012: Framing the way to relate climate extremes to climate change. *Climatic Change*, **115**, 283–290. doi: 10.1007/s10584-012-0441-5.

Trenberth, K. E., 2015: Has there been a hiatus? *Science*, **349**, 691–692. doi: 10.1126/science.aac9225.

Trenberth, K. E., and J. T. Fasullo, 2013: An apparent hiatus in global warming? *Earth's Future*, **1**, 19–32. doi: 10.002/2013EF000165.

Trenberth, K. E., J. T. Fasullo, G. Branstator, and A. S. Phillips, 2014: Seasonal aspects of the recent pause in surface warming. *Nature Climate Change*, **4**. doi: 10.1038/NCLIMATE2341.

Trenberth, K. E., J. T. Fasullo, and T. G. Shepherd, 2015: Attribution of climate extreme events. *Nature Climate Change*, **5**, 725–730. doi: 10.1038/NCLIMATE2657.

16 Prediction and Projection

Climate models will be essential for improved predictions of the future but have real shortcomings. The IPCC framework for assessments is outlined along with future emissions scenarios.

16.1 Using Models

Earth's energy imbalance is associated with the forcings of the climate system and how it responds, including all of the feedbacks. EEI is actually the net outcome, but climate models facilitate the understanding of the actual flows of energy through the climate system and in turn are a key step toward estimating the consequences. Hence, climate models are extensively used to encapsulate the knowledge and understanding of the climate system and how it works, and for making projections of the future. However, as noted in Chapter 1, the models are not perfect and involve approximations and assumptions. The IPCC approach has been very democratic in that all models have been treated equally even though some models have been shown to contain substantial errors. It makes sense to use only the models that perform well. Of course, all models contain errors, but some models actually violate physical principles, such as conservation of mass or energy, and these should not be used. Certainly, some models are a lot better than others.

Experience suggests that some models perform well for some variables but not others, and often this difference stems from how the models are constructed. In particular it may relate to how they get tuned to match the climate system observations. It has been a delicate and controversial topic to set some models aside, because that may have political implications for the associated modeling group in terms of future funding and credibility. Further, it is not right or fair to evaluate a model for a variable that it has been tuned to match.

Models are generally created with an emphasis on including the processes in all aspects of the climate system to the extent possible. But when this is done, the climate of the model may not look that much like the observed. Various tweaks are made to parameters within their recognized uncertainty range. But there is no standard set of variables used for this process. Usually, a final tweak is to obtain a balanced TOA radiation in the absence of changing forcing, and this can usually be done by adjusting aspects of the cloud parameters, particularly low cloud. Parameters often tuned may include cloud drop size, relative humidity for condensation, entrainment values in convection, autoconversion rates for how much condensate is converted to rainfall and how much ice goes to snow, and cloud extent and brightness.

Attention can be given to solar radiation and infrared radiation separately, as well as net radiation, usually mostly for the global mean, although meridional structure and seasonal variations may also be considered, especially when trying to reconcile a coupled model to one where the SSTs are specified. A lot of attention is paid to GMST and how it changes over the 20th century with estimated forcing. This also means paying attention to SST. Tropical mean values of radiation, cloud and precipitation may also receive scrutiny. Sea-level pressure fields are another favored metric. Some groups also pay attention to interannual variability associated with ENSO. Volcanic responses are also examined and tuned.

In terms of getting the observed 20th-century warming about right, it has been found that there is a strong trade-off between the model climate sensitivity and aerosol forcing specified. A model with high climate sensitivity, and thus large response to increasing carbon dioxide, can be reined in by having a larger aerosol forcing that produces cooling. The uncertainties in both of these quantities have allowed quite different models to fit the observations to a reasonable degree. But they have very diverse implications for the future. Models are useful for many purposes, but they can easily be abused, and they should not be used as black boxes (Fig. 16.1) without full understanding of the approximations, assumptions, limitations and strengths. Models are tools and can be extremely valuable if used appropriately.

The process of model development never ends. In general, each new generation of models does show improvements. Older versions of the models, which, it can be argued, are better evaluated in the literature and somewhat understood, are cast

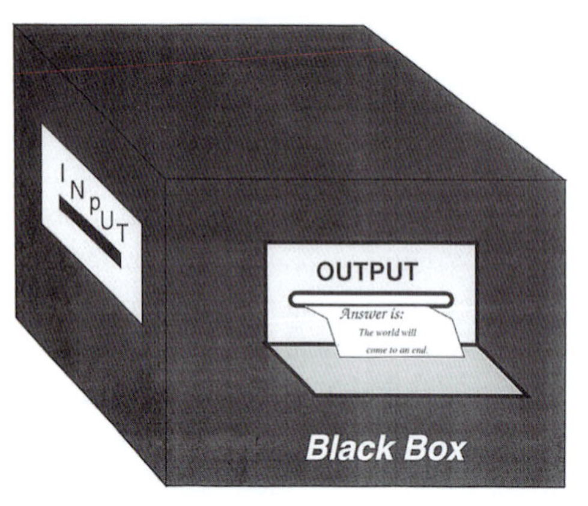

Fig. 16.1 Climate models are based on mathematical formulations of the laws of physics and thermodynamics. However, models are too often used as a "black box" and the output taken for granted without proper appreciation of the assumptions and approximations.

aside for the latest and greatest. While some IPCC models deliberately have modest evolutions, some are "bleeding edge" models that are not (yet) tried and tested. Transparency is a desirable goal but one that is easily undermined.

One way to test models is to simulate the paleoclimate where the past changes have been very large. However, as well as uncertainties in the climate, there are also considerable uncertainties in the forcings as time is extended backwards. Nevertheless, the paleoclimate reconstructions have been used as part of model and climate sensitivity evaluations. Moreover, the utility of models has been demonstrated by providing useful insights into past climate and interpretation of the proxy data.

The question is whether models are similar enough in relevant ways to the real world that it is possible to learn from the models and draw conclusions about the real world. The job of scientists is to find out where this is the case, and where it is not, and to quantify the uncertainties. For that reason, statements about future climate in IPCC always have a likelihood attached, and numbers have uncertainty ranges. All models contain flaws that are often difficult to pin down because of their complexity and sophistication, but they are so much better than any "back-of-the envelope" guesses, and the shortcomings and limitations are known.

These models can be used to guide decisions about what policies to follow, such as what areas or aspects are most vulnerable, how to reduce undesirable climate impacts and build resilience, and to plan for the future in the most cost-effective ways. Because human actions themselves are not predictable, these are not predictions, but rather they are called projections and depend on the nature of the "what if" question.

16.2 Climate Projections

With the increased recognition that Earth's climate is changing due to human activities comes a growing demand by decision-makers for reliable, quantitative climate information appropriate for use in the context of assessments for climate change adaptation, impacts and vulnerability. Climate models can be used for this purpose and, indeed, a goal is to be able to make actual climate predictions. This is extremely challenging, because it requires:

(i) a very good climate system model that is unbiased;
(ii) a complete knowledge of the observed current state of all aspects of the climate system to provide the starting point; and
(iii) specification or the ability to predict the forcings of the climate system, including human influences.

The example for prediction followed here is that in weather forecasting. Weather forecasts are issued daily for a few weeks and are routinely verified and assessed, so that biases and the nature of errors can be readily determined. Weather forecast numerical weather prediction models continue to improve, and that includes their resolution, which was down to about 10 km for the horizontal grid size in 2020. Commonly, they are very good at most things but may be prone to fail at other times. The identified biases can then be allowed for in translating the model output to an actual forecast for public use. However, this is not the case for climate forecasts of decades or more. Instead, the forecasts are conditional on certain assumptions and they are intended as decision-making tools but should not be taken literally.

Box 16.1: Models All models are wrong, some are useful

Often attributed to George Box.
The utility of models relates to how they are used and abused.

Unfortunately, it is not yet possible to fully specify the observed state of the climate at any time, as found in Chapter 14 in trying to determine Earth's Energy Imbalance. Ocean analyses have improved enormously for the top 2000 m of the ocean, but not yet to a point where different analyses agree adequately, especially on sub-annual timescales. Knowledge of the state of energy tied up in the land and ice systems is even worse.

On a 30-year time frame, global climate projections are simplified by being only weakly sensitive to different emissions scenarios for long-lived greenhouse gases, although how aerosols change is of concern. Aerosol emissions are usually assumed to continue as they have been. Emissions by humans vary enormously and capriciously and, as the lifetime of aerosols is only of the order of 1 week, the emissions matter. However, they are not predictable, and rapid large changes can occur, as experienced during the COVID-19 pandemic. Volcanic eruptions are not predictable, except that once one has occurred, the consequences for stratospheric aerosol may be predictable for a year or two. On longer time frames, it becomes essential to predict human behavior to determine the net emissions of all radiatively important substances, and this too is impossible. It would require predictions of election results, outcomes of conflicts, how many people there are and where they are, and their standard of living, among other things. Finally, climate models continue to improve but nonetheless contain substantial biases, simplifications, and errors.

Accordingly, for the short term, perhaps up to 30 years or so, reasonable assumptions and extrapolations can be used along with bias-corrected models to make predictions, and there is good reason to think that the initial state provides some predictability for maybe a decade or so. El Niño predictions have been operational with some success for up to about a year ahead, but predictability beyond about 18 months falls off.

More generally, instead of making *predictions*, climate *projections* are made. Usually these do not start from the observed state and instead, the forecast is of the changes that occur, rather than the absolute values, thereby removing systematic biases. In addition, instead of guessing at the forcings, a so-called "emissions scenario" is conjured up as a "what if" situation.

- What if the carbon dioxide in the atmosphere had not increased due to human activities?
- What if carbon dioxide continues to increase at 1% per year for 100 years?
- What if humans fail to rein in their consumption of fossil fuels and a "business as usual" scenario unfolds?
- What if, instead, all human net emissions are brought to zero in 50 years?
- What if the climate changes as projected in the model, then what would the impacts be on agriculture and society?
- What if those things happened, then what strategies might there be for coping with the changes?

These kinds of questions are well posed, and answers can be reasonably provided using climate models, impacts models, and Integrated Assessment Models (IAMs). They are all very legitimate questions for scientists to ask and address. The first set involve the physical climate system. The others involve biological and ecological scientists, and social scientists, and they may involve economists, as happens in a full IPCC assessment.

Sidebar 16.1: The Intergovernmental Panel on Climate Change (IPCC)

In 2007, the Fourth Assessment Report of the Intergovernmental Panel on Climate Change (IPCC), known as AR4, clearly stated that "***Warming of the climate system is unequivocal***" and it is "***very likely***" due to human activities. Later in 2007, the IPCC won the Nobel Peace Prize, jointly with Al Gore Jr., "for their efforts to build up and disseminate greater knowledge about man-made climate change, and to lay the foundations for the measures that are needed to counteract such change."

The IPCC is an international organization that includes a panel of governments and a body of scientists from around the world convened by the United Nations jointly under the United Nations Environment Programme (UNEP) and the World Meteorological Organization (WMO) and initiated in 1988. Its mandate is to provide policymakers with an objective assessment of the scientific and technical information available about climate change, its environmental and socioeconomic impacts, and possible response options. The IPCC reports on all aspects of the science of global climate and the effects of human activities on climate in particular. Major assessments were made in 1990, 1995, 2001, 2007 and 2013, and are called the First, Second and Third Assessment Reports (FAR, SAR and TAR) and AR4 and AR5 for the Fourth and Fifth Assessment Reports.

Each new IPCC report reviews all the published literature over the previous 5–7 years, and assesses the state of knowledge, while trying to reconcile disparate claims and resolve discrepancies, and document uncertainties. There is a *Technical Summary* and a short *Summary for Policy Makers* (SPM), and the volume from each of the three Working Groups (WGs) has tended to run to about 1000 pages or more; see Figs. 16.2 and 16.3.

WG I deals with how the climate has changed and the possible causes. It considers how the climate system responds to various agents of change and the ability to model the processes involved as well as the performance of the whole system. It further seeks to attribute recent changes to the possible various causes, including the human influences, and thus it goes on to make projections for the future (Fig. 16.3). WG II deals with impacts of climate change, vulnerability, and options for adaptation to such changes; and WG III deals with options for mitigating and slowing the climate change, including possible policy options (Fig. 16.3). Each WG has two co-chairs, one from a developing country and one from a developed one. Each WG is staffed by a small Technical Support Unit (TSU) that is hosted by one of the co-chairs of the WG.

The IPCC also includes a Task Force on National Greenhouse Gas Inventories to oversee the National Greenhouse Gas Inventories Programme, and to encourage its use by parties of the United Nations Framework Convention on Climate Change (UNFCCC). The Task Group on Data and Scenario Support for Impacts and Climate Analysis was established to facilitate cooperation between the climate modeling and climate impacts assessment communities by increasing the availability of climate change related data and scenarios for climate analysis and impacts, adaptation, vulnerability, and mitigation research. Along with comprehensive Assessment Reports, the IPCC has produced several Special Reports and Technical Papers on topics of interest, as well as Methodology Reports. Many of these reports are prepared in response to requests from the UNFCCC or from other international organizations and conventions.

The IPCC is an intergovernmental body. It is open to all member countries of the United Nations (UN) and WMO. Currently 194 countries are members of the IPCC. Governments participate in the review process

Sidebar 16.1: *(cont.)*

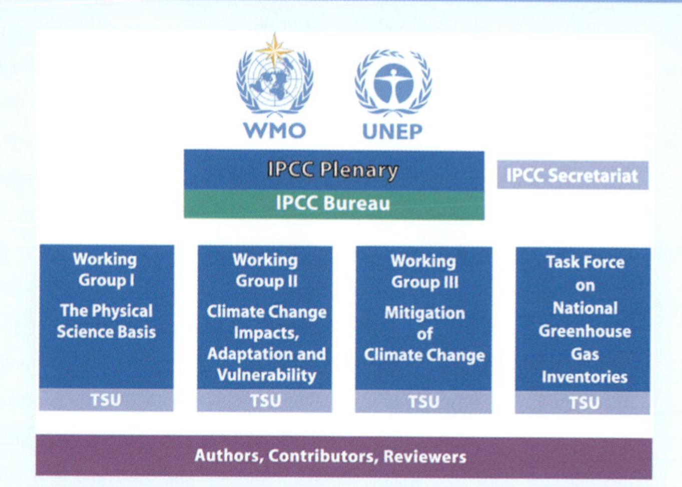

Fig. 16.2 The structure of IPCC under the World Meteorological Organization and the United Nations Environment Programme, with three WGs. Each has a Technical Support Unit (TSU).

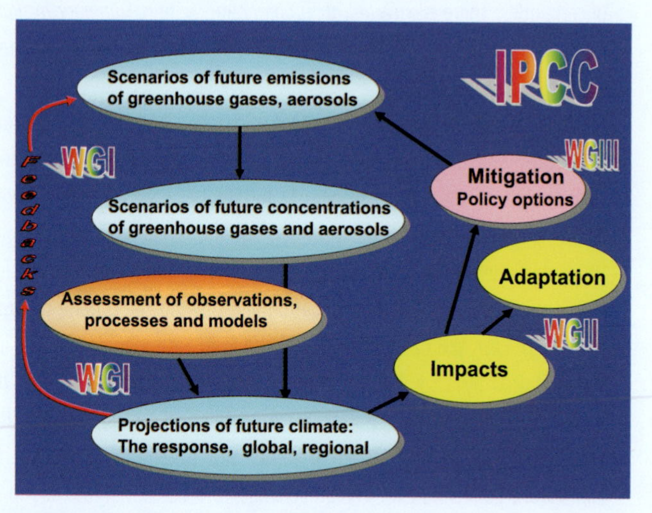

Fig. 16.3 The activities of WG I and the flow of information and feedbacks to WG II on impacts and adaptation, and WG III on mitigation and policy options.

and the plenary sessions, where main decisions about the IPCC work program are taken and reports are accepted, adopted, and approved. The IPCC Bureau Members, including the Chair, are also elected by governments during the plenary sessions. The IPCC Secretariat in Geneva coordinates all the IPCC work and liaises with governments. Thousands of scientists from all over the world contribute to the work of the IPCC on a voluntary basis.

Sidebar 16.1: *(cont.)*

The IPCC Bureau comprises the IPCC Chair, the IPCC Vice-Chairs, the Co-Chairs and Vice-Chairs of the Working Groups, and the Co-Chairs of the Task Force. The IPCC Bureau is chaired by the IPCC Chair. Coordinating Lead Authors (CLAs) and Lead Authors are selected by the relevant Working Group/Task Force Bureau from those experts cited in nominations provided by governments and participating organizations, and other experts as appropriate, known through their publications and works, and considering both geography and gender. None are paid by the IPCC. Coordinating Lead Authors take overall responsibility for coordinating a major section of a report and the chapter for which they are responsible. Lead Authors work in teams to produce the content of the chapter and may enlist Contributing Authors who provide additional technical information.

The IPCC starts a new assessment with a scoping process among experts which leads to the proposed general outline of a report, highlighting also new features and cross-cutting matters. Policymakers and other users of IPCC reports are consulted in order to identify the key policy-relevant issues along with experienced scientists who can attest to the state and capabilities of the science. The outlines are subject to formal approved by the Panel before work begins.

Each WG is made up of participants from the United Nations countries, and involves several hundred Lead Authors, many more Contributing Authors, and thousands of reviewers from over 130 countries who provided tens of thousands of review comments. The IPCC process is very open. Two major reviews are carried out in producing the report: a first review by experts and a second review by experts and governments. Climate "skeptics" can and do participate, some as authors. All comments are responded to in writing and result in many changes in the report. The process is overseen by two or more Review Editors for each chapter.

Summaries for Policymakers (SPMs) are prepared after the first expert review rather than concurrently with the main reports. They undergo one round of expert and government review. Each point undergoes careful scrutiny not only by the scientists, but also by government officials and nongovernmental organizations (NGOs). A second opportunity for comments by governments takes place as input to the approval process. The SPM is then approved line-by-line by governments in a major meeting, which takes place over 3 or 4 days, often well into the nights. Those participating include government representatives (typically more than 120 countries are present), several NGOs as observers, and typically 40–50 scientists. Simultaneous translation and interpretation occur throughout the meeting into English, French, Spanish, Russian, Chinese and Arabic, per UN practice.

The role of the IPCC is to provide policy-relevant but not policy-prescriptive scientific advice to policymakers and the general public. IPCC scientists, with all kinds of value systems, ethnic backgrounds, and from different countries, gather together to produce the best consensus description of what they jointly understand, with appropriate statements about confidence and uncertainty. Negotiations occur over wording to ensure accuracy, balance, clarity of message, and relevance to understanding and policy. The strength is that it is a consensus report, but the process also makes it a conservative report. The rationale is that the scientists determine what can be said, but the governments determine how it can best be said.

This sidebar is based upon Trenberth (2015).

Box 16.2: Integrated Assessment Models

IAMs are mathematical computer models based on explicit assumptions about how the coupled human and natural system behaves. The modules within the IAM are often greatly simplified, but there may be many modules describing the climate, carbon cycle, and various human dimensions including demographics (population), politics, economics, standards of living, and societal behavior. The strength of an IAM is its ability to calculate the consequences of different assumptions and to interrelate many factors simultaneously, but an IAM is constrained by the quality and character of the assumptions and data that underlie the model.

To explore how robust the results are, and to try to remove the dependence of results on a single model, intercomparison experiments have been set up under the auspices of the World Climate Research Programme through the Climate Model Intercomparison Project: CMIP. In turn the results have been used extensively by the IPCC for making future projections of climate under a number of specified emissions scenarios. The most recent CMIP is CMIP-5, and CMIP-6 is continuing in 2021.

16.3 Emissions Scenarios and Future Pathways

For the Third and Fourth IPCC Assessments the projections were based on a series of greenhouse gas emission storylines ("narratives"), from a Special Report on Emissions Scenarios (SRES), which made a range of assumptions about future energy consumption, technological innovations, land use allocations, and other human activities. These were used as input to Integrated Assessment Models. Starting from the forcings, and the emissions, this provided calculations of greenhouse gas concentrations and radiative forcings, allowing the climate modeling community to generate projections of global climate change. There are stresses other than climate change that must be considered when trying to understand vulnerabilities, which are a product of the sensitivity, anticipated exposure, and adaptive capacity of the system or region being impacted. Along with the climate fields, estimates of impacts were obtained, which, in turn, were used to identify adaptation and resilience options and to assess their effectiveness.

However, the emissions scenarios were not always consistent with those used by the impacts, adaptation and vulnerability communities that were based on a variety of socioeconomic and environmental (including climatic) baselines. A substantial problem that also emerged was the cascade of uncertainties and errors.

Box 16.3: SRES Storylines for Emissions Scenarios

A1: a future world of very rapid economic growth, global population that peaks in mid-century and declines thereafter, and rapid introduction of new and more efficient technologies.

A2: a very heterogeneous world with continuously increasing global population and regionally oriented economic growth that is more fragmented and slower than in other storylines.

B1: a convergent world with the same global population as in the A1 storyline but with rapid changes in economic structures toward a service and information economy, with reductions in material intensity, and the introduction of clean and resource-efficient technologies.

B2: a world in which the emphasis is on local solutions to economic, social, and environmental sustainability, with continuously increasing population (lower than A2) and intermediate economic development.

Quite aside from the highly idealized emissions scenario, assumptions, uncertainties and errors arise in generating concentrations of atmospheric gases, whether done by a comprehensive carbon system model or a much simpler model. In turn the concentrations were then fed into climate models to generate climate variables such as temperature and precipitation. These were further fed into impacts models to generate the effects on society and activities, such as farming, and those results were further fed into economic models, often in very simplified form (e.g., just using the changes in GMST). At every step along the way, the uncertainties were magnified to the point where the results could become meaningless. Of course, many uncertainties cancel because mass, energy, and other quantities are conserved, but it was difficult to build this into uncertainty analysis.

To remove some of the largest growth of uncertainties, it was subsequently decided that it was better to specify the concentrations of the atmospheric components and their changes over time, along with their radiative forcing and then work both backwards and forwards. Working backwards allowed the corresponding emissions to be computed along with the flexibility of redistributing the emissions by country. Working forwards, the *Representative Concentration Pathways* (RCPs), as they were called, pinned down a lot of aspects and allowed the corresponding climate change to be computed with reasonable limits. RCPs were also designed to provide new baselines to account for rapid growth in, for example, China and India, not previously accounted for.

The IPCC report in 2013 used a number of RCPs tied to possible radiative forcings in 2100. The values chosen for the RCPs range over 2.6, 4.5, 6.0, and

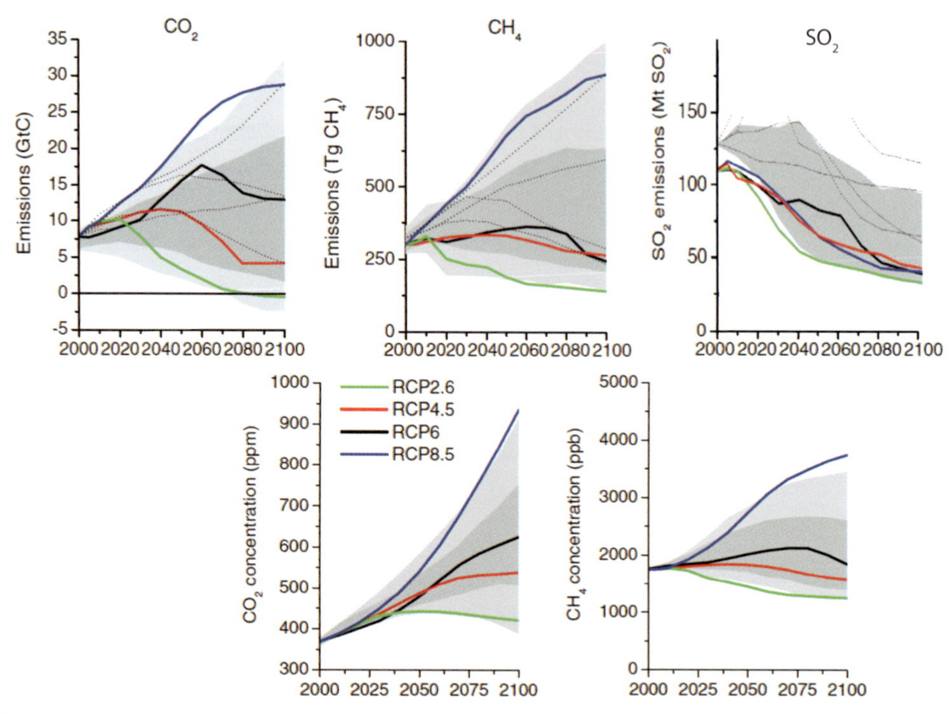

Fig. 16.4 Emissions of CO_2, CH_4, and SO_2 across the RCPs from 2000 to 2100 (top) and concentrations for CO_2 and CH_4 (below). The gray areas indicate values in the general literature (90th and 98th percentiles), and the light dotted lines relate to the SRES values. Adapted from van Vuuren et al. (2011)

8.5 watts per square meter in 2100, relative to 1750. These are transient, not equilibrium values, but may be compared with the 3.7 W m^{-2} that corresponds to the equilibrium forcing for a doubling of pre-industrial carbon dioxide concentrations (from 280 to 560 parts per million by volume [ppm]). Including also the prescribed concentrations of methane and nitrous oxide, the combined CO_2-equivalent concentrations in 2100 are 475 ppm (RCP2.6), 630 ppm (RCP4.5), 800 ppm (RCP6.0), and 1313 ppm (RCP8.5). Current CO_2-equivalent forcing from greenhouse gases is about 3.0 W m^{-2}, although, accounting also for aerosols, the net is about 2.3 W m^{-2}. Hence RCP2.6 is for very low emissions and very unlikely to be realized; while RCP8.5 is closer to business as usual and what has been occurring.

The possible associated emissions of carbon dioxide, methane, nitrous oxide and sulfur dioxide and nitrogen oxides (NOx) were computed and some of these are given in Fig. 16.4 along with the corresponding concentrations for carbon dioxide and methane.

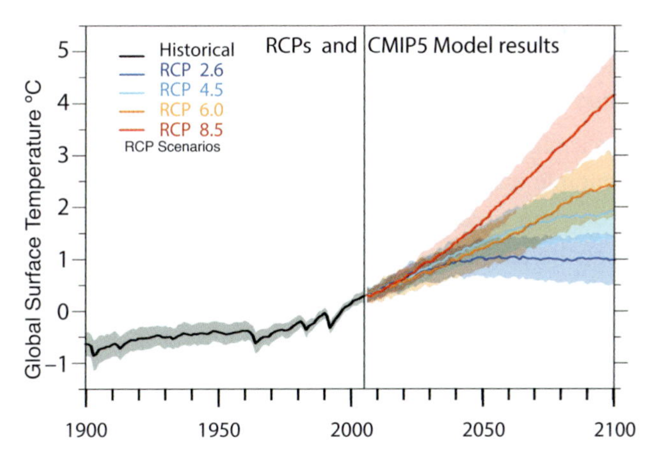

Fig. 16.5 Global mean surface temperature projections under different Representative Concentration Pathways (RCPs) based on Climate Model Intercomparison Project phase 5 (CMIP5) climate model simulations. Means are shown by solid lines, and one standard deviation by shading. Zero corresponds to 1986–2005. Mean temperatures from model historical runs are shown in black.

The CMIP5 models in IPCC AR5 were used to simulate GMST from 1900 until 2005, and then projections were made using the RCPs out to 2100 (Fig. 16.5). The biases were removed by using anomalies, and the spread around the mean model shows the natural variability. In Fig. 16.5 the short zigzags correspond to the responses to specified major volcanic eruptions, which each produced cooling for a few years. The model spread going into the future is greater, but by mid-21st century the different RCPs give distinctly different results. As of 2020, the RCP8.5 looks like it will be closest to reality, owing to the ongoing failure to rein in greenhouse gas emissions.

Note that the projections in Fig. 16.5 are all based on the assumed RCPs. None of them is a prediction or forecast. Yet all too often they have been misused as if the IPCC does offer predictions.

In the next IPCC report, using CMIP6 models along with RCPs, a new set of future scenarios called Shared Socioeconomic Pathways (SSPs) will be used to examine how global society, demographics and economics might change over the next century. These SSPs look at five different ways in which the world might evolve in the absence of climate policy and how different levels of climate change mitigation could be achieved when the mitigation targets of RCPs are combined with the SSPs. The SSPs define different baseline worlds that might occur in the absence of any concerted international effort to address climate change, beyond those already adopted by countries. The idea is that population, technological, and economic growth could lead to very different future emissions and warming outcomes, even without climate policy. They include: a world of sustainability-focused growth and equality (SSP1); a "middle of the road" world where trends broadly follow their historical patterns (SSP2); a fragmented world of "resurgent nationalism" (SSP3); a world of ever-increasing inequality (SSP4); and a world of rapid and unconstrained growth in economic output and energy use (SSP5). In turn, these have been characterized as: Taking the green road; Middle of the road; Regional rivalry and a rocky road; Inequality; and Fossil-fueled development. These narratives describe alternative pathways for future society; their characterization is given in italics, and several are quite similar to the SRES storylines.

Box 16.4: SSP Storylines for Emissions Scenarios

SSP1: a world of sustainability-focused growth and equality [*Taking the green road*];

SSP2: a "middle of the road" world where trends broadly follow their historical patterns [*Middle of the road*];

SSP3: a fragmented world of "resurgent nationalism" [*Regional rivalry and a rocky road*];

SSP4: a world of ever-increasing inequality [*Inequality*]; and

SSP5: a world of rapid and unconstrained growth in economic output and energy use [*Fossil-fueled development*].

Integrated Assessment Models were used to translate the socioeconomic conditions of the SSPs into estimates of future energy use characteristics and greenhouse gas emissions. The set of SSPs facilitate integrated research leading to a better understanding not only of the physical climate system consequences of these scenarios, but also of the climate impact on societies. They recognize that global radiative forcing levels can be achieved by different pathways of various greenhouse gases, aerosols, and land use; the set of SSPs therefore establishes a matrix of global forcing levels and socioeconomic storylines.

Simple models suggest the resulting increases in GMST range from about 3.0 to 5.0°C by 2100 relative to pre-industrial values. However, the SSPs then have to be integrated with the RCPs, producing a wide range of possible outcomes that should be useful for policy purposes. All scenarios in the SSP database that keep warming below 2°C incorporate some bioenergy with carbon capture and storage (BECCS). That is to say they overshoot the target and require net negative emissions at various points.

Although the developers of the SSPs make no claim as to the relative likelihood of any scenario, given the recent past, SSP2 is closest to business as usual and the warming by 2100 is likely to be about 4°C, unless substantive policy changes can be implemented globally. Other pathways are possible, but no assessment has been made as to how likely they could be.

References and Further Reading

Hausfather, Z., 2018: Explainer: How 'Shared Socioeconomic Pathways' explore future climate change. www.carbonbrief.org/explainer-how-shared-socioeconomic-pathways-explore-future-climate-change.

IPCC: Intergovernmental Panel on Climate Change, 2013: *Climate Change 2013. The Physical Science Basis*, ed. Stocker, T. F., et al. Cambridge: Cambridge University Press.

O'Neill, B. C., C. Tebaldi, D. P. van Vuuren, et al., 2016: The Scenario Model Intercomparison Project (ScenarioMIP) for CMIP6. Geoscience Model Development, **9**, 3461–3482. doi: 10.5194/gmd-9-3461-2016.

Trenberth, K. E., ed., 1992: *Climate System Modeling*. Cambridge University Press, Cambridge, 788pp.

Trenberth, K. E., 2015: Intergovernmental Panel on Climate Change. In: North, G. R. (ed.-in-chief), J. Pyle, and F. Zhang, eds., *Encyclopedia of Atmospheric Sciences*, 2nd ed., vol. 2. London: Academic Press, 90–94.

USGCRP, 2017: *Climate Science Special Report: Fourth National Climate Assessment, Vol. I*, edited by Wuebbles, D. J., D. W. Fahey, K. A. Hibbard, et al. Washington, DC: US Global Change Research Program, 470pp. https://science2017.globalchange.gov/.

van Vuuren, D. P., J. Edmonds, M. Kainuma, et al., 2011: The representative concentration pathways: an overview. *Climatic Change*, **109**,5 5–31. doi: 10.1007/s10584-011-0148-z.

17 Emissions and Information

The carbon emissions by countries depend on energy intensity, energy efficiency, standards of living and population. Four countries are used for illustration. An informed response in terms of mitigation and adaptation is needed.

17.1 Carbon Emissions

This book has addressed the energy flows through the climate system. It has made the point that the total energy generation by humans is relatively small compared with natural flows, and the main way humans cause climate change is by interfering with the natural energy flows. Ironically, that interference comes mainly from our use of energy and the associated emissions of greenhouse gases and aerosols. In this case, "energy" encapsulates electricity, heat, transport, industrial, and agricultural activities. The latter include changes in land use. The primary source of the errant emissions is the burning of fossil fuels, generating carbon dioxide. A key issue then is how to decarbonize energy systems.

The emissions are tracked according to country, or groups of countries such as the European Union (Fig. 17.1). The EU-28 includes the EU members as of 2019 including the United Kingdom. Emissions declined in Europe after 1990 or so (and this affects the curves for the United States and other regions tracked in Fig 17.1). Emissions leveled off after about 1990 for many other countries, but not for much of Asia, including China and India. The increases in the latter countries stem from extensive use of fossil fuels in order to raise living standards and make electricity more generally available, especially to enable refrigeration.

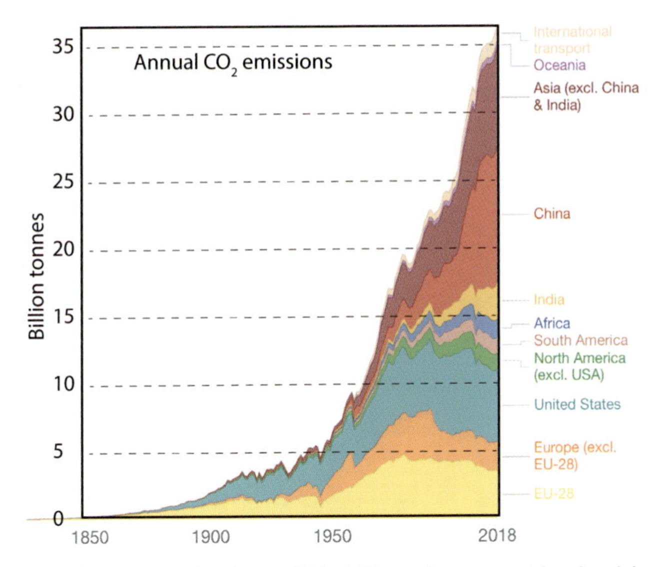

Fig. 17.1 Estimated carbon dioxide emissions for the world in billions of tonnes, with a breakdown by region. Adapted from Our World in Data, downloaded July 24, 2020. Data from Carbon Dioxide Information Analysis Center (CDIAC) and the Global Carbon Project. Adapted from https://ourworldindata.org/.

The year 2020 is highly anomalous owing to the COVID-19 pandemic, and lockdowns appear to have caused reductions in greenhouse gas emissions, of up to 30% for nitrogen oxides. However, preliminary information suggests that sulfate emission declines have offset the associated radiative forcing changes affecting global warming. In addition, the huge increase in carbon dioxide from extensive wildfires, especially in western North America, is not yet accounted for.

Rather than emissions, some political arguments refer to emissions per capita and claim that this should be one metric of allocation of responsibility; but the atmosphere cares not about emissions per capita, only about total emissions. This implies that population should also be a factor. The population and its standard of living relate directly to the demands on precious natural resources that are inherently limited. In this sense, climate change is but part of the major issue of sustainability. Far too many things being done and exploited by humans are simply not sustainable, and it is easy to argue that the world is already overpopulated if the elimination of poverty and better standards of living for all are to be achieved. In the following, a framework is put forward to bring the various factors together.

A number of metrics have been developed to provide insights into the current status of these issues and how they have changed over the years. One metric used is the "*energy intensity*" or *energy efficiency*, which is usually expressed as the energy per GDP (Gross Domestic Product). Similarly, by tracking emissions, another metric is the "*carbon intensity*," expressed as the emission rate per energy

Table 17.1 For 1990, given for the countries at left are: total emissions, carbon intensity, emissions per capita, energy per GDP, electricity generation, GDP per capita, and the population in millions

	Emissions	Carbon intensity	Emissions per capita	Energy per GDP	Electricity generation	GDP per capita	Population
1990	Mt	kg/$	t	kWh/$	TWh	US$ (2011)	Millions
China	2 420	0.87	2.1	2.85	621	2 460	1 180
Germany	1 050	0.53	13.3	2.09	550	25 054	79
NZ	25	0.37	7.5	2.51	32	19 955	3.4
USA	5 120	0.55	20.3	2.42	3 233	36 982	252

unit. The *standard of living* is often expressed as the GDP per capita. Then, of course, how many people and consumers there are also matters.

Accordingly, it is possible to break down the emissions (Fig. 17.1) in several ways beyond the breakdown by country, and one way is as follows:

$$Emissions = \frac{Emissions}{Energy} \cdot \frac{Energy}{GDP} \cdot \frac{GDP}{Capita} \cdot Population$$

The first ratio on the right-hand side is the carbon intensity, the second is the energy intensity, and the third crudely depicts the standard of living. We could also write this as

$$Emissions = \frac{Emissions}{Capita} \cdot Population$$

Perhaps surprisingly, numbers are available for all of these, often over much of the 20th century and beyond, from https://ourworldindata.org/, although the degree to which they are current differs.

Here 1990 is selected as a base value and then 2016 is selected for a more recent value, although in one case 2018 values are included (Tables 17.1 and 17.2). Metrics from four nations have been selected, and the purpose here is to be illustrative, not comprehensive.

Earlier, it was argued that the most pertinent number is actually the accumulated emissions of carbon dioxide, because of its long lifetime. The USA leads by far with 399.4 billion tonnes (bt) through 2017. China was at 200.1 bt, Germany 90.5 bt, and New Zealand (NZ) 1.8 bt. China is included because it now leads the world in total annual emissions. Germany represents Europe as a region that led the way in the Kyoto Protocol (Chapter 18) which was designed to reduce emissions, and New Zealand is included as a small nation.

Table 17.2 For 2016, given for the countries at left are: total emissions, carbon intensity, emissions per capita, energy per GDP, electricity generation in terawatt-hours for 2018, the 2016 GDP per capita in 2011 US$, and the population in millions

	Emissions	Carbon intensity	Emissions per capita	Energy per GDP	Electricity generation	GDP per capita	Population
2016	Mt	kg/$	t	kWh/$	TWh 2018	US$ (2011)	Millions
China	9 700	0.57	6.9	2.10	7 111	12 320	1 379
Germany	802	0.20	9.8	0.97	649	46 841	82
NZ	34	0.22	7.4	1.58	44	34 040	4.7
USA	5 310	0.31	16.4	1.50	4 461	53 015	323

China (28%), the USA (15%), the European Union (10%), and India (7%) were responsible for 60% of global carbon dioxide emissions in 2018. In Tables 17.1 and 17.2, it can be seen that only Germany reduced emissions from 1990 to 2016. A big factor was the drop in carbon intensity, although the energy intensity also declined substantially, and the population increase was modest. Standards of living increased in each country, but by less than a factor of 2, except that the increase in China was a factor of over 5 and accompanied by an order of magnitude increase in electricity generation. Although improved, carbon intensity and energy intensity are high among these nations. The USA stands out for the highest emissions per capita, and its carbon intensity is relatively high compared with Germany and NZ.

Germany's emissions declined by 35.7% from 1990 to 2019. Germany was given a head start in 1990 when, following the fall of the Berlin Wall and reunification, the decline of the East German industrial and power sectors meant automatic CO_2 reductions. Germany is bound by the European Union's Emissions Trading System and operates with a "cap and trade" approach. It had set a goal of cutting emissions by 40% by 2020 and was very successful in renewable energy (40% of power consumption was renewable as of 2019) and improving efficiency. It has been less successful in transportation, although train use has increased. Economic growth has been largely decoupled from emissions.

New Zealand has an unusual greenhouse gas emissions profile for a developed country. Nearly half (48%) of NZ's gross emissions in 2018 (almost all methane and nitrous oxide) came from agriculture (includes animal husbandry). NZ is also unusual in that it generates a large proportion (84% in 2018) of electricity renewably, mainly through hydroelectricity. Nevertheless, NZ had carbon dioxide emissions of 7.9 tonnes per person in 2017, close to average for a developed country. This reflects the continued large carbon dioxide emissions from other sectors such as transport, manufacturing, and construction.

In a broad sense, economic growth is historically linked to growing energy and CO_2 emissions, but countries have different levels of CO_2 emissions per capita despite similar GDP per capita levels. Energy efficiency is related to productivity and technology efficiency, but can also be related to the type of economic activity. The latter reflects the degree to which an economy is manufacturing- versus service-based, as the latter requires less energy. However, carbon efficiency relates to the energy mix, such as fossil fuel versus renewable energy. Global carbon dioxide intensity has been steadily falling since 1990, partly because of increased use of improved technology. On average, carbon intensity is low for low GDP, rises as countries transition from low to middle incomes often in rapidly growing industrial economies, and then, at higher incomes, carbon intensity falls again.

A large amount of carbon dioxide comes from traded goods, so that some countries, like China, produce more carbon dioxide than is generated by the goods they consume, versus other countries, like the United States, that import goods that required energy to produce. In recent years (2015–2018), production-based emissions in China of order 7.0 t have exceeded consumption-based emissions (6.0 t), by 15–20%. In the USA, consumption-based emissions (5.7 t) exceeded production-based (5.3 t). This means assigning blame for emissions can be fraught with complications. However, it is plain that there are huge inequalities and that the world's poorest nations have contributed very little to the carbon dioxide emissions, yet they are often the most vulnerable to climate change.

The goal should be to decarbonize the economy as fast as possible in an economic fashion bearing in mind the health and safety of the processes. Health costs, for instance, can be greatly reduced by cutting down on fossil fuel burning for energy because of the particulate pollution and air quality issues. Hence there is a huge double benefit. At the same time, it is important to recognize the continuing increasing demand for electricity and energy. Note that some industrial plants, e.g., blast furnaces, require high temperatures that cannot be satisfied by electricity but might be satisfied by nuclear energy.

Another important way to decarbonize is by catching the carbon dioxide generated and storing/disposing of it in depleted underground fossil fuel caverns or deep-sea formations. Called "*carbon capture and storage*" (CCS), this is a stopgap measure, as the storage capacity is finite, but is a necessary part of the toolkit. Otherwise, fossil fuel must be switched for either renewable technologies (including bioenergy, hydropower, solar, wind, geothermal, and marine energy) or nuclear energy. Over 2005–2015 the share of renewables in the electricity mix has increased by approximately 5–6%. However, the share from nuclear production has decreased by a similar amount.

Progress on electricity decarbonization has been stalled over the last decade as a result of a growing – often irrational – aversion to nuclear energy. New nuclear technology is changing the face of nuclear power, small modular reactors (SMRs) are advocated by the industry, and the modular aspect means that power plants could be built incrementally, adding more modules as needed. SMRs are smaller, cheaper, and safer than conventional nuclear plants, but still have high up-front costs and suffer from regulations. Safety regulations have made nuclear power more expensive, though plant safety and waste transport and disposal have been partly solved. Even new breeder reactors still produce radioactive waste, although less than their predecessors. Nuclear waste presents both an engineering problem and a social problem, because most people want nuclear waste to go elsewhere. Consequently, unresolved waste, economic, national security, and sociopolitical concerns need to be resolved. However, nuclear power is thriving in Russia, China, and South Korea, where nuclear power plants cost a lot less than in the West. Proper accounting is also necessary for renewables. For instance, solar photovoltaic panels typically have a lifetime of 20 years or so, but contain toxic heavy metals, such as cadmium, that should be recycled or properly disposed of.

As well as the development of renewables, another priority is to shift sectors such as transport towards electricity, because it is otherwise difficult to decarbonize. Electric vehicles may then be able to use renewable energy. Two other outstanding issues concern the intermittency of some forms of energy, such as wind and solar, thus requiring improved methods of electricity storage, such as batteries, and the need for greatly improved low-cost (in terms of both power and expense) transmission of electricity from where it is generated to where it is needed. The storage problem becomes much easier if the intermittent sources are coupled with water and hydro power, where the storage is of water behind a dam. However, this has often been difficult to achieve owing to fragmentation of the sources, distributors, and users.

Clearly, improvements in efficiency matter a great deal, and this extends from generation of power to improved production of crops, dairy, and food. The latter may include changes in diet and reduction in waste.

17.2 Adaptation and Mitigation

While it has been stated that "*the science is settled*" in regard to climate change, there remain many uncertainties. For example, the warming is not uniform; land is generally warming more than the oceans – but not everywhere. Moreover, associated changes in precipitation, winds, storms, sea-level rise, and many other important aspects of weather and climate are far from clear and settled. What is settled is that global warming is real; the rate is substantial and faster than humans have ever experienced, and it is caused by humans.

One facet of how to deal with climate change has been outlined by Trenberth et al. (2016). The main responses come under the headings of "*mitigation*," which refers to slowing or stopping the emissions of greenhouse gases and therefore limiting the nature of future changes, or "*adaptation*," which refers to assessing the impacts, building resilience to them, and planning sensibly for the expected changes. But what are we adapting to? In fact, we humans will adapt in some form, either autonomously or through planning, building resiliency, and coping with the changes, or by suffering the consequences. Accordingly, a third vital component is to build an *information* system to tell us what is happening and why, what the prospects are for the future on different time horizons, and thus what we must adapt to. As a whole, not enough is being done of any of these.

There are several steps under the adaptation heading. These include assessing the impacts of the projected climate change effects on various regions and sectors, assessing vulnerability to the impacts, making plans to reduce the vulnerability and build resiliency, and generally cope with the expected changes, including extremes. The longer society delays steps to cut the release of planet-warming greenhouse gases, the more severe and widespread the harm will be, according to the IPCC. Global warming threatens food and water supplies, security, and economic growth, and will worsen many existing problems, including hunger, drought, flooding, wildfires, poverty, and war. The IPCC WG II has emphasized eight major climate risks:

1. Death or harm from coastal flooding.
2. Harm or economic losses from inland flooding.
3. Extreme weather disrupting electrical, emergency, or other systems.
4. Extreme heat, especially for the urban and rural poor.
5. Food insecurity linked to warming, drought, or flooding.
6. Water shortages causing agricultural or economic losses.
7. Loss of marine ecosystems essential to fishing and other communities.
8. Loss of terrestrial and inland water ecosystems.

Given the pandemic in 2020, human health and wellbeing should be added, because it is not understood how climate change may affect transmission and emergence of new disease vectors. A case can be made that many of the biggest potential issues arise in association with water availability owing to increasing demand and effects of climate change, especially the extremes of drought and flooding. Extreme heat and drought also relate to risk of wildfire.

According to IPCC WG II, global adaptation cost estimates are substantially greater than current adaptation funding and investment, particularly in developing countries, suggesting a funding gap and a growing adaptation deficit. Global adaptation cost estimates suggest a range from 70 to 100 US$ billion per year in developing countries to 2050. The IPCC concludes that the world's poorest people will suffer the most as temperatures rise, with many of them already contending with food and water shortages, higher rates of disease and premature death, and the violent conflicts that result from those problems.

For mitigation, many good things are happening at the level of towns, cities, states, and some countries, which responsibly attempt to limit their carbon footprint. However, in general the national and international framework is inadequate, though it is essential. If one region implements a carbon tax, for example, some companies and even industries threaten to move to the next town or state, or even to the next country.

17.3 Geoengineering

For the most part, humans are responsible for the current climate change, but it is a side effect of human activities, it is not intentional. With recognition of the consequences, there have been multiple proposals to deliberately modify the climate and "fix" it. The essence of geoengineering is not to attempt to directly forestall the problem but rather to implement alternative devices to deal with the symptoms. The difficulty with this approach is that even when it proves possible to alleviate the immediate predicament, there are often unintended consequences and side effects that could prove even worse than the original problem. Geoengineering involves considerable costs that may be better spent on adaptation. It also raises major ethical questions concerning who is entitled to make the decision (on behalf of all humanity) to intentionally change the climate. Moreover, once begun, if ever stopped, the results could be extremely abrupt climate change with highly undesirable consequences.

One approach is to remove carbon dioxide from the atmosphere in various ways, which may work but could also be quite costly, and then what is to be done with the carbon dioxide? Another proposal is euphemistically called "solar radiation management" (SRM). At the present time, such plans are all either in the proof-of-concept phase or face quite challenging technological and practical challenges. It is often assumed that it is not only known what would otherwise happen (perfect prediction) but also that the solution works as advertised; both assumptions are questionable and introduce major uncertainties.

For instance, based on the temperature decreases observed following major volcanic eruptions that inject sulfate aerosols into the stratosphere, SRM proposes that a possible partial solution to global warming may be to emulate these effects. However, global warming is not caused by increased sunshine, rather it arises from the increased greenhouse effect. Geoengineering by blocking the Sun addresses neither the central problem of climate change nor acidification of the oceans. Instead, adverse effects on the hydrological cycle may result (Chapter 10) and the outcome is apt to reduce precipitation and change atmospheric heating patterns. Creating a risk of widespread drought and reduced freshwater resources for the world to cut down on global warming does not seem like an appropriate fix.

17.4 Climate Information

The climate-observing system is in decay, continuity of satellite observations has been in jeopardy, and climate models must continue to improve. Building climate services is a priority of the World Meteorological Organization, but is a struggle in some countries. In particular, many more data on social science aspects are needed to properly enable adaptation. This includes, for example, information about infrastructure and its vulnerability to such things as sea-level rise, or more severe flooding; or agricultural systems and their vulnerability; or water supplies; or ... basically all of the non-natural systems constructed by humans to support their lifestyle. It is worth noting that the benefits of building a climate information system occur regardless of whether or not climate change occurs.

The many challenges encountered in making, interpreting, and acting on climate analyses, predictions, and projections point to the need for much better climate information: a robust *climate information system* should serve as the third leg of a "*climate change stool*" (Fig. 17.2). Climate information complements and rounds out adaptation and mitigation.

In particular, such a system could provide clarity regarding the uncertainties in climate predictions, and allow development of sound risk management strategies. The climate information system would also enable and support *climate services*, which involve the production, translation, transfer, and use of climate knowledge and information in climate-informed decision making and climate-smart policy and planning.

Chapter 16 addressed climate change *projections*, where the goal is to discern the average future climate in response to the external influences on the climate system including human influences, as well as influences from natural sources such as changes in solar output or in atmospheric composition such as from volcanoes. The attribution of cause is important (Chapter 15) as it better informs communities and decision-makers what aspects of climate change to take account of and plan for, what parts are natural variability, and how they interact. Current climate projections are conflated with natural variability, which in part accounts for diversity of outcomes among the different projections, as different possible realizations

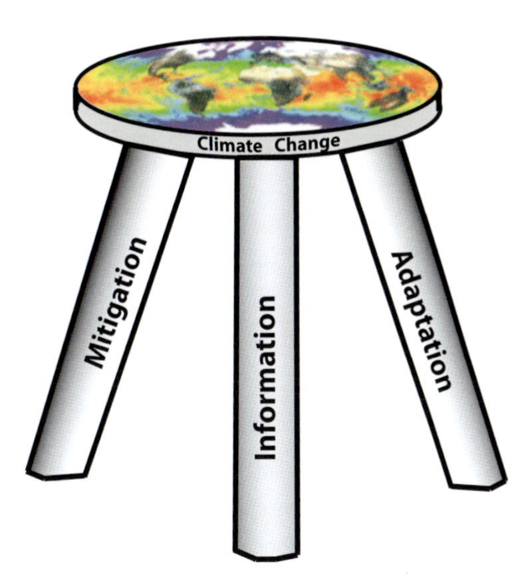

Fig. 17.2 Schematic of the "three-legged stool" response to climate change, with satellite imagery of the planet on top and the legs labeled as adaptation, mitigation and information. Adapted from Trenberth et al. (2016)

of climate variability are being sampled. Differences in model performance are an additional factor.

17.5 What Is a Climate Information System?

A "system" needs to be "end-to-end." There are many pieces of a climate information system already in place, but they are neither coherently managed nor responsive to needs. Overall there is a need for observations (that satisfy the climate-observing principles); a performance tracking system; the archiving and stewardship of data; access to data, including data management and integration; analysis and reanalysis of the observations (the delayed reprocessing and quality control of all data, analyzed with a comprehensive system that remains constant over time to minimize spurious changes) and products that communicate the results; assessment of what has happened and why (attribution), including likely impacts on humans and ecosystems; prediction of near-term climate change from seasons to decades; dissemination of information; and responsiveness to decision-makers and users.

Many aspects require establishing relationships among physical, environmental and social impact variables, understanding and coping with uncertainties, regionalizing results, and helping users understand the information. The proposed climate information system must include the links from basic research to applied research and transfer to operations of new technology, methods and products, and the

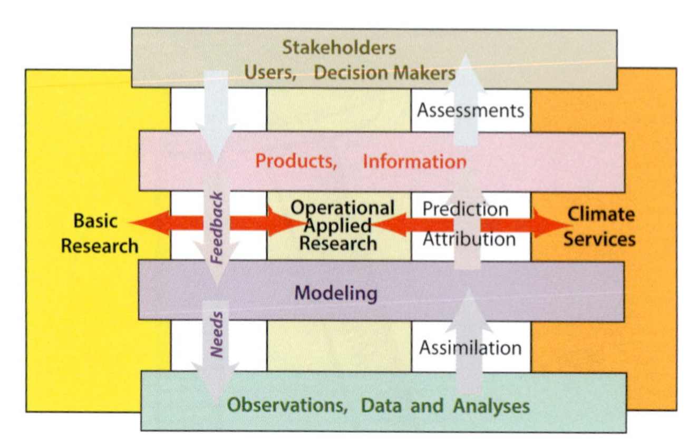

Fig. 17.3 A schematic of the flow of the climate information system, as basic research feeds into applied and operational research and the development of climate services. The system is built on the climate-observing system that includes the analysis and assimilation of data using models to produce analyses and fields for initializing models; and the use of models for attribution and prediction, with all the information assessed and assembled into products and information that are disseminated to users. The users in turn provide feedback on their needs and how to improve information.

development of more effective climate services (Fig. 17.3). It will require high-quality observations and analysis of the determinants of social and ecosystem vulnerability, resilience, and adaptive capacity. It involves academia, the private sector, business and investment communities as much as the public sector.

Attribution needs to become real-time and routine, but also to be much improved. The attribution of extreme climate events is a highly visible topic, featured frequently in the media after just about every major climate event, whether a flood, drought, wildfire, hurricane, or snowstorm. Yet it has been difficult to achieve scientifically robust quantification of the relative contributions of human-induced climate change and natural variability because both are factors in any extreme event; see Chapter 15.

In the United States, various assessments, analyses and short-term predictions are assembled monthly and seasonally, and special issues of the *Bulletin of the American Meteorological Society* are produced annually. However, there is a need for much better integration of the different timescales, and in building the knowledge from research to applications and services. The monthly and annual assessments need to be extended to encompass decadal and longer variations, and should properly account for human-caused climate change. While valuable contributions are made that inform decision-makers, gaps and shortcomings need to be addressed. Not the least of these is the interpretation and translation of information into forms useful to decision-makers, and to advancing risk management strategies. The proposed Climate Information System should serve as a near real-time version of IPCC Assessment Reports, probably produced at quarterly frequency, and complemented by the various national assessments.

The proposed Climate Information System will provide data and services not only for understanding, attributing, and predicting climate changes, but also for informing mitigation and adaptation to climate changes. For instance, transitioning from the current fossil-fuel based energy system to one based on renewable sources, such as wind and solar power, is a key component of any climate-change mitigation strategy. The characteristics of wind and solar resources for generating power, especially with regard to their intermittency, are different from those of fossil fuels. Although the impacts of weather and climate on energy demand have long been considered by the energy sector, their effects on power generation and all the resultant implications have not been adequately incorporated into it. Foundational data sets and projections of wind and solar resources across time-scales are examples of information needed for wind and solar plant siting, and for operating power generation and transmission.

It has been noted that climate services are crucial for successful adaptation to occur. But good observational data are critical to climate services. Observations provide the context relative to the past; allowing assessment of model performance and initialization for predictions. They provide essential information for assessing resilience, and any predictive capacity adds to their value. In addition, society needs translation of information to manage the risks and opportunities of climate change and variability. Upsides include such items as reduced need for heating in colder climates, new agricultural opportunities with longer growing seasons, fewer crop losses due to frost, etc.

Nevertheless, there is a continuing lament that current climate information "may not be certain enough to motivate its use." "Mounting evidence shows, however, that climate information can improve behaviors and outcomes when appropriately incorporated within the decision context" (Trenberth et al., 2016). Many examples exist, such as for farmers who utilize drought outlooks, or successful interventions from health warning systems such as from dengue fever risks. The climate information provided for the 2015–16 El Niño event was used much more effectively than during the 1997–98 event.

There is abundant evidence that decision-makers need and want help in understanding the complicated climate/society interface in ways that facilitate better outcomes within their communities and businesses. Indeed, there is a surging interest from corporations related to how their supply chains and business model may be affected by and vulnerable to climate change. In light of the increasingly expensive and devastating impacts of climate-related extreme events, it is now critical to build an integrated knowledge system that includes public and private partners. Firmly establishing the third leg of the stool in addressing climate change is essential. Such a climate information system would be invaluable for preparing for the challenges ahead. It would provide comprehensive climate information to communities and decision-makers concerned with impacts and adaptation at regional scales. It could *usher in an era of climate services akin to the weather enterprise of the past 50 years.*

References and Further Reading

Goddard, L., 2016: From science to service. *Science*, **353**, 1366–1367. doi: 10.1126/science.aag3087.

Ritchie, H., and M. Roser, 2017: CO_2 and Greenhouse Gas Emissions. Published online at OurWorldInData.org. Retrieved from: https://ourworldindata.org/co2-and-other-greenhouse-gas-emissions.

Solomon, S., G.-K. Plattner, R. Knutti, and P. Friedlingstein, 2009: Irreversible climate change due to carbon dioxide emissions *Proceedings of the National Academy of Sciences USA*, **106**, 1704–1709. doi: 10.1073/pnas.0812721106.

Trenberth, K. E., 2008: Observational needs for climate prediction and adaptation. *WMO Bulletin*, **57**(1), 17–21.

Trenberth, K. E., R. A. Anthes, A. Belward, et al., 2013: Challenges of a sustained climate observing system. In: Asrar, G. R., and J. W. Hurrell, eds., *Climate Science for Serving Society: Research, Modelling and Prediction Priorities.* Dordrecht: Springer, 13–50.

Trenberth, K. E., M. Marquis, and S. Zebiak, 2016: The vital need for a climate information system. *Nature Climate Change*, **6**, 1057–1059. doi: 10.1038/NCLIM-16101680.

18 Climate Change and Environmental Issues

Following a brief outline of the main global political agreements related to climate change, a critical appraisal of economists' projections is offered, along with the importance of nonscientific factors in making decisions. Value systems matter. There is no Planet B and we are all together on Spaceship Earth.

18.1 Politics of Climate Change

The United Nations Framework Convention on Climate Change (UNFCCC) is an international environmental treaty adopted on 9 May 1992 and taken up at the Earth Summit in Rio de Janeiro in June 1992. The UNFCCC objective is to "stabilize greenhouse gas concentrations in the atmosphere at a level that would prevent dangerous anthropogenic interference with the climate system." It set up important annual reporting requirements, including annual meetings of the Conference of the Parties (COP) to assess and promote progress.

The Kyoto Protocol was initiated by the UNFCCC at COP-3 in 1997. On February 16, 2005, the Kyoto Protocol was ratified by 164 countries, but it did not include Australia and the USA. Australia ratified it much later in December 2007 and it has been ratified by 192 parties but not by the United States. The Kyoto Protocol was designed to limit carbon dioxide emissions and those of other greenhouse gases from developed countries, but did not impose restrictions on developing countries. Some good progress was achieved in Europe, but undermined by tremendous industrialization and emissions on the part of China, in particular, and other developing countries.

Great hopes for a further agreement occurred after the IPCC AR4 report in 2007, which announced that "global warming was unequivocal," and the IPCC shared the Nobel Peace Prize with Al Gore in 2007. These hopes were carried forward into COP-15 in Copenhagen in 2009, but failed to be realized. One factor was the development of so-called "*Climategate*" whereby a large number of emails were stolen from the University of East Anglia server, and cherry-picked, distorted and abused by climate change deniers to carry out malicious attacks on some scientists who participated in the IPCC report and thereby undermine the scientific basis for the agreements. Although there was no basis for these claims, they appeared to achieve their purpose. Seven major investigations of the scientists involved in the hacked emails showed some minor violations of Freedom of Information Acts but complete vindication of all other aspects.

Belatedly, 197 countries around the world, recognizing the need to respond in various ways, enabled the Paris Agreement to reduce greenhouse gas emissions to be put in place in December 2015, and it was adopted less than a year later. This agreement sets emissions reduction targets and requires progress towards meeting those targets to be tracked and reported to the international community. This includes the goal of "*holding the increase in the global average temperature to well below 2°C above pre-industrial levels and pursuing efforts to limit the temperature increase to 1.5 °C.*" The Paris Agreement also commits to enhancing adaptive capacity and resilience to the impacts of climate change. Leadership from the United States under President Obama played a substantial role. Unfortunately, the advent of the Trump administration with a Republican Party majority in the US Senate, deniers of human-caused climate change, meant that progress stalled as the USA announced withdrawal from the Paris Agreement, to take effect in November 2020. In 2021, the new Biden administration restored the USA in the Paris Agreement.

The UN framework requires a unanimous consensus. The Agreement works largely through peer pressure and goodwill. It is voluntary. It is toothless, as there are no penalties. There is no compensation for loss and damage. Unless substantive steps forward are taken, climate change is apt to continue. Adaptation plans then become essential. A Green Fund was established under the Paris Agreement to enable resilience to be built in developing nations. However, funding is voluntary.

Four years after the Paris Agreement, most nations are failing to meet their commitments and emissions of greenhouse gases continue without diminution. While many countries have significantly added renewable energy to their mix, demand for energy continues to rise.

18.2 Winners and Losers

The climate has always varied on multiple timescales, but now humans are the main agents of change and are likely to remain so for the next few centuries. Climate change is already affecting every continent and ocean, posing immediate and growing risks to people. Global warming threatens food and water supplies, security and economic growth, and will worsen many existing problems, including hunger, drought, flooding, wildfires, poverty, and war. Scientists have long been aware of this and it has been spelled out in the many IPCC assessment reports since the second in 1995, and in domestic reports, such as the US National Climate Assessments.

There is no doubt that there are winners and losers as the climate changes, and some regions can benefit from climate change, for instance from a longer growing season. Moreover, climate change is not necessarily bad – after all, climate has always varied – but rapid climate change is always disruptive. And the climate is changing at unprecedented rates. It may well be that the climate locally changes to be one that is better in some respect, but it will not stay that way because it keeps changing, and changing, and changing. Even short-term benefits sooner or later become negatives as the climate continues to change. So, a key point of climate change is the "change" part. No sooner has the climate changed to be nicer than it changes again. It behooves us to greatly slow the pace of climate change in order to provide future generations with a manageable and livable planet.

A major concern of scientists, not adequately appreciated by the public and politicians, is that evidence of dangers warranting policy responses may be delayed or muted by the tremendous inertia in climate change, so that by the time problems are abundantly clear, it may be too late to do anything about it. The longer society delays taking steps to cut the emissions of planet-warming greenhouse gases, the more severe and widespread the harm will be.

The inertia and long time-horizons come about from several aspects:

(i) the long atmospheric life of carbon dioxide (centuries) as well as other heat-trapping greenhouse gases;
(ii) the very slow response of the climate system to the changes because of the slow warming of the oceans and melting of glacier ice, which result in a lag on the order of 20 years;
(iii) the complexity of current infrastructures, particularly energy systems, whose transformation would require on the order of two or more decades.

All of which means that actions taken now have their main effects 50 years from now and beyond. Some of the projected changes in climate are essentially irreversible, at least on the timescale of human lifetimes. While climate change has always been present, it is estimated that the current rates of change are a factor of 100 or greater more rapid than natural changes. It is the rates of change that are very disruptive, not only to natural ecosystems and biodiversity, but also to society.

Already the costs are substantial every year from drought, wildfires, floods, heatwaves, storm surges, and strife. Although the climate change factor may only be 5–20% in terms of an increase in precipitation, say, records are broken, thresholds are crossed, things break, or burn, floods occur, and people die. The Actuaries Climate Risk Index (ACRI) suggests for 1990–2016 that US property losses were about $1 billion per year higher than expected based on 1961–90 (ACRI 2020). Note that this does not include the devastating 2017 year (hurricanes Harvey, Irma, Maria, etc.), and it does not include loss of life or disruption. Hence, there are extreme nonlinearities associated with such events. Instead of $1 billion in damage the damage may be at least $10 billion and perhaps $100 billion. Damage from such extremes and nonlinearity is greatly underestimated by economists. This is "*The straw that breaks the camel's back*" syndrome. Economic assessments of the potential future risks of climate change have omitted or grossly underestimated many of the most serious consequences for lives and livelihoods because these risks are difficult to quantify precisely and lie outside of human experience. For instance, they fail to take account of the potential for large concurrent impacts across the world that would cause mass migration, displacement and conflict, with huge loss of life.

The climate events that cause the damage are isolated, regional in nature, and affect but few at a time. The public does not see an integrated view. A major IPCC report comes out and it is a headline for at most one day. But the problem continues, and in fact gets worse every day. Yet it is no longer news because it remains the same problem, although the problem has not been solved. It is easy for the public to set it aside.

18.3 Values

Science knowledge has been going in one direction while some political views are going in another. Jørgen Randers, a professor of climate strategy at the Norwegian Business School, in January 2015 concluded that "*it is profitable to let the world go to hell*," and so the private sector will not solve the problem without strong incentives. Pope Francis' 2015 encyclical "Laudato Si" (Praise Be to You) highlights the ethical nature of the problem. *"Climate change is a global problem with grave implications: environmental, social, economic, political and for the distribution of goods. It represents one of the principal challenges facing humanity in our day."*

The report "Risky Business: The Economic Risks of Climate Change in the United States," co-chaired by Henry Paulsen and Michael Bloomberg, concluded in July 2014 that *"taking a cautious approach, waiting for more information, a business-as-usual approach, is actually radical risk taking."* But *"if we act now the U.S. can still avoid most of the worst impacts and significantly reduce the odds of costly climate outcomes – but only if we start changing our business and public policy practices today . . ." "The longer we wait to adapt to and mitigate climate change, the more devastating the economic impacts will be."*

Patrick Daniel Moynihan famously said, *"Everyone is entitled to his own opinion, but not his own facts."* The observations and data – the facts – are of mixed quality and duration, but together tell a compelling story that leaves no doubt about the human role in climate change. Observed changes in some phenomena, such as hurricanes and tornadoes, are confounded by large variability, the changing observing system and shortness of reliable records. But the absence of evidence is not evidence of absence of important changes, and physical understanding and climate modeling can fill the gaps.

Box 18.1: Yogi Bera (the famous baseball player; 1925–2015) said,

"Baseball is 90% mental and the other half is physical."

On climate change he might have said:

"Climate change is 90% ethical and the other half is science."

There are many facts related to climate to demonstrate conclusively that the problem is real. The observational evidence combined with physical understanding based on well-established physical principles, as detailed in this book, makes this abundantly clear. However, the facts are not enough. Climate change is happening because of human activities, but what we do about it involves *value systems* and politics. The role of scientists is to lay out the facts, evidence, uncertainties, prospects, and consequences, but the decisions on what to do about them reside in the realm of politics and should involve all of society.

18.4 Spaceship Earth

Climate change is inherently an intergenerational problem. What kind of a planet are we leaving our grandchildren? It is also a problem of equity among nations. Small island states and developing countries have not contributed much to the problem but are greatly affected by it. Costs of climate change and air pollution are not borne by those who cause these problems. There are uncertainties associated with exactly in what form and where climate change effects will be felt, but the risks are growing. A normal way society deals with risk is by building resilience and taking out insurance. The precautionary principle should come into play. But society is not doing enough to mitigate the problem or plan for the consequences. On the other hand, the world as a whole has been getting more prosperous and resilient, and perhaps the best example is the warning and evacuation systems and shelters that have been adopted in Bangladesh and India, where tropical cyclones that would have killed tens of thousands of people years ago kill only hundreds or fewer today. In many areas, improved building codes and early warning systems have also reduced damage and loss of life in spite of increasing extremes. More generally, adaptation efforts have the potential to pay off in major ways by investing in and strengthening early warning systems, making new infrastructure more resilient, protecting mangroves, and making water resources more resilient. The Global Commission on Adaptation estimates that US$1.8 trillion invested by 2030 could pay off with net benefits of over US$7 trillion.

Box 18.2: Al Gore (former Vice President of the United States) from the Democratic National Convention on August 28, 2008:

"Inconvenient truths must be acknowledged if we are to have wise governance."

The issue of global climate change really boils down to one of the "*tragedy of the commons*." The oceans are one major commons, and there is very limited protection of the oceans from the Law of the Sea. The atmosphere is the other major commons. Air over China one day is over North America 5 days later, and over Europe in another 5 days or so. It is in everyone's interest to exploit the atmosphere and use it as a convenient dumping ground for pollutants and emissions. This applies to individuals, companies, industries, cities, counties, states, and nations. But there are major costs, in terms of air quality, climate change, acidification of the ocean, environmental degradation, water pollution, lung diseases caused by bad air quality, oil supply disruptions, patrolling the world's oil shipping lanes for years, and so on, that are not borne by the users. This raises the question about whether there ought to be a principle of "user

Fig. 18.1 There is no Planet B. Imagery courtesy of NOAA GOES 17. www.flickr.com/photos/noaasatellites/48127127191/sizes/l/.

pays," in which case there is a great need for a price on carbon that is universal. It must account for all the externalities and downstream effects. This might be implemented in many ways, through cap-and-trade schemes, fees or a carbon tax, combined with tariffs for international trade involving noncompliant countries.

Economists generally agree that users should pay: there must be an appropriate price on carbon and a removal of the substantial subsidies for fossil fuels that currently exist. The US federal government subsidizes fossil fuels by providing access to public lands, plus a variety of government services, and tax breaks have been estimated as having a value of $20 billion to $39 billion per year in recent times. Indirect costs are estimated to increase costs by at least an order of magnitude. Opponents of taking action often point to the cost of taking action but fail to account for the high cost of not taking action. In fact, there are countless benefits to taking action, but therein lies the problem: they are countless. Quality of life, greater security, fewer refugees, biodiversity, and so forth, have no dollar figure attached.

The atmosphere is a global commons (and dumping ground), and the solution to the global warming climate change is that people of the world must work together to save us on our "Spaceship Earth" (Fig. 18.1). There is no vaccine for climate change! Yet there is much that can be done, and with the right incentives in place the private sector would become fully engaged in the solutions.

References and Further Reading

ACRI (Actuaries Climate Risk Index) 2020: www.actuary.org/sites/default/files/2020-01/ACRI.pdf.

Bast, E., 2015: Empty promises: G20 subsidies to oil, gas and coal production http://priceofoil.org/2015/11/11/empty-promises-g20-subsidies-to-oil-gas-and-coal-production/.

Blunden, J., and D. S. Arndt (eds.), 2020: State of the climate in 2019. *Bulletin of the American Meteorological Society*, **101**(8), Si–S429. doi: 10.1175/2020BAMSStateofthe Climate.1.

Global Commission on Adaptation, 2019: https://gca.org/global-commission-on-adaptation/report.

IPCC, Fifth Assessment Reports (AR5): www.ipcc.ch/report/ar5/; and reports from Working Groups 1, 2 and 3: www.ipcc.ch/report/ar5/wg1/; www.ipcc.ch/report/ar5/wg2/; www.ipcc.ch/report/ar5/wg3/.

Keen, S., 2020: The appallingly bad neoclassical economics of climate change. *Globalizations*. doi: 10.1080/14747731.2020.1807856.

Kyoto Protocol: https://unfccc.int/kyoto_protocol/items/2830.php.

Mann, M. E., 2012: *The Hockey Stick and the Climate Wars*. Columbia University Press, New York. 448 pp.

McKibben, W., 2018: *How extreme weather is shrinking the planet*. *New Yorker*, November 26, 2018. www.newyorker.com/magazine/2018/11/26/how-extreme-weather-is-shrinking-the-planet?utm_medium=email&utm_source=actionkit.

Moynihan, P. D., 1983: https://quoteinvestigator.com/2020/03/17/own-facts/#return-note-437589-9.

Paris Agreement: http://unfccc.int/paris_agreement/items/9485.php.

Paulsen, H., and M. Bloomberg, 2014: Risky Business: The Economic Risks of Climate Change in the United States. http://riskybusiness.org/report/national/.

Pope Francis, 2015: Laudato Si. https://laudatosi.com/watch.

Randers, J.: https://www.theguardian.com/sustainable-business/2015/jan/19/davos-climate-action-democracy-failure-jorgen-randers.

UNFCCC: https://unfccc.int/2860.php.

Glossary

Absorbed solar radiation: Net incoming radiation from the Sun in the shortwave part of the spectrum at the top of the atmosphere that is absorbed (after reflections are accounted for).

Acidification: Ocean acidification refers to a reduction in the pH of the ocean over an extended period, typically decades or longer.

Advection: Transport of water or air along with its properties (e.g., temperature, chemical tracers) by winds or currents.

Aerosol: Microscopic particles suspended in the atmosphere, originating from either a natural source (e.g., volcanoes) or human activity (e.g., coal burning).

Albedo: The reflectivity of the Earth.

Altimetry: A technique for measuring the height of the Earth's surface with respect to the geocentre of the Earth within a defined terrestrial reference frame.

Anthropogenic climate change: Climate change arising from human influences.

Anticyclone: A high-pressure weather system. The wind rotates around these in a clockwise sense in the northern hemisphere and counterclockwise in the southern hemisphere. They usually give rise to fine settled weather.

Atmosphere: The air above Earth's surface, made up of a mixture of gases, predominantly nitrogen and oxygen.

Attribution: The process of evaluating the relative contributions of multiple causal factors to a change or event.

Baroclinic instability: Arises from horizontal temperature gradients in a stable atmosphere to produce down-gradient heat transports.

Black body: All objects with a temperature above absolute zero (0 K) emit electromagnetic radiation. A black body absorbs all radiation falling on it, and does not reflect any.

Blocking: A blocking event refers to the development of a strong slow-moving anticyclone displaced into mid to high latitudes. It serves to block the normal eastward progression of weather systems.

Chaos: A dynamical system such as the climate system may exhibit chaos with erratic chaotic behavior in the sense that very small changes in the initial state of the system in time lead to large and apparently unpredictable changes in its temporal evolution.

Climate: The time mean and variability of the climate system.

Climate feedback: An interaction in which a perturbation in one climate quantity causes a change in a second, and the change in the second quantity ultimately leads to an additional change in the first. A negative feedback is one in which the initial perturbation is weakened by the changes it causes; a positive feedback is one in which the initial perturbation is enhanced.

Climategate: The episode where a large number of emails were stolen from the University of East Anglia server in 2009, and misused to undermine the COP proceedings.

Climate model: A numerical representation of the climate system typically solved in a supercomputer. The model is based on the physical, chemical, and biological properties of its components, their interactions and feedback processes. The climate system can be represented by models of varying complexity. Earth system models that include coupled atmosphere–ocean general circulation models provide a representation of the climate system that is most comprehensive.

Climate prediction: A climate forecast starting from a particular state of the climate system to estimate the actual evolution of the climate in the future.

Climate projection: The simulated response of the climate system to a scenario of future emission or concentration of greenhouse gases and aerosols, depicted as an expected change.

Climate sensitivity: Equilibrium climate sensitivity (units: °C) refers to the equilibrium (steady-state) change in the annual global mean surface temperature following a doubling of the atmospheric carbon dioxide concentration. The **transient climate response** (units: °C) is the change in the global mean surface temperature, averaged over a 20-year period, centered at the time of atmospheric carbon dioxide doubling, in which CO_2 increases at 1% $year^{-1}$.

Climate system: The atmosphere, oceans, land, biosphere, and cryosphere, plus all the hydrology and vegetation.

Clausius–Clapeyron (C-C): The Clausius–Clapeyron relation, named after Rudolf Clausius and Benoît Paul Émile Clapeyron, characterizes the atmospheric vapor pressure as a function of temperature and pressure.

Convection: In weather, the process of warm air rising rapidly while cooler air usually more gradually subsides over broader regions elsewhere to take its place. This process often produces cumulus clouds and may result in rain.

COP: Conference of the Parties under the United Nations Framework Convention on Climate Change (UNFCCC), held annually.

Coriolis effect: An effect whereby a mass moving in a rotating system (the Earth) experiences a force (the Coriolis force) perpendicular to the direction of motion and to the axis of rotation. On Earth, the effect deflects moving objects to the right in the northern hemisphere and to the left in the southern hemisphere.

Cryosphere: Sea ice, snow cover, and land ice (including the semi-permanent ice sheets of Antarctica and Greenland, and glaciers).

Cyclone: A low-pressure weather system. The wind rotates around cyclones in a counterclockwise direction in the northern hemisphere and clockwise in the southern hemisphere. Cyclones are usually associated with rainy, unsettled weather and may include warm and cold fronts.

Dry static energy: The sum of the atmospheric sensible heat and potential energy.

Ekman transport: The total transport resulting from a balance between the Coriolis force and the frictional stress due to the action of the wind on the ocean surface.

El Niño: The occasional warming of the tropical Pacific Ocean from the west coast of South America to the central Pacific that typically lasts a year or so and alters weather patterns around the world.

El Niño–Southern Oscillation (ENSO): Consists of El Niño or La Niña and the Southern Oscillation together.

Ensemble: In numerical weather prediction and climate, instead of making a single forecast or simulation, a large number (ensemble) of runs of a model with small perturbations is produced as a form of Monte Carlo analysis to give not only the mean but also the spread or uncertainty in outcome.

Enthalpy: The heat content of a substance per unit mass. Used to refer to sensible heat in atmospheric science (as opposed to latent heat).

Equilibrium Climate Sensitivity (ECS): The net Global Mean Surface Temperature (GMST) change in response to doubling carbon dioxide concentrations at equilibrium.

Evapotranspiration: The combined processes of evaporation from the surface and transpiration from vegetation.

External forcings: Influences from outside of the climate system, including the Sun, internal upheavals from tectonics and volcanoes, and humans.

Extratropical cyclone: A large-scale (thousands of km) storm in the middle or high latitudes having low central pressure and cold and warm fronts with strong horizontal gradients in temperature and humidity.

Faculae: Bright patches on the Sun.

Feedback: Amplification or reduction in response to a forcing from internal processes.

Forcing: An external influence on the climate system.

Geostrophic balance: The balance between the Coriolis force and pressure gradient force. The geostrophic flow is parallel to the isobars.

Glacial: An interval of time within an ice age where a large portion of Earth's surface is cold and covered in ice.

Greenhouse effect: The effect produced as certain atmospheric gases allow incoming solar radiation to pass through to the Earth's surface, but reduce the outgoing (infrared) radiation, which is reradiated from Earth, escaping into outer space. The effect is responsible for warming the planet.

Greenhouse gas: Any gas that absorbs infrared radiation in the atmosphere.

Group velocity: Applies to the modulation and overall envelope shape of the amplitude of a group of waves as they propagate through space. It determines the direction and speed of energy propagation.

Hadley Circulation: The large-scale meridional overturning in the tropical atmosphere.

Heat capacity: Depends on the mass involved as well as its capacity for holding heat, as measured by the specific heat of each substance.

Hurricane: Tropical cyclone with sustained winds over 74 mph or 119 km per hour.

Hydrological cycle: The cycle of water evaporating from the oceans and the land surface, transporting the moisture elsewhere and especially onto land as water vapor, condensing to form clouds, precipitating over ocean and land as rain or snow, running off on the land surface, infiltrating into soils, recharging groundwater, discharging into streams and rivers, and flowing back out into the oceans.

Hydrostatic balance: The state in which the vertical pressure gradient is determined by gravity, so that the pressure at the bottom of a column of fluid (air or water) is determined by the weight of fluid above.

Hysteresis: The state of a system depends on its history. It comes about because of slow reactions to forces. Consequently, the evolution is often not reversible and bistable states can be created.

Ice: Often a shorthand for the cryosphere, consisting of all forms of ice, sea ice, glaciers, ice caps, ice sheets, and ice precipitation (snow, hail graupel, etc.).

Insolation: The amount of solar radiation reaching Earth, measured in W m^{-2}.

Interglacials: The warm periods between ice age glaciations.

Inter-Tropical Convergence Zone (ITCZ): A near equatorial zonal belt of low pressure, strong convection, and heavy precipitation, where the northeast trade winds meet the southeast trade winds.

Jet stream: Relatively narrow meandering bands of strong wind in the upper levels of the troposphere.

Kinetic energy: The energy associated with movement. It depends on the mass and velocity.

La Niña: The occasional cooling of the tropical Pacific Ocean from the west coast of South America to the central Pacific, in contrast to El Niño.

Land: The solid surface of Earth, including all of its topography and variety, not covered in water and making up some 29.2% in terms of area.

Latent heat: The heat required to convert a solid into a liquid, or a liquid into a vapor (change phase), without change of temperature. Conversion in the opposite direction returns the latent energy.

Longwave radiation: Infrared radiation in the longwave part of the electromagnetic spectrum, corresponding to wavelengths of 0.8 microns to 1000 microns. For Earth, it also corresponds to the wavelengths of thermal emitted radiation.

Mean sea level: The surface level of the ocean at a particular point averaged over an extended period of time.

Meridional Overturning Circulation (MOC): Meridional (north–south) overturning circulation in the ocean quantified by zonal mean (east–west) mass transports.

Mesoscale Convective System (MCS): Cluster of thunderstorms organized on a scale larger than the individual thunderstorms but smaller than extratropical cyclones; also lasts longer than a thunderstorm and normally persists for several hours or more.

Mixing ratio: The ratio of the mass of a substance to the total mass (usually applied to the atmosphere).

Moist static energy: The sum of the dry static energy plus latent energy.

Monsoon: A tropical and subtropical seasonal reversal in both the surface winds and associated precipitation, caused by differential heating between a continental-scale land mass and the adjacent ocean. Monsoon rains occur over land in summer.

Ocean: The main saline water-covered area of Earth, making up most of Earth's hydrosphere, also called the sea. The main oceans are the Pacific, Atlantic, Indian, Southern, and Arctic.

Outgoing longwave radiation: Net outgoing radiation in the infrared part of the spectrum at the top of the atmosphere.

Paleoclimate: Climate for which only proxy climate records are available, including historic and geologic time.

Parameterization: Parametric representation of processes that cannot be explicitly resolved at the spatial or temporal resolution of a model (sub-grid scale processes) by relationships with variables that are resolved.

Paris Agreement: The UNFCCC agreement of 187 countries to reduce greenhouse gas emissions at the COP 21 in Paris in December 2015.

pH: A dimensionless measure of the acidity of water; a pH decrease of 1 unit corresponds to a 10-fold increase in the concentration of hydrogen ions, or acidity.

Photosynthesis: The process by which plants take carbon dioxide from the air (or bicarbonate in water) to build plant material (carbohydrates), releasing oxygen in the process.

Precipitable water: The total air column water vapor amount converted to the depth of liquid water content.

Proxy: A record that is interpreted using physical and biophysical principles to represent climate-related variations back in time. Examples include data from pollens, tree rings, speleothems, corals, marine sediments, and ice cores.

Radiative forcing: The influence a given climatic factor has on the amount of downward-directed radiant energy impinging upon Earth.

Reanalysis: Gridded estimates of historical atmospheric or oceanographic variables, such as temperature and wind, created by reprocessing past meteorological or oceanographic data using fixed state-of-the-art weather forecasting or ocean circulation models with data assimilation techniques.

Relative humidity: Usually expressed as a percentage, is the fraction of water vapor relative to the amount needed for saturation at the same temperature.

Respiration: The process in living organisms involving the production of energy, with the intake of oxygen and the release of carbon dioxide from the oxidation of complex organic substances.

Sensible heat: see Enthalpy.

Shortwave radiation: Radiation from the Sun, most of which occurs at wavelengths shorter than the infrared.

Southern Oscillation: A global-scale variation in the atmosphere associated with El Niño events.

South Pacific Convergence Zone (SPCZ): A band of low-level convergence, cloudiness and precipitation extending from the west Pacific Warm Pool southeastwards towards French Polynesia.

Specific humidity: Mass of water vapor in a unit mass of moist air.

Stream-function: Lines depicting the rotational (nondivergent) part of the flow fields.

Thermocline: The region of vertical temperature gradient in the oceans lying between the deep abyssal waters and the surface mixed layer.

Total solar irradiance: The solar radiation received at the top of the Earth's atmosphere on a surface oriented perpendicular to the incoming radiation and at the mean distance of Earth from the Sun.

Transient climate response (TCR): The value for GMST increase at the time of doubling CO_2 with increases of 1% $year^{-1}$.

Transpiration: The process of water movement through a plant and its evaporation from leaves, stems, and flowers.

Troposphere: The part of the atmosphere in which we live, ascending to about 15 km above Earth's surface; in which temperatures generally decrease with height. The atmospheric dynamics known as "weather" take place within the troposphere.

Typhoon: The same phenomenon as a hurricane but occurring in the northwest Pacific.

Urban heat island: The region of warm air over built-up cities associated with the presence of city structures, roads, etc.

Vorticity: A measure of the local rotational character of the fluid and its spin about a vertical axis.

Walker Circulation: Direct thermally driven zonal east–west overturning circulation in the atmosphere over the tropical Pacific Ocean, with rising air in the west and sinking air in the east Pacific.

Warm Pool: The region of high sea surface temperatures and deep thermocline in the tropical western Pacific in the vicinity of Indonesia.

Weather: The events in the atmosphere related to atmospheric flow or motion occurring on a variety of timescales and involving a rich collage of phenomena.

Acronyms

AAAS	American Association for the Advancement of Science
ACRI	Actuaries Climate Risk Index
AMO	Atlantic Multidecadal Oscillation
AMOC	Atlantic Meridional Overturning Circulation
AR	Assessment Report
ASR	absorbed solar radiation
AVISO	Archiving, Validation and Interpretation of Satellite Oceanographic data
BECCS	bioenergy with carbon capture and storage
C-C	Clausius–Clapeyron relationship
CCS	carbon capture and storage
CCSM	Community Climate System Model
CDIAC	Carbon Dioxide Information Analysis Center
CERES	Clouds and the Earth's Radiant Energy System
CFCs	chlorofluorocarbons
CMIP	Climate Model Intercomparison Project
COP	Conference of the Parties
DJF	December–January–February average
DMS	dimethyl sulfide
DSE	Dry Static Energy
EBAF	Energy Balanced and Filled
ECMWF	European Centre for Medium-range Weather Forecasts
EEI	Earth's Energy Imbalance
ENSO	El Niño–Southern Oscillation
EOF	empirical orthogonal function
EPA	Environmental Protection Agency (US)
ERA	ECMWF Reanalysis
ET	evapotranspiration
GDP	Gross Domestic Product
GHG	greenhouse gas
GMST	global mean surface temperature
GPCP	Global Precipitation Climatology Project

hPa	hectoPascal
IAM	Integrated Assessment Model
IAP	Institute of Atmospheric Physics (Beijing)
IOD	Indian Ocean Dipole
IPCC	Intergovernmental Panel on Climate Change
IPO	Inter-decadal Pacific Oscillation
ITCZ	Inter-Tropical Convergence Zone
ITF	Indonesian ThroughFlow
JJA	June–July–August average
LULCC	land-use and land-cover change
MEI	Multivariate ENSO Index
MERRA	Modern-Era Retrospective analysis for Research and Applications
MHT	meridional heat transport
MJO	Madden–Julian Oscillation
MSLP	mean sea-level pressure
NAM	Northern Annular Mode
NAO	North Atlantic Oscillation
NASA	National Aeronautics and Space Administration (US)
NCAR	National Center for Atmospheric Research (US)
NEC	North Equatorial Current
NECC	North Equatorial Countercurrent
NGO	nongovernmental organization
NOAA	National Oceanographic and Atmospheric Administration (US)
NPI	North Pacific Index
NRC	National Research Council (US)
NSIDC	National Snow and Ice Data Center (US)
NWP	numerical weather prediction
OHC	ocean heat content
OLR	outgoing longwave radiation
ONI	Oceanic Niño Index
ORAS5	Ocean Reanalysis System 5
PDO	Pacific Decadal Oscillation
PDSI	Palmer Drought Severity Index
PIOMAS	Pan-Arctic Ice Ocean Modeling and Assimilation System
PNA	Pacific–North American pattern
ppb	parts per billion by volume
ppm	parts per million by volume
RCP	Representative Concentration Pathway
SAM	Southern Annular Mode
SEC	South Equatorial Current

SLP	sea-level pressure
SMR	small modular reactor
SOI	Southern Oscillation Index
SPCZ	South Pacific Convergence Zone
SPM	Summary for Policymakers
SRES	Special Report on Emissions Scenarios
SRM	solar radiation management
SSP	Shared Socioeconomic Pathway
SST	sea surface temperature
TCWV	total column water vapor
TNI	Trans Niño Index
TOA	top-of-atmosphere
TSI	total solar irradiance
TSU	Technical Support Unit
UNEP	United Nations Environment Programme
UNFCCC	United Nations Framework Convention on Climate Change
UV	ultraviolet
WCRP	World Climate Research Programme
WG	Working Group
WMO	World Meteorological Organization
WSI	Water Stress Index
XBT	expendable bathythermograph

Bibliography

At the end of each entry, in red, the chapter(s) where the material in the reference was used is given.

Abraham, J. P., M. Baringer, N. L. Bindoff, et al., 2013: A review of global ocean temperature observations: implications for ocean heat content estimates and climate change. *Reviews of Geophysics*, **51**, 450–483. https://doi.org/10.1002/rog.20022. [14]

Allan, R. P., C. Liu, M. Zahn, et al., 2014: Physically consistent responses of the global atmospheric hydrological cycle in models and observations. *Surveys in Geophysics*, **35**, 533–552. https://doi.org/10.1007/s10712–012-9213-z. [10]

Balmaseda, M. A., K. E. Trenberth, and E. Källén, 2013: Distinctive climate signals in reanalysis of global ocean heat content, *Geophysical Research Letters*, **40**, 1754–1759. https://doi.org/10.1002/grl.50382. [14]

Beltrami, H., 2002: Climate from borehole data: energy fluxes and temperatures since 1500. *Geophysical Research Letters*, **29**, 2111. https://doi.org/10.1029/2002GL015702. [14]

Biskaborn, B. K., S. L. Smith, J. Noetzli, et al., 2019: Permafrost is warming at a global scale. *Nature Communications*, **10**, 264. https://doi.org/10.1038/s41467–018-08240-4. [14]

Blunden, J. and D. S. Arndt, eds., 2020: State of the climate in 2019. *Bulletin of the American Meteorological Society*, **101**, Si–S429 https://doi.org/10.1175/2020BAMSStateoftheClimate.1 [2] [18]

Bosson, J.-B, M. Huss, and E. Osipova, 2019: Disappearing World Heritage glaciers as a keystone of nature conservation in a changing climate. *Earth's Future*, **7**. https://doi.org/10.1029/2018EF001139. [14]

Brodribb, T. J., J. Powers, H. Cochard, and B. Choat, 2020: Hanging by a thread? Forests and drought. *Science*, **368**, 261–266. https://doi.org/10.1126/science.aat7631. [2]

Broecker, W., 1975: Climatic change: are we on the brink of a pronounced global warming? *Science*, **189**, 460–463. [2]

Bronseleiaer, B., and L. Zanna, 2020: Heat and carbon coupling reveals ocean warming due to circulation changes. *Nature*, **584**, 227–233. https://doi.org/10.1038/s41586-020-2573-5. [8]

Carbone, R. E., and J. D. Tuttle, 2008: Rainfall occurrence in the U.S. warm season: the diurnal cycle. *Journal of Climate*, **21**, 4132–4146. https://doi.org/10.1175/2008JCLI2275.1. [7]

Carbone, R. E., J. W. Wilson, T. D. Keenan, and J. M. Hacker, 2000: Tropical island convection in the absence of significant topography. Pt I: Lifecycle of diurnally forced convection. *Monthly Weather Review*, **128**, 3459–3480. https://doi.org/10.1175/1520-0493(2000)128<3459:TICITA>2.0.CO;2. [7]

Cazenave, A., H.-B. Dieng, B. Meyssignac, K. von Shuckmann, B. Decharme, and E. Berthier, 2014: The rate of sea-level rise. *Nature Climate Change*, **4**, 358–361. https://doi.org/10.1038/nclimate2159. [14]

Cazenave, A., B. Meyssignac, M. Ablain, et al., 2018: Global sea-level budget 1993–present. *Earth System Science Data*, **10**, 1551–1590. https://doi.org/10.5194/essd-10-1551-2018. [14]

Charney, J. G., B. Stevens, I. H. Held, et al., 1979: *Carbon Dioxide and Climate: A Scientific Assessment*. Washington, DC: US National Academy of Sciences. [13]

Cheng, L., and J. Zhu, 2014: Uncertainties of the Ocean Heat Content estimation induced by insufficient vertical resolution of historical ocean subsurface observations. *Journal of Atmospheric and Oceanic Technology*, **31**(6), 1383–1396. https://doi.org/10.1175/JTECH-D-13-00220.1. [6]

Cheng, L., J. Zhu, and R. L. Sriver, 2015: Global representation of tropical cyclone-induced short-term ocean thermal changes using Argo data. *Ocean Science*, **11**, 719–741. https://doi.org/10.5194/os-11-719-2015. [7]

Cheng, L., K. Trenberth, J. Fasullo, T. Boyer, J. Abraham, and J. Zhu, 2017: Improved estimates of ocean heat content from 1960–2015. *Science Advances*, **3** (3), e1601545. https://doi.org/10.1126/sciadv.1601545. http://advances.sciencemag.org/content/3/3/e1601545. [8] [11] [12] [14]

Cheng, L., J. Abraham, Z. Hausfather, and K. E. Trenberth, 2019a: How fast are the oceans warming? Observational records of ocean heat content show that ocean warming is accelerating. *Science*, **363**, 128–129. https://doi.org/10.1126/science.aav7619. [14]

Cheng, L., K. E. Trenberth, J. Fasullo, M. Mayer, M. Balmaseda, and J. Zhu, 2019b: Evolution of ocean heat content related to ENSO. *Journal of Climate*, **32**, 3529–3556. https://doi.org/10.1175/JCLI-D-18-0607.1. [12]

Cheng, L., K. E. Trenberth, N. Gruber, et al., 2020a: Improved estimates of changes in upper ocean salinity and the hydrological cycle. *Journal of Climate*, **33**. https://doi.org/10.1175/JCLI-D-20-0366.1. [8] [10] [14]

Cheng, L., J. P. Abraham, J. Zhu, et al., 2020b: Record-setting ocean warmth continued in 2019. *Advances in Atmospheric Science*, **37**, 137–142. https://doi.org/10.1007/s00376-020-9283-7. [14]

Church, J. A., and N. J. White, 2011: Sea-level rise from the late 19th to the early 21st Century. *Surveys in Geophysics*, **32**(4–5), 585–602. http://doi.org/10.1007/s10712–011-9119-1. [14]

Coddington, O., J. L. Lean, P. Pilewskie, M. Snow, and D. Lindholm, 2016: A solar irradiance climate data record. *Bulletin of the American Meteorological Society*, **97**, 1265–1282. https://doi.org/10.1175/BAMS-D-14-00265.1. [4]

Cook, E. R., R. Seager, M. A. Cane, and D. W. Stahle, 2007: North American drought: reconstructions, causes, and consequences. *Earth Science Reviews*, **81**, 93–134. https://doi.org/10.1016/j.earsciReview2006.12.002. [10]

Cornwall, W., 2019: In hot water. *Science*, **363**, 442–445. https://doi.org/10.1126/science.363.6426.442. [8]

Covey, C., P. J. Gleckler, C. Doutriaux, et al., 2016: Metrics for the diurnal cycle of precipitation: toward routine benchmarks for climate models. *Journal of Climate*, **29**, 4461–4471. https://doi.org/10.1175/JCLI-D-15-0664.1. [7]

Cowtan, K., Z. Hausfather, E. Hawkins, et al., 2015: Robust comparison of climate models with observations using blended land air and ocean sea surface temperatures. *Geophysical Research Letters*, **42**, 6526–6534. https://doi.org/10.1002/2015GL064888. [2]

Dai, A., 2011a: Characteristics and trends in various forms of the Palmer Drought Severity Index (PDSI) during 1900–2008, *Journal of Geophysical Research*, **116**, D12115. https://doi.org/10.1029/2010JD015541. [10]

Dai, A., 2011b: Drought under global warming: a review. *WIREs Climate Change*, **2**, 45–65. https://doi.org/10.1002/wcc.81. [10]

Dai, A., T. Qian, K. E. Trenberth, and J. D. Milliman, 2009: Changes in continental freshwater discharge from 1949–2004. *Journal of Climate*, **22**, 2773–2791. [10]

de Boisseson, E., M. Balmaseda, and M. Mayer, 2018: Ocean heat content variability in an ensemble of twentieth century ocean reanalyses. *Climate Dynamics*, **50**, 3783–3798. https://doi.org/10.1007/s00382-017-3845-0. [14]

Deser, C., A. S. Phillips, and J. W. Hurrell, 2004: Pacific interdecadal climate variability: linkages between the tropics and the north Pacific during boreal winter since 1900. *Journal of Climate*, **17**, 3109–3124. [11] [12]

Deser, C., I. R. Simpson, K. A. McKinnon, and A. S. Phillips, 2017: The Northern Hemisphere extratropical atmospheric circulation response to ENSO: how well do we know it and how do we evaluate models accordingly? *Journal of Climate*, **30**, 5059–5082. [12]

Dessler, A. E., 2020: Potential problems measuring climate sensitivity from the historical record. *Journal of Climate*, **33**, 2237–2248. [13]

Dessler, A. E., and P. M. Forster, 2018: An estimate of equilibrium climate sensitivity from interannual variability. *Journal of Geophysical Research: Atmospheres*, **123**, 8634–8645. https://doi.org/10.1029/2018JD028481. [13]

Dessler, A. E., T. Mauritsen, and B. Stevens, 2018: The influence of internal variability on Earth's energy balance framework and implications for estimating climate sensitivity. *Atmospheric Chemistry and Physics*, **18**, 5147–5155. https://doi.org/10.5194/acp-18-5147-2018. [13]

Emanuel, K., 2005: Increasing destructiveness of tropical cyclones over the past 30 years. *Nature*, **436**, 686. https://doi.org/10.1038/nature03906. [7]

Ezer, T., L. P. Atkinson, W. B. Corlett, and J. L. Blanco, 2013: Gulf Stream's induced sea level rise and variability along the U.S. mid-Atlantic coast. *Journal of Geophysical Research: Oceans*, **118**, 685–697. https://doi.org/10.1002/jgrc.20091. [8]

Frederikse, T., F. Landerer, L. Caron, et al., 2020: The causes of sea-level rise since 1900. *Nature*, **584**, 393–397. https://doi.org.cuucar.idm.oclc.org/10.1038/s41586–020-2591-3. [14]

Friedlingstein, P., M. W. Jones, M. O'Sullivan, et al., 2019: Global carbon budget 2019. *Earth System Science Data*, **11**, 1783–1838. https://doi.org/10.5194/essd-11-1783-2019. [2]

Goddard, L., 2016: From science to service. *Science*, **353**, 1366–1367. https://doi.org/10.1126/science.aag3087. [17]

Gregory, J. M., and T. Andrews, 2016: Variation in climate sensitivity and feedback parameters during the historical period. *Geophysical Research Letters*, 43, 3911–3920. [13]

Horel, J. D., and J. M. Wallace, 1981: Planetary-scale atmospheric phenomena associated with the Southern Oscillation. *Monthly Weather Review*, **109**, 813–829. [11]

Hoskins, B. J., and D. J. Karoly, 1981: The steady linear response of a spherical atmosphere to thermal and orographic forcing. *Journal of the Atmospheric Sciences*, **38**, 1179–1196. [11]

Hu, D., L. Wu, W. Cai, A. S. Gupta, et al., 2015: Pacific western boundary currents and their roles in climate. *Nature*, **522**, 299–308. https://doi.org/10.1038/nature14504. [11]

Huang, S., 2006: 1851–2004 annual heat budget of the continental landmasses. *Geophysical Research Letters*, **33**, L04707. https://doi.org/10.1029/2005GL025300. [14]

Huffman, G. J., R. F. Adler, D. T. Bolvin, and G. Gu, 2009: Improving the global precipitation record: GPCP version 2.1. *Geophysical Research Letters*, **36**, L17808. https://doi.org/10.1029/2009GL040000. [5] [10]

Hurrell, J. W., 1995: Decadal trends in the North Atlantic Oscillation and relationships to regional temperature and precipitation. *Science*, **269**, 676–679. [11]

Hurrell, J. W., and C. Deser, 2009: Atlantic climate variability. *Journal of Marine Systems*, **78**, 28–41. https://doi.org/10.1016/j.jmarsys.2008.11.026. [11]

Hurrell, J. W., Y. Kushnir, G. Ottersen, and M. Visbeck, 2003: An overview of the North Atlantic Oscillation. In: Hurrell, J. W., Y. Kushnir, G. Ottersen, and M. Visbeck, eds. *The North Atlantic Oscillation: Climatic Significance and Environmental Impact.* Geophysical Monograph **134**, 1–35. Washington, DC: American Geophysical Union. [11]

IMBIE Team, 2018: Mass balance of the Antarctic Ice Sheet from 1992 to 2017. *Nature*, **558**, 219–222. [14]

IMBIE Team, 2020: Mass balance of the Greenland Ice Sheet from 1992 to 2018. *Nature*, **579**, 233–239. [14]

IPCC: Intergovernmental Panel on Climate Change, 2013 : *Climate Change 2013. The Physical Science Basis*, ed. Stocker,T.F., et al. Cambridge:Cambridge University Press. [1] [11] [14] [16]

Jones, P. D., M. New, D. E. Parker, S. Martin, and I. G. Rigor, 1999: Surface air temperature and its changes over the past 150 years. *Reviews in Geophysics*, **37**, 173–199. [5]

Karl, T. R., and K. E. Trenberth, 2003: Modern global climate change. *Science*, **302**, 1719–1723. https://doi.org/10.1126/science.1090228. [1]

Kennedy, J. J., N. A. Rayner, C. P. Atkinson, and R. E. Killick, 2019: An ensemble data set of sea surface temperature change from 1850: the Met Office Hadley Centre HadSST.4.0.0.0 data set. *Journal of Geophysical Research: Atmospheres*, **124**. https://doi.org/10.1029/2018JD029867. [2]

Kiehl, J. T., and K. E. Trenberth, 1997: Earth's annual global mean energy budget. *Bulletin of the American Meteorological Society*, **78**, 197–208. [3]

Kopp, G., 2016: Magnitudes and timescales of total solar irradiance variability. *Journal of Space Weather and Space Climate*, **6**, A30. https://doi.org/10.1051/swsc/2016025. [4]

Kosaka, Y., and S-P. Xie, 2016: The tropical Pacific as a key pacemaker of the variable rates of global warming. *Nature Geoscience*, **9**, 669–673. https://doi.org/10.1038/NGEO2770. [12]

Kwok, R., 2018: Arctic sea ice thickness, volume, and multiyear ice coverage: losses and coupled variability (1958–2018). *Environmental Research Letters*, **13**, 105005. https://doi.org/10.1088/1748-9326/aae3ec. [14]

Kwok, R., and J. C. Comiso, 2002: Spatial patterns of variability in Antarctic surface temperature: connections to the Southern Hemisphere Annular Mode and the Southern Oscillation. *Geophysical Research Letters*, **29**, 1705. https://doi.org/10.1029/2002GL015415. [11]

Lacis, A. A., G. A. Schmidt, D. Rind, and R. A. Ruedy, 2010: Atmospheric CO_2: principal control knob governing Earth's temperature. *Science*, **330**, 356–359. www.sciencemag.org/cgi/content/full/330/6002/356/DC1. [3]

Lee, S.-K., W. Park, M. O. Baringer, A. L. Gordon, B. Huber, and Y. Liu, 2015: Pacific origin of the abrupt increase in Indian Ocean heat content during the warming hiatus. *Nature Geoscience*, **8**, 445–449. https://doi.org/10.1038/ngeo2438. [12]

Li, G., L. Cheng, J. Zhu, J. Abraham, K. E. Trenberth and M. E. Mann, 2020: Increasing ocean stratification over the past half century. *Nature Climate Change*, **10**. https://doi.org/10.1038/s41558–020-00918-2. [8] [14]

Li, Y., W. Han, A. Hu, G. A. Meehl, and F. Wang, 2018: Multidecadal changes of the upper Indian Ocean heat content during 1965–2016. *Journal of Climate*, **31**, 7863–7884. https://doi.org/10.1175/JCLI-D-18-0116.1. [11]

Loeb, N. G., B. A. Wielicki, D. R. Doelling, et al., 2009: Toward optimal closure of the earth's top-of-atmosphere radiation budget. *Journal of Climate*, **22**, 748–766. https://doi.org/10.1175/2008jcli2637.1. [4]

Lorenz, E. N., 1967: *The Nature and Theory of the General Circulation of the Atmosphere*. Vol **218**. Geneva: World Meteorological Organization.161pp. [1]

Lumpkin, R., and G. C. Johnson, 2013: Global ocean surface velocities from drifters: mean, variance, El Nino–Southern Oscillation response, and seasonal cycle, *Journal of Geophysical Research: Oceans*, **118**, 2992–3006. https://doi.org/10.1002/jgrc.20210. [8]

Mann, M. E., 2012: *The Hockey Stick and the Climate Wars*. New York: Columbia University Press, 448pp. [18]

Mantua, N. J., S. Hare, Y. Zhang, et al., 1997: A Pacific interdecadal climate oscillation with impacts on salmon production. *Bulletin of the American. Meteorological Society*, **78**, 1069–1079. [11]

Marshall, G. J., 2003: Trends in the Southern Annular Mode from observations and reanalyses. *Journal of Climate*, **16**, 4134–4143. https://doi.org/10.1175/1520-0442 (2003)016<4134:TITSAM>2.0.CO;2. [11]

Marzeion, B., A. H. Jarosch, and M. Hofer, 2012: Past and future sea-level change from the surface mass balance of glaciers. *The Cryosphere*, **6**, 1295–1322. https://doi.org/10.5194/tc-6-1295-2012. [14]

Marzeion, B., P. W. Leclercq, J. G. Cogley, and A. H. Jarosch, 2015: Global reconstructions of glacier mass change during the 20th century are consistent. *The Cryosphere*, **9**(6), 2399–2404. [14]

Mayer, M., L. Haimberger, and M. A. Balmaseda, 2014: On the energy exchange between tropical ocean basins related to ENSO. *Journal of Climate*, **27**, 6393–6403. https://doi.org/10.1175/JCLI-D-14-00123.1. [12]

McCoy, I. L., D. T. McCoy, R. Wood, et al., 2020: The hemispheric contrast in cloud microphysical properties constrains aerosol forcing. *Proceedings of the National Academy of Sciences USA*, 117(32),18998–19006. https://doi.org/10.1073/pnas.1922502117. [13]

McKibben, W., 2018: How extreme weather is shrinking the planet. *New Yorker*, November 26, 2018. www.newyorker.com/magazine/2018/11/26/how-extreme-weather-is-shrinking-the-planet?utm_medium=email&utm_source=actionkit [18]

McPhaden, M. J., 2012: A 21st century shift in the relationship between ENSO SST and warm water volume anomalies. *Geophysical Research Letters*, **39**, 9706. https://doi.org/10.1029/2012GL051826. [12]

McPhaden, M. J., T. Lee, and D. McClurg, 2011: El Niño and its relationship to changing background conditions in the tropical Pacific Ocean. *Geophysical Research Letters*, **38**, L15709. https://doi.org/10.1029/2011GL048275. [12]

Meehl, G. A., J. Arblaster, J. Fasullo, A. Hu, and K. Trenberth, 2011: Model-based evidence of deep-ocean heat uptake during surface temperature hiatus periods. *Nature Climate Change*. **1**, 360–364. https://doi.org/10.1038/nclimate1229. [15]

Meehl, G. A., A. Hu, J. Arblaster, J. T. Fasullo, and K. E. Trenberth, 2013: Externally forced and internally generated decadal climate variability in the Pacific. *Journal of Climate*, **26**, 7298–7310. https://doi.org/10.1175/JCLI-D-12-00548.1. [15]

Meehl, G. A., C. A. Senior, V. Eyring, et al., 2020: Context for interpreting equilibrium climate sensitivity and transient climate. *Science Advances*, **6**. https://doi.org/10.1126/sciadv.aba1981. [13]

Meyssignac, B., T. Boyer, Z. Zhao, et al., 2019: Measuring global ocean heat content to estimate the Earth energy imbalance. *Frontiers in Marine Science*, **6**, 432. https://doi.org/10.3389/fmars.2019.00432. [14]

National Academy of Sciences and The Royal Society, 2020: *Climate Change: Evidence and Causes: Update 2020*. 36pp. https://doi.org/10.17226/25733. [1] [17] [18]

National Research Council, 2015a: *Climate Intervention: Reflecting Sunlight to Cool Earth*. Washington, DC: The National Academies Press. https://doi.org/10.17226/18988. [17]

National Research Council, 2015b: *Climate Intervention: Carbon Dioxide Removal and Reliable Sequestration*. Washington, DC: The National Academies Press. https://doi.org/10.17226/18805. [17]

National Research Council, 2016: *Attribution of Extreme Weather Events in the Context of Climate Change*. Washington, DC: The National Academies Press, 165pp. https://doi.org/10.17226/21852. [15]

NeremR. S., B. D. Beckley, J. T. Fasullo, et al., 2018: Climate-change–driven accelerated sea-level rise detected in the altimeter era. *Proceedings of the National Academy of Sciences USA*, **115**, 2022–2025. [14]

Newman, M., M. A. Alexander, T. R. Ault, et al., 2016: The Pacific Decadal Oscillation, revisited. *Journal of Climate*, **29**, 4399–4427. https://doi.org/10.1175/JCLI-D-15-0508.1. [11] [12]

Newman, M., A. T. Wittenberg, L. Cheng, G. P. Compo, and C. A. Smith, 2018: The extreme 2015/16 El Niño, in the context of historical climate variability and change. *Bulletin of the American Meteorological Society*, **99**, S16–S20. [12]

O'Neill, B. C., C. Tebaldi, D. P. van Vuuren, et al., 2016: The Scenario Model Intercomparison Project (ScenarioMIP) for CMIP6. *Geoscience Model Development*, **9**, 3461–3482. https://doi.org/10.5194/gmd-9-3461-2016. [16]

Oort, A. H., 1971: The observed annual cycle in the meridional transport of atmospheric energy. *Journal of the Atmospheric Sciences*, **28**, 325–339. [9]

Oort, A. H, and T. Vonder Haar, 1976: On the observed annual cycle in the ocean–atmosphere heat balance over the Northern Hemisphere. *Journal of Physical Oceanography*, **6**, 781–800. [9]

Oreskes, N., and E. M. Conway, 2010: *Merchants of Doubt*. London: Bloomsbury Press, 355pp. [1]

Pall, P., C. M. Patricola, M. Wehner, et al., 2017: Diagnosing conditional anthropogenic contributions to heavy Colorado rainfall in September 2013. *Weather Climate Extremes*, **17**, 1–6. https://doi.org/10.1016/j.wace.2017.03.004. [10] [15]

Palmer, W. C., 1965: Meteorological drought. Report 45, US Department of Commerce, Washington, DC, 58pp. www.ncdc.noaa.gov/oa/climate/research/drought/palmer.pdf. [10]

Paolo, F., H. Fricker, and L. Padman, 2015: Volume loss from Antarctic ice shelves is accelerating. *Science*, **348**, 327–331. https://doi.org/10.1126/science.aaa0940. [14]

Parkinson, C., 2019: A 40-y record reveals gradual Antarctic sea ice increases followed by decreases at rates far exceeding the rates seen in the Arctic. *Proceedings of the National Academy of Sciences USA*, **116**, 14414–14423. www.pnas.org/cgi/doi/10.1073/pnas.1906556116. [14]

Peixoto, J. P., and A. H. Oort, 1992: *Physics of Climate*.New York: American Institute of Physics, 520pp. [9] [10]

Rabett, E., 2017: The Green Plate Effect. http://rabett.blogspot.com/2017/10/an-evergreen-of-denial-is-that-colder.html. [3]

Riser, S. C., H. J. Freeland, D. Roemmich, et al., 2016: Fifteen years of ocean observations with the global Argo array. *Nature Climate Change*, **6**(2), 145–153. https://doi.org/10.1038/nclimate2872. [14]

Ruppert, J. H., Jr., and X. Chen, 2020: Island rainfall enhancement in the maritime continent. *Geophysical Research Letters*, **47**, e2019GL086545. https://doi.org/10.1029/2019GL086545. [7]

Santoso, A., M. J. McPhaden, and W. Cai, 2017: The defining characteristics of ENSO extremes and the strong 2015/16 El Niño. *Review of Geophysics*, **55**, 1079–1129. https://doi.org/10.1002/2017RG000560. [12]

Schär, C., A. Arteaga, C.C.N Ban, et al., 2020: Kilometer-scale climate models: prospects and challenges. *Bulletin of the American Meteorological Society*, **101**. https://doi.org/10.1175/BAMS-D-18-0167.2. [1]

Schlosser, C. A., K. Strzepek, X. Gao, et al., 2014: The future of global water stress: an integrated assessment. *Earth's Future*, **2**, 341–361. https://doi.org/10.1002/2014EF000238. [10]

Schmidt, A., M. J. Mills, S. Ghan, et al., 2018: Volcanic radiative forcing from 1979 to 2015. *Journal of Geophysical Research: Atmospheres*.**123**, 12491-412508. [14]

Schmidt, G. A., R. A. Ruedy, R. L. Miller, and A. A. Lacis, 2010: Attribution of the present-day total greenhouse effect. *Journal of Geophysical Research*, **115**, D20106. https://doi.org/10.1029/2010JD014287. [3]

Schneider, A. M., A.Friedl, and D. Potere, 2009: A new map of global urban extent from MODIS satellite data. *Environmental Research Letters*, **4**, 044003. https://doi.org/10.1088/1748-9326/4/4/044003. [6]

Schweiger, A., R. Lindsay, J. Zhang, M. Steele, H. Stern, and R. Kwok, 2011: Uncertainty in modeled Arctic sea ice volume. *Journal of Geophysical Research*, **116**, C00D06. https://doi.org/10.1029/ 2011JC007084. [14]

Shepherd, A., L. Gilbert, A. S. Muir, et al., 2019: Trends in Antarctic Ice Sheet elevation and mass. *Geophysical Research Letters*, **46**. https://doi.org/10.1029/2019GL082182. [14]

Shepherd, T. G., E. Boyd, R. A. Calel, et al., 2018: Storylines: an alternative approach to representing uncertainty in climate change. *Climatic Change*, **151**, 555–571. https://doi10.1007/s10584–018-2317-9. [15]

Sherwood, S., M. J. Webb, J. D. Annan, et al., 2020: An assessment of Earth's climate sensitivity using multiple lines of evidence. *Review of Geophysics*. https://doi.org/10.1029/2019RG000678. [13]

Shindell, D. T., G. Faluvegi, and G. A. Schmidt, 2020: Influences of solar forcing at ultraviolet and longer wavelengths on climate. *Journal of Geophysical Research: Atmospheres*, **124**. https://doi.org/10.1029/2019JD031640. [4]

Silverman, J., 2013: *Opening Heaven's Floodgates: The Genesis Flood Narrative, Its Context, and Reception*. Piscataway, NJ: Gorgias Press, 548pp. [5]

Smith, B., H. A. Fricker, A. S. Gardner, et al., 2020: Pervasive ice sheet mass loss reflects competing ocean and atmosphere processes. *Science*, **368**, 1239–1242. https://doi.org/10.1126/science.aaz5845. [14]

Solomon, S., G.-K. Plattner, R. Knutti, and P. Friedlingstein, 2009: Irreversible climate change due to carbon dioxide emissions. *Proceedings of the National Academy of Sciences USA*, **106**, 1704–1709. https://doi.org/10.1073/pnas.0812721106. [17]

Song. X.-P., M. C. Hansen, S. V. Stehman, et al., 2018: Global land change from 1982 to 2016. *Nature*, **560**, 639–643. https://doi.org/10.1038/s41586-018-0411-9. [14]

Stoerk, T., G. Wagner, and R. E. T. Ward, 2018: Policy brief. Recommendations for improving the treatment of risk and uncertainty in economic estimates of climate impacts in the Sixth Intergovernmental Panel on Climate Change Assessment Report. *Reviews of Environmental Economics Policy*, **12**, 371–376. https://doi.org/10.1093/reep/rey005. [18]

Storto, A., A. Alvera-Azcárate, M. Balmaseda, et al., 2019: Ocean reanalyses: recent advances and unsolved challenges. *Frontiers in Marine Science*, **6**, 418. https://doi.org/10.3389/fmars.2019.00418. [14]

Sun, Q., C. Miao, Q. Duan, et al., 2018: A review of global precipitation data sets: data sources, estimation, and intercomparisons. *Reviews of Geophysics*, **56**, 79–107. https://doi.org/10.1002/. [5]

Swart, N. C., J. C. Fyfe, N. Gillett, and G. J. Marshall, 2015: Comparing trends in the Southern Annular Mode and surface westerly jet. *Journal of Climate*, **28**, 8840–8855. https://doi.org/10.1175/JCLI-D-15-0334.1. [11]

Thompson, D. W. J., and J. M. Wallace, 2000: Annular modes in the extratropical circulation, Pt I: Month-to-month variability. *Journal of Climate*, **13**, 1000–1016. [11]

Thompson, D. W. J., J. M. Wallace, and G. C. Hegerl, 2000: Annular modes in the extratropical circulation. Part II: Trends. *Journal of Climate*, **13**, 1018–1036. [11]

Thompson, D. W. J., M. P. Baldwin, and J. M. Wallace, 2002: Stratospheric connection to Northern Hemisphere wintertime weather: implications for prediction. *Journal of Climate*, **15**, 1421–1428. [11]

Tollefson, J., 2020: Can the world slow global warming.? *Nature*, **573**, 324–327. https://nature.us17.list-manage.com/track/click?u=2c6057c528fdc6f73fa196d9d&id=731f36c0aa&e=3cdb1f115b. [17]

Trenberth, K. E., 1976: Spatial and temporal variations of the Southern Oscillation. *Quarterly Journal of the Royal Meteorological Society*, **102**, 639–653. [12]

Trenberth, K. E., 1983: What are the seasons? *Bulletin of the American Meteorological Society*, **64**, 1276–1282. [5] [6]

Trenberth, K. E., 1984: Signal versus noise in the Southern Oscillation. *Monthly Weather Review*, **112**, 326–332. [12]

Trenberth, K. E., 1990: Recent observed interdecadal climate changes in the Northern Hemisphere. *Bulletin of the American Meteorological Society*, **71**, 988–993. [11] [12]

Trenberth, K. E., ed., 1992: *Climate System Modeling*. Cambridge: Cambridge University Press, 788pp. [1][16]

Trenberth, K. E., 1994: The different flavors of El Niño. 18th Annual Climate Diagnostics Workshop, November 1–5, 1993, Boulder, CO, 50–53. [12]

Trenberth, K. E., 1997a: The definition of El Niño. *Bulletin of the American Meteorological Society*, **78**, 2771–2777. [12]

Trenberth, K. E., 1997b: The use and abuse of climate models in climate change research. *Nature*, **386**, 131–133. [1] [2]

Trenberth, K. E., 1998: Atmospheric moisture residence times and cycling: implications for rainfall rates with climate change. *Climatic Change*, **39**, 667–694. [5] [10]

Trenberth, K. E., 1999a: Conceptual framework for changes of extremes of the hydrological cycle with climate change. *Climatic Change*, **42**, 327–339. [5]

Trenberth, K. E., 1999b: Atmospheric moisture recycling: role of advection and local evaporation. *Journal of Climate*, **12**, 1368–1381. [10]

Trenberth, K. E., 2001: Stronger evidence for human influences on climate: the 2001 IPCC Assessment. *Environment*, **43**(4), 8–19. [1]

Trenberth, K. E., 2007: Warmer oceans, stronger hurricanes. *Scientific American*, July, 45–51. [7]

Trenberth, K. E., 2008: Observational needs for climate prediction and adaptation. *WMO Bulletin*, **57** (1), 17–21. [17]

Trenberth, K. E., 2009: An imperative for adapting to climate change: tracking Earth's global energy. *Current Opinion on Environmental Sustainability*, **1**, 19–27. https://doi.org/10.1016/j.cosust.2009.06.001. [14]

Trenberth, K. E., 2011a: Attribution of climate variations and trends to human influences and natural variability. *WIREs Climate Change*, **2**, 925–930. https://doi.org/10.1002/wcc.142;http://doi.wiley.com/10.1002/wcc.142. [15]

Trenberth, K. E., 2011b: Changes in precipitation with climate change. *Climate Research*, **47**, 123–138. https://doi.org/10.3354/cr00953. [5] [10]

Trenberth, K. E., 2012: Framing the way to relate climate extremes to climate change. *Climatic Change*, **115**, 283–290, https://doi.org/10.1007/s10584-012-0441-5. [15]

Trenberth, K. E., 2015a: Has there been a hiatus? *Science*, **349**(2649), 691–692. https://doi.org/10.1126/science.aac9225. [11] [15]

Trenberth, K. E., 2015b: Intergovernmental Panel on Climate Change. In: North, G. R. (ed.-in-chief), J. Pyle, and F. Zhang, eds., *Encyclopedia of Atmospheric Sciences*, 2nd ed., vol. 2. London: Academic Press, 90–94. [16]

Trenberth, K. E., 2018: Climate change caused by human activities is happening and it already has major consequences. *Journal of Energy Water Resources Law*, **36**, 463–481. https://doi.org/10.1080/02646811.2018.1450895. [1]

Trenberth, K. E., 2020: Understanding climate change through Earth's Energy Flows. *Journal of the Royal Society New Zealand*, **50**, 331–347. NZJR-2020-0003, https://doi.org/10.1080/03036758.2020.1741404. [1]

Trenberth, K. E., and J. M. Caron, 2000: The Southern Oscillation revisited: sea level pressures, surface temperatures and precipitation. *Journal of Climate*, **13**, 4358–4365. [12]

Trenberth, K. E., and J. M. Caron, 2001: Estimates of meridional atmosphere and ocean heat transports. *Journal of Climate*, **14**, 3433–3443. [9]

Trenberth, K. E., and A. Dai, 2007: Effects of Mount Pinatubo volcanic eruption on the hydrological cycle as an analog of geoengineering. *Geophysical Research Letters*, **34**, L15702. https://doi.org/10.1029/2007GL030524. [10]

Trenberth, K. E., and J. Fasullo, 2007: Water and energy budgets of hurricanes and implications for climate change. *Journal of Geophysical Research*, **112**, D23107. https://doi.org/10.1029/2006JD008304. [7] [10]

Trenberth, K. E., and J. Fasullo, 2008: Energy budgets of Atlantic hurricanes and changes from 1970. *Geochemistry, Geophysics, Geosystems*, **9**, Q09V08. https://doi.org/10.1029/2007GC001847. [7]

Trenberth, K. E., and J. T. Fasullo, 2010: Tracking Earth's energy. *Science*, **328**, 316–317. [14]

Trenberth, K. E., and J. T. Fasullo, 2011: Tracking Earth's energy: from El Niño to global warming. *Surveys in Geophysics*, https://doi.org/10.1007/s10712-011-9150-2. [3]

Trenberth, K. E., and J. T. Fasullo, 2012: Climate extremes and climate change: the Russian heat wave and other climate extremes of 2010. *Journal of Geophysical Research*, **117**, D17103. https://doi.org/10.1029/2012JD018020. [5]

Trenberth, K. E., and J. T. Fasullo, 2013: An apparent hiatus in global warming? *Earth's Future*. **1**, 19–32. https://doi.org/10.002/2013EF000165. [15]

Trenberth, K. E., and T. J. Hoar, 1996: The 1990–1995 El Niño–Southern Oscillation Event: longest on record. *Geophysical Research Letters*, **23**, 57–60. [12]

Trenberth, K. E., and T. J. Hoar, 1997: El Niño and climate change. *Geophysical Research Letters*, **24**, 3057–3060. [12]

Trenberth, K. E., and J. W. Hurrell, 1994: Decadal atmosphere–ocean variations in the Pacific. *Climate Dynamics*, **9**, 303–319. [11]

Trenberth, K. E., and J. W. Hurrell, 2019: Climate change. In: Dunn, P. O., and A. P. Møller, eds., *The Effects of Climate Change on Birds*, 2nd ed. Oxford: Oxford University Press, 5–25. [11] [12]

Trenberth, K. E., and B. L. Otto-Bliesner, 2003: Toward integrated reconstructions of past climates. *Science*, **300**, 589–591. [4]

Trenberth, K. E., and D. J. Shea, 1987: On the evolution of the Southern Oscillation. *Monthly Weather Review*, **115**, 3078–3096. [12]

Trenberth, K. E., and D. J. Shea, 2005: Relationships between precipitation and surface temperature. *Geophysical Research Letters*, **32**, L14703. https://doi.org/10.1029/2005GL022760. [5]

Trenberth, K. E., and D. J. Shea, 2006: Atlantic hurricanes and natural variability in 2005. *Geophysical Research Letters*, **33**, L12704. https://doi.org/10.1029/2006GL026894. [11]

Trenberth, K. E., and L. Smith, 2005: The mass of the atmosphere: a constraint on global analyses. *Journal of Climate*, **18**, 864–875. [6]

Trenberth, K. E., and D. P. Stepaniak, 2001: Indices of El Niño evolution. *Journal of Climate*, **14**, 1697–1701. [12]

Trenberth, K. E., and D. P. Stepaniak, 2003a: Co-variability of components of poleward atmospheric energy transports on seasonal and interannual timescales. *Journal of Climate*, **16**, 3691–3705. https://doi.org/10.1175/1520-0442(2003)016,3691:COCOPA.2.0.CO;2. [7] [9]

Trenberth, K. E., and D. P. Stepaniak, 2003b: Seamless poleward atmospheric energy transports and implications for the Hadley circulation. *Journal of Climate*, **16**, 3706–3722. https://doi.org/10.1175/1520-0442(2003)016,3706: SPAETA.2.0.CO;2. [7] [9]

Trenberth, K. E., and D. P. Stepaniak, 2004: The flow of energy through the Earth's climate system. *Quarterly Journal of the Royal Meteorological Society*, **130**, 2677–2701. https://doi.org/10.1256/qj.04.83. [6]

Trenberth, K. E., and Y. Zhang, 2018a: How often does it really rain? *Bulletin of the American Meteorological Society*, **99**, 289–298. https://doi.org/10.1175/BAMS-D-17-0107.1. [5] [7] [10]

Trenberth, K. E. and Y. Zhang, 2018b: Near global covariability of hourly precipitation in space and time. *Journal of Hydrometeorology*, **19**, 695–713. https://doi.org/10.1175/JHM-D-17-0238.1. [5] [7] [10]

Trenberth, K. E., and Y. Zhang, 2019: Observed inter-hemispheric meridional heat transports and the role of the Indonesian ThroughFlow in the Pacific Ocean. *Journal of Climate*, **32**, 8523–8536. https://journals.ametsoc.org/doi/pdf/10.1175/JCLI-D-19-0465.1. [9] [14]

Trenberth, K. E., J. T. Houghton, and L. G. Meira Filho. 1996: The climate system: an overview. In: Houghton, J. T., L. G. Meira Filho, B. Callander, et al., eds., *Climate Change 1995. The Science of Climate Change*. Second Assessment Report of the Intergovernmental Panel on Climate Change. Cambridge: Cambridge University Press, 51–64. [1]

Trenberth, K. E., G. W. Branstator, D. Karoly, A. Kumar, N-C. Lau, and C. Ropelewski, 1998: Progress during TOGA in understanding and modeling global teleconnections associated with tropical sea surface temperatures. *Journal of Geophysical Research*, **103**, 14291–14324. [11] [12]

Trenberth, K. E., D. P. Stepaniak, and J. M. Caron, 2000: The global monsoon as seen through the divergent atmospheric circulation. *Journal of Climate*, **13**, 3969–3993. [7]

Trenberth, K. E., J. M. Caron, D. P. Stepaniak, and S. Worley, 2002: Evolution of El Niño Southern Oscillation and global atmospheric surface temperatures. *Journal of Geophysical Research*, **107**(D8), 4065. https://doi.org/10.1029/2000JD000298. [12]

Trenberth, K. E., A. Dai, R. M. Rasmussen, and D. B. Parsons, 2003: The changing character of precipitation. *Bulletin of the American Meteorological Society*, **84**, 1205–1217. https://doi.org/10.1175/bams-84-9-1205. [5] [10]

Trenberth, K. E., L. Smith, T. Qian, A. Dai, and J. Fasullo, 2007: Estimates of the global water budget and its annual cycle using observational and model data. *Journal of Hydrometeorology*, **8**, 758–769. https://doi.org/10.1175/JHM600.1. [10]

Trenberth, K. E., P. D. Jones, P. Ambenje, et al., 2007a: Observations: surface and atmospheric climate change. In: Solomon, S.,D.Qin, M.Manning, et al., eds., *Climate Change 2007. The Physical Science Basis*. Fourth Assessment Report of the Intergovernmental Panel on Climate Change. Cambridge: Cambridge University Press, 235–336. [11] [12]

Trenberth, K. E., C. A. Davis, and J. Fasullo, 2007b: Water and energy budgets of hurricanes: case studies of Ivan and Katrina. *Journal of Geophysical Research*, **112**, D23106. https://doi.org/10.1029/2006JD008303. [7]

Trenberth, K. E., J. T. Fasullo, and J. Kiehl, 2009: Earth's global energy budget. *Bulletin of the American Meteorological Society*, **90**, 311–323. https://doi.org/10.1175/2008BAMS2634.1. [3]

Trenberth, K. E., R. A. Anthes, A. Belward, et al., 2013: Challenges of a sustained climate observing system. In: Asrar, G. R., and J. W. Hurrell, eds., *Climate Science for Serving Society: Research, Modelling and Prediction Priorities*. Dordrecht: Springer, 13–50. [17]

Trenberth, K. E., J. T. Fasullo, and M. A. Balmaseda, 2014a: Earth's energy imbalance. *Journal of Climate*, **27**, 3129–3144. https://doi.org/10.1175/JCLI-D-13-00294. [14]

Trenberth, K. E., A. Dai, G. van der Schrier, et al., 2014b: Global warming and changes in drought. *Nature Climate Change*, **4**, 17–22. https://doi.org/10.1038/NCLIMATE2067. [10] [12]

Trenberth, K. E., J. T. Fasullo, G. Branstator, and A. S. Phillips, 2014c: Seasonal aspects of the recent pause in surface warming. *Nature Climate Change*, **4**, 911–916. https://doi.org/10.1038/NCLIMATE2341. http://rdcu.be/o7wB [15]

Trenberth, K. E., J. T. Fasullo, and T. G. Shepherd, 2015a: Attribution of climate extreme events. *Nature Climate Change*, **5**, 725–730. https://doi.org/10.1038/NCLIMATE2657. [15]

Trenberth, K. E., Y. Zhang, J. T. Fasullo, and S. Taguchi, 2015b: Climate variability and relationships between top-of-atmosphere radiation and temperatures on Earth. *Journal of Geophysical Research*, **120**, 3642–3659. https://doi.org/10.1002/2014JD022887. [13]

Trenberth, K. E., Y. Zhang, and J. T. Fasullo, 2015c: Relationships among top-of-atmosphere radiation and atmospheric state variables in observations and CESM. *Journal of Geophysical Research*, **120**, 10,074–10,090. https://doi.org/10.1002/2015JD023381. [13]

Trenberth, K. E., J. T. Fasullo, K. von Schuckmann and L. Cheng, 2016a: Insights into Earth's energy imbalance from multiple sources. *Journal of Climate*, **29**, 7495–7505. http://dx.doi.org/10.1175/JCLI-D-16-0339.1. [14]

Trenberth, K. E., M. Marquis, and S. Zebiak, 2016b: The vital need for a climate information system. *Nature Climate Change*, **6**, 1057–1059. https://doi.org/10.1038/NCLIM-16101680. [17]

Trenberth, K. E., Y. Zhang, and M. Gehne, 2017: Intermittency in precipitation: duration, frequency, intensity, and amounts using hourly data. *Journal of Hydrometeorology*, **18**, 1393–1412. https://doi.org/10.1175/JHM-D-16-0263. [5] [10]

Trenberth, K. E., L. Cheng, P. Jacobs, Y. Zhang, and J. Fasullo, 2018: Hurricane Harvey links to ocean heat content. *Earth's Future*, **6**, 730–744. https://doi.org/10.1029/2018EF000825. [7] [10] [12]

Trenberth, K. E., Y. Zhang, J. T. Fasullo, and L. Cheng, 2019: Observation-based estimates of global and basin ocean meridional heat transport time series. *Journal of Climate*, **32**, 4567–4583. https://doi.org/10.1175/JCLI-D-18-0872.1. [8] [9] [10]

Turner, J., T. Phillips, G. J. Marshall, et al., 2017: Unprecedented springtime retreat of Antarctic sea ice in 2016. *Geophysical Research Letters*, **44**, 6868–6875. https://doi.org/10.1002/2017GL073656. [11] [14]

USGCRP, 2017: *Climate Science Special Report: Fourth National Climate Assessment, Vol. I*, edited byWuebbles, D.J., D.W. Fahey, K.A. Hibbard, et al. Washington, DC: US Global Change Research Program, 470pp. https://science2017.globalchange.gov/. [16] [17]

USGCRP, 2018: *Impacts, Risks, and Adaptation in the United States: Fourth National Climate Assessment, Vol. II*, edited by Reidmiller, D. R., C. W.

Avery, D. R. Easterling, et al. Washington, DC: US Global Change Research Program, 1515 pp. https://doi.org/10.7930/NCA4.2018. [17]

Vanderkelen, I., N. P. M. van Lipzig, D. M. Lawrence, et al., 2020. Global heat uptake by inland waters. *Geophysical Research Letters*, **47**. https://doi.org/10.1029/2020GL087867. [14]

van Vuuren, D. P., J. Edmonds, M. Kainuma, et al., 2011: The representative concentration pathways: an overview. *Climatic Change*, **109**, 5–31. https://doi.org/10.1007/s10584–011-0148-z. [16]

Vecchi, G. A., and B. J. Soden, 2007: Effect of remote sea surface temperature change on tropical cyclone potential intensity. *Nature*, **450**, 1066–1070. https://doi.org/10.1038/nature06423. [7]

von Schuckmann, K., M. D. Palmer, K. E. Trenberth, et al., 2016: Earth's energy imbalance: an imperative for monitoring. *Nature Climate Change*, **6**, 138–144. https://doi.org/10.1038/NCLIM-15030445C. www.nature.com/nclimate/journal/v6/n2/full/nclimate2876.html. [14]

Wells, N., S. Goddard, and M. J. Hayes, 2004: A self-calibrating Palmer Drought Severity Index. *Journal of Climate*, **17**, 2335–2351. https://doi.org/10.1175/1520-0442(2004)017<2335:ASPDSI>2.0.CO;2. [10]

Wild, M., D. Folini, M. Z. Hakuba, et al., 2015: The energy balance over land and oceans: an assessment based on direct observations and CMIP5 climate models. *Climate Dynamics*, **44**, 3393–3429. [3]

Williams, A. P., J. T. Abatzoglou, A. Gershunov, et al., 2019: Observed impacts of anthropogenic climate change on wildfire in California. *Earth's Future*, **7**, 892–910. https://doi.org/10.1029/2019EF001210. [10]

Williams, I. N., and C. M. Patricola, 2018: Diversity of ENSO events unified by convective threshold sea surface temperature: a nonlinear ENSO index. *Geophysical Research Letters*, **45**. 9236–9244. https://doi.org/10.1029/2018GL079203. [12]

Wolter, K., and M. S. Timlin, 2011: El Niño/Southern Oscillation behaviour since 1871 as diagnosed in an extended multivariate ENSO index (MEI.ext). *International Journal of Climatology*, **31**, 1074–1087. https://doi.org/10.1002/joc.2336. [12]

Xie, P., R. Joyce, S. Wu, et al., 2017: Reprocessed, bias-corrected CMORPH global high-resolution precipitation estimates. *Journal of Hydrometeorology*, **18**, 1617–1641. https://doi.org/10.1175/JHM-D-16-0168.1 [5] [7]

Yang, C.; S. Masina, and A. Storto, 2017: Historical ocean reanalyses (1900–2010) using different data assimilation strategies. *Quarterly Journal of the Royal Meteorological Society*, **143**, 479–493. https://doi.org/10.1002/qj.2936. [14]

Zanna, L.,S. Khatiwala, J. M. Gregory, J. Ison, and P. Heimbach, 2019: Global reconstruction of historical ocean heat storage and transport. *Proceedings of the*

National Academy of Sciences USA, **116**, 1126–1131. https://doi.org/10.1073/pnas.1808838115. [14]

Zelinka, M. D., T. A. Myers, D. T. McCoy, et al., 2020: Causes of higher climate sensitivity in CMIP6 models. *Geophysical Research Letters*, **47**, e2019GL085782. [13]

Zhu, J., C. J. Poulsen, and B. L. Otto-Bliesner, 2020: High climate sensitivity in CMIP6 model not supported by paleoclimate. *Geophysical Research Letters*, **47**. https://doi.org/10.1038/s41558-020-0764-6. [13]

OTHER RESOURCES: WEBSITES

ACRI(Actuaries Climate Risk Index), 2020: www.actuary.org/sites/default/files/2020-01/ACRI.pdf [18]

Amazon fires: https://edition.cnn.com/2020/08/07/americas/brazil-bolsonaro-amazon-fires-intl/index.html [10]

AVISO: Archiving, Validation and Interpretation of Satellite Oceanographic data: www.aviso.altimetry.fr/en/home.html [14]

Bast, E., 2015: Empty promises: G20 subsidies to oil, gas and coal production. http://priceofoil.org/2015/11/11/empty-promises-g20-subsidies-to-oil-gas-and-coal-production/ [18]

Climate services: www.climate-services.org/ [17]

EPA (Environmental Protection Agency),2020: www.epa.gov/climate-indicators/climate-change-indicators-atmospheric-concentrations-greenhouse-gases [6]

First Street Foundation, 2020: First national flood risk assessment. https://assets.firststreet.org/uploads/2020/06/first_street_foundation__first_national_flood_risk_assessment.pdf [14]

Flood, 2020: https://edition.cnn.com/2020/01/06/asia/jakarta-floods-intl-hnk/index.htmlhttps://earthobservatory.nasa.gov/images/146113/torrential-rains-flood-indonesia?src=eoa-iotd [10]

Geothermal heat: www.builditsolar.com/Projects/Cooling/EarthTemperatures.htm [6]

Global Commission on Adaptation, 2019: https://gca.org/global-commission-on-adaptation/report [18]

Hausfather, Z., 2018: Explainer: How 'Shared Socioeconomic Pathways' explore future climate change. www.carbonbrief.org/explainer-how-shared-socioeconomic-pathways-explore-future-climate-change [16]

Hausfather, Z., and R. Betts, 2020: Importance of carbon-cycle feedback uncertainties. www.carbonbrief.org/analysis-how-carbon-cycle-feedbacks-could-make-global-warming-worse [2]

IPCC (Intergovernmental Panel on Climate Change), Fifth Assessment Reports (AR5) www.ipcc.ch/report/ar5/ and reports from Working Groups 1, 2 and 3: www.ipcc.ch/report/ar5/wg1/; www.ipcc.ch/report/ar5/wg2/; www.ipcc.ch/report/ar5/wg3/ [17] [18]

Keen, S., 2020: The appallingly bad neoclassical economics of climate change. *Globalizations*, https://doi.org/10.1080/14747731.2020.1807856 [18]

Kyoto Protocol: https://unfccc.int/kyoto_protocol/items/2830.php [18]

Moynihan, P. D., 1983: https://quoteinvestigator.com/2020/03/17/own-facts/#return-note-437589-9 [18]

NASA albedo: https://earthobservatory.nasa.gov/images/2599/global-albedo [6]

NOAA (National Oceanic and Atmospheric Administration), 2020: www.climate.gov/ [14]

NRC (National Research Council), 2010: America's Climate Choices: Advancing the Science of Climate Change. www.nap.edu/catalog.php?record_id=12782 [18]

NRC, 2011: Climate Stabilization Targets: Emissions, Concentrations, and Impacts over Decades to Millennia. www.nap.edu/catalog.php?record_id=12877 [18]

NRC, 2013: Abrupt Impacts of Climate Change: Anticipating Surprise. www.nap.edu/catalog.php?record_id=18373 [18]

NRC, 2014, and Royal Society, 2014: *Climate Change: Evidence and Causes*, Washington, DC: National Academy Press, 33pp. www.nap.edu/catalog.php?record_id=18730 [18]

NSIDC (National Snow and Ice Data Center): https://nsidc.org/cryosphere/sotc/sea_level.html [6]. http://nsidc.org/greenland-today/ [14]

Paris Agreement: http://unfccc.int/paris_agreement/items/9485.php [18]

Paulsen,H. and M. Bloomberg, 2014: Risky Business: The Economic Risks of Climate Change in the United States. http://riskybusiness.org/report/national/ [18]

Pope Francis, 2015: Laudato Si. https://laudatosi.com/watch [18]

Randers, J., 2015: www.theguardian.com/sustainable-business/2015/jan/19/davos-climate-action-democracy-failure-jorgen-randers [18]

Ritchie, H., and M. Roser, 2017: CO_2 and Greenhouse Gas Emissions. Published online at OurWorldInData.org. Retrieved from: https://ourworldindata.org/co2-and-other-greenhouse-gas-emissions. [17]

Ritchie, H., and M. Roser, 2020: *Land Use*. Published online at OurWorldInData.org. Retrieved from: https://ourworldindata.org/land-use, 26 April 2020. [6]

Satellite Imagery: www.flickr.com/photos/noaasatellites/48127127191/sizes/l/ [7]

Science settled: www.skepticalscience.com/print.php?r=222 or www.wsj.com/articles/climate-science-is-not-settled-1411143565 [17]

SolarCycleScience.com: http://solarcyclescience.com/solarcycle.html [4]

UNFCCC(United Nations Framework Convention on Climate Change): https://unfccc.int/2860.php [18]

World Wildlife Fund, 2020: www.wwf.fr/sites/default/files/doc-2020-08/20200827_Report_Fires-forests-and-the-future_WWF-min.pdf [10]

Index